Sustainable Development and Biodiversity

Volume 3

Series editor
Kishan Gopal Ramawat
M.L. Sukhadia University Botany Department, Udaipur, Rajasthan, India

This book series provides complete, comprehensive and broad subject based reviews about existing biodiversity of different habitats and conservation strategies in the framework of different technologies, ecosystem diversity, and genetic diversity. The ways by which these resources are used with sustainable management and replenishment are also dealt with. The topics of interest include but are not restricted only to sustainable development of various ecosystems and conservation of hotspots, traditional methods and role of local people, threatened and endangered species, global climate change and effect on biodiversity, invasive species, impact of various activities on biodiversity, biodiversity conservation in sustaining livelihoods and reducing poverty, and technologies available and required. The books in this series will be useful to botanists, environmentalists, marine biologists, policy makers, conservationists, and NGOs working for environment protection.

More information about this series at http://www.springer.com/series/11920

Dinesh K. Maheshwari
Editor

Composting for Sustainable Agriculture

Springer

Editor
Dinesh K. Maheshwari
Dept. of Botany and Microbiology
Gurukul Kangri University
Haridwar (Uttarakhand)
India

ISSN 2352-474X ISSN 2352-4758 (electronic)
ISBN 978-3-319-08003-1 ISBN 978-3-319-08004-8 (eBook)
DOI 10.1007/978-3-319-08004-8
Springer Cham Heidelberg New York Dordrecht London

Library of Congress Control Number: 2014952660

Printed on acid-free paper

Springer is part of Springer Science+Business Media (www.springer.com)

Preface

The global consensus to reduce inputs of agrochemicals, which are perceived as being hazardous in nature, has provided opportunity for the development of novel benign sustainable crop management strategies. One of the strategies is the application of effective microbial product in the form of 'Compost', beneficial for both farmers and ecosystem.

Microorganisms are able to degrade solid waste organic material into compost, which is a mixture of decayed organic matter, manure etc. Incomplete microbial degradation of organic waste involving both aerobic and anaerobic process lead to compost formation. If such products is incorporated in to soil, increases soil fertility and enhances plant growth and development. The beneficial activities bestowed upon plants by compost utilization are multifaceted, hence most promising alternatives for achieving sustainable agricultural production.

The present book entitled "Composting for Sustainable Agriculture" comprises 13 chapters contributed by leading experts having authoritative experience both in teaching and research on fundamental and applied aspects of compost science. The intensification through nutrient cycling, aerobic-anaerobic processing of organic waste, lignocellulosic bioconversion including both terrestrial and aquatic biomass residue into compost, its amendment into soil ensure farmers to obtain better crop productivity are suitably described. A due account is provided with respect to physio-chemical and biological parameters and their analysis in mature compost for quality assessment. The application of metabolites enzymes of cellulolytic thermophiles has also been focused. Compost tea is a watery extract of microorganisms and nutrients acts as potential source for the management of foliar and fruit diseases besides municipal solid waste, oil palm waste. Compost proved efficient in improvement of agricultural soil fertility have also been included.

The book provides adequate new insights to students, teachers, NGO's and other professionals interested to enrich the subject of knowledge of compost process, analysis and application particularly in the context of Environmental studies, Biotechnology, Microbiology, Agriculture, Plant protection, Agronomy and field practices in crop ecosystem.

I would like to express my sincere thanks to all the contributors for their contribution for mutual co-operation of scientific benefits. I acknowledge with thanks the assistance rendered by my research students Mohit and Shrivardhan. I owe my sincere thanks to Dr. Valeria Rinaudo and Dr. Ineke from the publisher Springer for their valuable support. Last but not the least, I owe thanks to my wife Dr. Sadhana and my children Dr. Charul and Ashish for taking care of me during this project.

Uttarakhand, India Dinesh K. Maheshwari

Contents

Contributors

Nuha Abdalla Department of Soil Science, CCS Haryana Agricultural University, Hisar, India

Hamada M. Abdelrahman Soil Science Dept., Faculty of Agriculture, Cairo University, Giza, Egypt

Hassen Abdennaceur Laboratoire Traitement et Recyclage des Eaux, Centre de Recherches et des Technologies des Eaux (CERTE), Tunis, Tunisie

Mohit Agarwal Department of Botany and Microbiology, Gurukul Kangri University, Haridwar, India

Rajinder Singh Antil Department of Soil Science, CCS Haryana Agricultural University, Hisar, India

Abhishek Bhattacharya Department of Biochemistry, Microbiology and Biotechnology, Rhodes University, Grahamstown, South Africa

Bougnom Blaise Pascal Faculty of Science, Department of Microbiology, University of Yaounde 1, Yaounde, Cameroon

Francesco G. Ceglie Organic Farming Dept., Mediterranean Agronomic Institute of Bari—CIHEAM-IAMB, Valenzano, Italy

K. Yin Chan Formerly NSW Department of Primary Industries, Richmond, NSW, Australia

Shrivardhan Dheeman Department of Botany and Microbiology, Gurukul Kangri University, Haridwar, India

Nwaga Dieudonné Faculty of Science, Department of Microbiology, University of Yaounde 1, Yaounde, Cameroon

Nerida J. Donovan Elizabeth Macarthur Agricultural Institute, NSW Department of Primary Industries, Menangle, NSW, Australia

Simon M. Eldridge Wollongbar Primary Industries Institute, NSW Department of Primary Industries, Wollongbar, NSW, Australia

Katherine J. Evans Perennial Horticulture Centre, Tasmanian Institute of Agriculture, University of Tasmania, New Town, TAS, Australia

Etoa François Xavier Faculty of Science, Department of Microbiology, University of Yaounde 1, Yaounde, Cameroon

A. W. Gandahi Department of Soil Science, Faculty of Crop Production, Sindh Agriculture University, Tandojam, Sindh, Pakistan

M. M. Hanafi Laboratory of Plantation Crops, Institute of Tropical Agriculture/ Department of Land Management, Faculty of Agriculture, Universiti Putra Malaysia, Serdang, Selangor, Malaysia

Martin A. Hubbe Dept. of Forest Biomaterials, North Carolina State University, Raleigh, NC, USA

Kazuyuki Inubushi graduate School of Horticulture, Chiba University, Matsudo, Chiba, Japan

A. N. Ivankin Moscow State University of Forest, Mytishchi, Moscow oblast, Russia

Essia Ngang Jean Justin Faculty of Science, Department of Microbiology, University of Yaounde 1, Yaounde, Cameroon

Dinesh Kumar Maheshwari Department of Botany and Microbiology, Gurukul Kangri University, Haridwar, India

Jedidi Naceur Laboratoire Traitement et Recyclage des Eaux, Centre de Recherches et des Technologies des Eaux (CERTE), Tunis, Tunisie

Bouzaiane Olfa Laboratoire Traitement et Recyclage des Eaux, Centre de Recherches et des Technologies des Eaux (CERTE), Tunis, Tunisie

Boyomo Onana Faculty of Science, Department of Microbiology, University of Yaounde 1, Yaounde, Cameroon

Urja Pandya Department of Microbiology and Biotechnology, University School of Sciences, Gujarat University, Ahmedabad, Gujarat, India

Catello Massimo PaneZaccardelli Consiglio per la Ricerca e la Sperimentazione in Agricoltura, Centro di Ricerca per l'Orticoltura, Pontecagnano, SA, Italy

Alice K. Percy Sprout Tasmania, Howrah, TAS, Australia

Brett I. Pletschke Department of Biochemistry, Microbiology and Biotechnology, Rhodes University, Grahamstown, South Africa

Dev Raj Department of Soil Science, CCS Haryana Agricultural University, Hisar, India

Meenu Saraf Department of Microbiology and Biotechnology, University School of Sciences, Gujarat University, Ahmedabad, Gujarat, India

Chapter 1
Ecological Intensification through Nutrients Recycling and Composting in Organic Farming

Francesco G. Ceglie and Hamada M. Abdelrahman

Abstract In organic agriculture fertilizers are permitted in organic forms, as defined by regulation. Mineralization of organic fertilizers is a biological decomposition that release plants' available nutrients; hence soil microbial communities are vital in the organic cropping systems. Composting microorganisms can work for the farmer's benefit recycling agricultural organic wastes into materials that contribute to healthy and biologically active soil. Composting process has been deeply described to highlight the link among starting mixture, process factors and final resulting compost. Composting and crop residues incorporation are fundamental to recycle resources at farm level to improve the nutrients use efficiency and to decrease the off-farm input needs. In the organic farming a balanced combination of compost application and crop residues incorporation increases the microbial carbon use efficiency, which regulates the soil organic matter decomposition and nutrients mineralization resulting both to increase the yield and to decrease the negative impact on the environment.

Keywords Crop residues recycling · Microbial C · Nutrients use efficiency · On-farm input · C/N ratio

1.1 Introduction

In organic farming systems crop rotation, cover crops, livestock integration and organic amendment are the pillars for sustainable soil fertility management. Composting is recommended as a tool to recycle inputs (biomass and nutrients) available in the farm and to reduce off-farm inputs. Compost application has been widely

F. G. Ceglie (✉)
Organic Farming Dept., Mediterranean Agronomic Institute of Bari—CIHEAM-IAMB, Via Ceglie 9, 70010 Valenzano, Italy
e-mail: ceglie@iamb.it

H. M. Abdelrahman
Soil Science Dept., Faculty of Agriculture, Cairo University, Gamma St., Giza 12613, Egypt
e-mail: hamada@agr.cu.edu.eg

D. K. Maheshwari (ed.), *Composting for Sustainable Agriculture,* Sustainable Development and Biodiversity 3, DOI 10.1007/978-3-319-08004-8_1

studied for its beneficial effects on: (i) soil physical properties such as water holding capacity and soil structure; (ii) soil chemical characteristics as cation exchange capacity, soil organic matter (SOM) quantity and quality (Oades 1984; Rivero et al. 2004), and (iii) soil biochemical and biological indicators such as global microbial biomass and soil enzyme activities, which are potentially involved in biogeochemical cycles and can exert a great influence on plant productivity parameters (Albiach et al. 2000; Adani et al. 2009; Fernández et al. 2009). Compost affects soil microbial communities (Debosz et al. 1999; Perucci et al. 2000; Debosz et al. 2002) increasing the competition among soil native microorganisms and the compost ones, which seems to lower the soil-borne pathogens load on plants (Höper and Alabouvette 1996). Moreover, compost effects on decreasing the severity of soil-borne diseases were reported (Chen et al. 1992; Abawi and Widmer 2000). The reasons mentioned above resemble the basis of the importance of compost and composting in organic agriculture. Compost could be locally produced, it complies with the ecological principles of recycling and contributes to the ecosystem viability by reducing wastes size that could disrupt the ecosystem. On the other hand, compost cannot be the solution for all constraints in organic farming and the other side of the coin should be considered in terms of: (i) compost quality level, (ii) heavy metals and toxic compounds content and availability, (iii) nutrients immobilization or leaching potentiality and iv) economical cost/benefit analysis. To navigate this precarious waters there is no universal solution; a deep knowledge of composting science linked to practical experiences identified at the local field conditions is the main method to maximize the compost use benefits.

The philosophy of organic farming is to feed the soil rather than the plant as the plants will be fed indirectly by microorganism activity. It is possible to feed microorganisms under controlled conditions by composting process and then supply compost and its microbial communities to the soil. Composting microorganisms might work for farmers if the farmers are enough expert to formulate a good mixture and if they can keep the biomass under composting in the proper conditions of temperature, moisture and gas exchange to facilitate the roles that the microorganism perform in the decomposition and maturation of organic matter. Moreover, farm residues could be recycled through composting or through direct incorporation into soil. Basically, the decomposition mechanism and dynamics are the same in both cases but the characteristics of the starting substrate and the process conditions are not controllable in case of crop residues incorporation.

1.2 Compost and Composting in Organic Farming

In organic agriculture fertilizers are permitted in organic forms, as defined by regulation (Luttikholt 2007; Huber et al. 2009). Mineralization of organic fertilizers is a biological decomposition that release plants' available nutrients; hence soil microbial communities are vital in the organic cropping systems. One of the most used organic fertilizers is compost, which plays a double role as both fertilizer and soil improver.

1.2.1 Composting: Microbial Transformation of Raw Materials

Compost and composting process definition have been extensively reported in the last decades of scientific literature (Feinstein and Morris 1975; Haug 1993; Kutzner 2001). Composting is “sensu stricto, a self-heating, aerobic solid phase biodegradative process of organic materials under controlled conditions, which distinguishes it from natural rotting or putrefaction” (Ryckeboer et al. 2003). Insam and de Bertoldi (2007) also defined the composting process as “a biodegradation process of a mixture of substrates carried out by a microbial community composed of various populations in aerobic condition in the solid state”. Composting transforms raw organic materials into biologically stable, humic-like substances, easier-to-handle materials. In this framework, we highlight in our own definition the functional role of the raw materials as microorganisms' feedstock and the importance of the ‘time factor’ in the never-ending process of organic matter maturation. Compost is a result of microbiological process under human control: the composting. It is a function of (i) the starting substrates physical, chemical and biological characteristics, (ii) the percentage of each raw material in the starting mixture, (iii) the process conditions in terms of water and oxygen range in the core of the mass under composting, and (iv) time (days of process after the starting mixture formulation). The biological decomposition of organic substrates and organic matter maturation are the two opposed paths which drive the process all through different phases distinguished by time and temperature values. A thermophilic period is representative of the biodegradation processes as a result of biologically produced heat (Haug 1993), while the maturation has been underlined through the organic matter evolution from the simple and the easy degradable forms in the raw materials to the complex and humic-like substances of the compost (Lhadi et al. 2006). The interactions and the dynamics among these biological forces both of degradation and stabilization result in a final product sufficiently stable for storage and soil application without adverse environmental effects.

During the composting process, microbial communities follow one another in a predictable succession pattern without breaks (Goyal 2005). Each community growth changes tremendously the composting mass's parameters (that are the environmental conditions for the subsequent microorganisms) in terms of temperature, organic substrate compositions, pH and bulk density. This process goes ahead until one microbial community is not any more able to survive in the new environmental conditions created by its own metabolism and catabolites. At that time one other community take place carrying on the composting process. As compost temperature increase, mesophilic zymogenous organism groups leave the place to thermophilic microorganisms, while late in the process, mesophilic microorganisms reemerge but at the same time they are basically autochthonous groups (Hermann and Shann 1997).

The composting process causes the loss of the more-degradable organic materials and the concentration of the less-degradable ones resulting in biomass enrichment of the humified materials. Composting is a process where humic substances are built up in a technological process that produces compost having humic-like materials. Humification of biomolecules provides environmental benefits through carbon (C) sequestration and reduction of CO_2 released through mineralization (Lhadi et al. 2006). Compost is usually used as a soil improver through the enrichment of stable organic matter; compost application into soil is invaluable measure for C sequestration in soil as it contains stable organic matter that helps accumulate soil organic C (SOC) in short term (Adani et al. 2009). It is worth to note that soil degradation by organic matter depletion together with land-use change are the major reasons for CO_2 emission from agricultural activities that contributed 33% of CO_2 increase among the other anthropogenic emissions (about 28% increase) (IPCC 2000, 2001).

1.2.2 A Plethora of Compost

Compost is one term that is used to describe millions of composts that are different in quality parameters, starting mixture and composting technological solutions. So far, it is possible to use either the term 'compost' or to address the term 'composting' at an aerobic maturation process of organic substrates in a plenty of cases and circumstances: (i) different raw materials available to start the process, (ii) their state of conservation; (iii) their pre-treatments e.g., shredding, size and age, (iv) their relative percentages in the starting mixture, (v) inclusion or exclusion of inocula, (vi) the structuring material (source, size and amount), (vii) the technologies used to mix, to aerate, to keep the right moisture content, (viii) the days of process, and (ix) the control procedures, the most important variable, which are able to affect in a relevant way the characteristics of the final products of any composting process. At the end, the product will be named compost in any case but its final characteristics (physical, chemical and biological) will vary in a vast range. Materials used for composting varies largely and consequently their microbial loads and their changes during the composting process, which leads to different final ready-to-use composts. Judging these final products through unique parameter might not give the best indication or characterization but a set of representative parameters should be used to characterize and categorize final compost. These parameters might include total organic C, pH, total nitrogen (N) and germination index (Campitelli and Ceppi 2008).

1.2.3 Regulatory Framework for Industrial Compost

At industrial facilities level, the compost product can go on the market as commercial 'compost' only under the observance of national law. The situation differs from country to another but it is possible to summarize it into two sets of parameters: (i) in case of heavy metals (Cu, Zn, Hg, Cd, Pb, Cr, Ni), pH, unfavorable bacteria (Enterobacteriaceae, Faecal Coliform, *Streptococcus* and *Salmonella*), Protozoa, eggs of Platyhelminthes, eggs of Nematodes and inertia (plastic, glass, stone), these parameters must be below specific thresholds fixed for each parameter; (ii) in case of organic matter content, humic and fulvic acids and organic N percentage over total N amount they must be above the fixed thresholds. Nevertheless, some countries regulations include parameters related to the process monitoring that have to be respected. Among the others, the most used parameter is the number of days over a threshold temperature (55, 60 or 65 °C) achieved in the composting mass.

According to EU regulations (EC 889/07; 834/08) in organic farming one can use every kind of compost excluding the sewage sludge based compost. Compost from municipal solid waste are permitted in any case, however, its quality increases in case compost was prepared from the organic separated collection of the urban wastes especially when the organic fraction is source-separated or comes from door-to-door collection (Richard and Woodbury 1992). National regulations or private standards can be more restrictive or specific in terms of compost definition and its use in organic farming. It is worth noting that the composting business is based on waste recycling much more than compost commercialization. The raw materials' rate in the starting mixture is decided according to the economical values of each waste that could be accepted as input to the facility (usually those are in order: sewage sludge, other sludge, municipal solid waste, other urban waste like public green pruning). The most profitable waste amount in the starting mixture is set to the maximum rate, which could warranty final compost acceptable under the law limits.

1.2.4 On-farm Composting: An Agroecological Practice

On contrary to commercially produced composts, the on-farm compost possibly presents higher organic C content and high level of humus-like substances compared to the mean values generally reported on commercial compost's labels. Unfortunately, this is accompanied with a slower mineralization rate of the on-farm compost in comparison with those of commercial products due to the high content of lingo-cellulosic residues, which are generally used in the on-farm composting preparation. At farm level any products resulted from the on-farm composting process could be used as soil amendment inside the same farm area without any law constrictions. This does not guarantee that compost produced by the farmer will have better characteristics than commercial compost available on the market. Specifically, in organic farms it is possible to assume that the raw materials available presents lower content of heavy metals, antibiotics and pesticides residues than the raw materials used in commercial composting plants. Again, this does not mean that the farmers' know-how and the composting facilities available in loco will be enough to prepare the appropriate mixture and to manage the composting process at the best.

1.3 Composting: Feed Microorganism to Produce Compost

The composting process includes four phases: (i) initial mesophilic phase (up to 42 °C), which lasts for few hours after the pile formation and it is strongly influenced by raw materials characteristics and storage conditions; (ii) thermophilic phase (45–68 °C) that might last several days to several weeks basically depending on the nature of the C compounds in the composted materials; (iii) second mesophilic phase takes place when mesophilic microorganisms recolonize the substrate and lasts a couple of months, and (iv) the maturation or curing phase, which might extend to several months and actually leads the never-ending maturation of organic matter (Insam and de Bertoldi 2007). The length of any composting phase depends on the type of feedstock and on the process efficiency, which is controlled by several factors such as moisture, aeration frequency and composting technology (Ryckeboer et al. 2003). Microbial communities present in the compost are the result of dynamic complex interactions between the microorganisms and their environment during each different composting phase. Starting from the first mesophilic phase through the whole process time, the self-selection process of composting microbial communities take place by the continuous increasing of the autochthones microorganism to the detriment of the exogenous one (Hermann and Shann 1997). Incessantly and in a very short lapse of time, compost material conditions change and the succession of bacterial communities follow it subsequently (Yamamoto et al. 2009). Zymogenous microorganism present in the raw materials rapidly decomposes soluble and easily degradable substrates, which results in production of organic acids that are responsible for decreasing pH to acidic values during the first days of the composting process (Beffa et al. 1996). Fungi and yeasts take advantage of this environmental conditions until the ammonification process increases the pH promoting bacterial metabolism. The mean generation time is shorter for bacteria than for fungi; it gives a great advantage to bacteria, which could better adapt to the rapidly changing environment than fungi. As a result, bacteria are responsible for most of the initial decomposition and, therefore, for the compost heat production (Ryckeboer et al. 2003). As compost temperature increases, thermophilic microorganisms take over and the process passes to the second phase. The temperature in the compost pile typically increases rapidly to 55–65 °C within 24–72 h of pile formation: it is the composting active phase. In the active 'thermophilic' phase, temperatures are enough high to kill pathogens, devitalize weed seeds and break down phytotoxic substances.

Temperature is the major selective factor for microbial populations among pH, moisture and C/N ratio (Rebollido et al. 2008). High temperature during the composting active phase of is a result of the microbial activity, which in the same time affects the composition of microbial communities (Fuchs 2010). The presence of very specific flora dominated by actinobacteria is important for compost hygienization through the production of microbial antibiotics. During this phase, oxygen must be replenished through passive or forced aeration, or turning the compost pile (Cooperband 2002). The high temperature in this stage accelerates the breakdown of proteins, fats, and complex carbohydrates, e.g., cellulose and hemicelluloses, but it is important that temperature should not exceed 70 °C as at this temperature only few species resists and thus mass microbial recolonization is retarded (Insam and de Bertoldi 2007). After most of the degradation has taken place the temperature decreases and the mesophilic phase starts. Mesophilic microorganisms reemerge in the process and take over the last stage, which is the maturation or curing stage of compost (Garcia-Prendes 2001).

As temperature gradually declines to around 40 °C the mesophilic microorganisms recolonize the pile and the composted materials enter the curing phase. The rate of oxygen consumption declines to the point where compost can be stockpiled without turning. During this phase, organic materials continue to decompose and are converted to biologically-stable humic-like substances. A long curing phase is needed if the compost is unfinished or immature; this can happen if the pile has received too little oxygen or too little/much moisture. There is no clearly defined time for the curing phase but common practices in commercial composting operations range from one to four months (Cooperband 2002). Maturation phase could be controlled by windrow turning and moisture check for obtaining stabilized organic matter in good conditions especially for the compost screened into smaller sizes.

Compost is considered finished when the raw feedstock is no longer actively decomposing and are biologically and chemically stable. When the temperature at the center of the pile returns to near-ambient levels and oxygen concentration in the middle of the pile remain greater than 10–15 % for several days, compost is considered stable or finished. These measurements should be taken when the compost pile has at least 50 % moisture content by weight (Cooperband 2002). In general, the composting process could be completed in a 3- to 4-month period. The literature provides different scientific works that discuss methods to determine compost stability or parameters to be considered to assess when composting process is finished. Compost stability assessment through CO_2 production or molecular oxygen demand indices are much more important in the industrial processes monitoring than in the on-farm composting as they indicate the minimum period to finish a process and to start a new one. During on-farm composting the temperature trend is the most practical way to decide when a composting process is finished, however, a germination assay should be carried out especially if compost will be used as a nursery substrate component.

1.3.1 Starting Mix C/N Ratio: The Compost Secret

Decomposing microorganisms: bacteria (including actinomycetes) and fungi are the main actors in the composting process, they require beside C and N other macro- and micronutrients for their growth. Carbon compounds provide energy for the metabolism and in most of the cases it is in excess while the N is the critical limiting element for microbial growth: if it is supplied in an insufficient amount the decomposition process will slow down, if it supplied in excess N will be volatilized as ammonia or leached as nitrate. It can be assumed that a mixture of plant materials should contain the essential level of P, K, S, and other trace elements, however, type and source of input feedstock affect the microbial community during the composting process according to their starting C/N ratio.

The aerobic metabolism of a generic decomposer microorganism growing on glucose substrate might have a net yield of 0.1–0.2 g of new microbial mass per each gram of substrate (Haug 1993). Based on this assumption, it has been estimated that the new microbial biomass synthesis uses 0.16–0.26 M NH_3 as N source for cell protein per 1 M glucose ($C_6H_{12}O_6$), which results in a C/N ratio range of 32–15 parts of C per each part of N. A similar calculation has been done by Ryckeboer et al. (2003) assuming a microbial yield coefficient of 30 % and an average microbial C/N content of 10, they suggested a theoretical optimum C/N ratio of 30 for a composting starting mix substrate while other works (Larsen and McCartney 2000; Tuomela et al. 2000) have confirmed the optimum C/N ratio in the range between 25 and 35.

The C/N ratio of compost starting mix is the leading parameter when setting up a new composting process. However, the C/N should not be used as absolute parameter as it is important to identify the nature of C in the composted materials. The complexity of the C compounds affects the rate at which organic wastes are broken down. The ease with which compounds degrade generally follows the order: car bohydrates>hemicelluloses>cellulose=chitin>lignin. Naturally, the decomposition rate of organic matter continues at very slow rate since there is a slow release of C. In contrast, fruit and vegetable wastes are easily degraded as they contain mostly simple carbohydrates (sugars and starches) while, leaves, stems, nutshells, bark and trees decompose more slowly as they contain cellulose, hemicelluloses and lignin (Epstein 1997). Only the biodegradable fraction of a substrate, in terms of volatile solids (VS), is available to decomposers and so it takes part to the process. The practical formula to calculate this fraction is: biodegradable fraction $= 0.83 - 0.028 \times$ lignin content of VS (Chandler et al. 1980); it means that 83 % of a substrate is the maximum percentage of degradable fraction of VS. However, this is true in case of lignin-free starting mix but there is no lignin-free plant residue. On the other hand, lignin plays a very important role in composting process being a precursor of the humic-like substances. The microbial degradation of the mass results in a precise trend of the C/N ratio during the composting process while the ash content per unit of dry mass under composting is a constant value for the whole process (Breitenbeck and Schellinger 2004), which means that the C/N decreases by the on-going process. Microbial respiration releases CO_2, which reduces the C content, but the C/N will decrease only if the diminishing of C is greater than that of N, which is possible only in case of negligible N rate in the leachate and of low ammonia volatilization.

1.3.2 Microorganisms' Environment in Composting

As discussed earlier, the composting process includes numerous microorganisms and it is the most important natural process of organic waste recycling. Compost microorganisms include useful microorganisms that perform the composting process and others that are potentially harmful for human, animal, plant and/or the environment. One of the benefits of the composting process is the "inactivation" of the harmful microorganisms and the development of beneficial microbial community. Thus, composters have to provide favorable conditions of oxygen, moisture and proper C/N ratio for microbial growth (Fuchs 2010).

Many factors affect the composting process, some of these factors have an important role in the process and others can influence its direction. The major factors affecting the composting process are oxygen, moisture and bulk density. Temperature must be monitored carefully because it is a function of the process wellness; its values drive the control procedures of the composting process. Once the composting process starts, the aerobic transformation takes place causing heat production with consequent oxygen consumption and water evaporation.

Daily temperature values of the process are the most important index of the composting status and it is the result of the microbial activity. It can be regulated within certain ranges only by indirect methods, decreasing temperature can be done through: (i) heat removing by turning, (ii) heat decreasing by wetting, while increasing temperature can be done through (a) oxygen supply by turning, (b) water supply by wetting. Other important factors that could limit the composting process are nutrients and pH. Nutrients especially C and N play an important role in the process as they are essential for microorganisms growth (Epstein 1997). These factors cannot be regulated during the process and can be managed only by an appropriate starting mixture.

Decomposition occurs mainly in the thin biofilm on the organic residues surface, which requires proper moisture content. Moisture is a key element affecting the composting process in a dynamic composting system where biological drying, metabolic water production and changes in compaction and porosity are all occurring over time (Richard et al. 2002). Compost mass humidity is measureable and can be regulated through watering.

The bulk density increases during the composting process due to decreasing particle size of the composting biomass as a result of decomposition. Bulk density affects the natural airflow from and to the composting pile, which in turn affects the heat dissipation form the pile's core to surrounding environment, which is important to determine the frequency of turning process or of the aeration flux. Bulking agents should be included in the starting mixture, e.g., different types of barks or wood chips. Larsen and McCartney (2000) reported 20–35 % of free air space in the composting mass as optimum range during the whole process. However, the effect of composting itself on the bulking agent could decrease its size, therefore, increasing mass compaction and limiting oxygen transfer.

It is worth noting that at the bottom of the pile, especially in the case of turning windrows, mostly used at farm scale, some niches (microenvironments) of anaerobiosis may occur even under optimal process management (Ryckeboer et al. 2003). Yamamoto et al. (2009) reported the evidence of Clostridia not only in the non-well aerated compost samples, as expected, but also in the well-managed process. *Clostridium* sp. converts organic compounds to sugar, acids and alcohol (Wiegel et al. 2006) playing an important role in compost maturation, i.e., anaerobic communities carry on the decomposition after that aerobic bacteria had consumed oxygen in the areas, or in the periods, when gas exchange is very slow. It has been estimated that 1 % of all the bacteria found in municipal solid waste compost are anaerobic (Atkinson et al. 1996). Anaerobic activities have been revealed by CH_4 release from composting pile (Fig. 1.1). The anaerobic microbial communities take parts in the decomposition process increasing its complexity and microbial biodiversity.

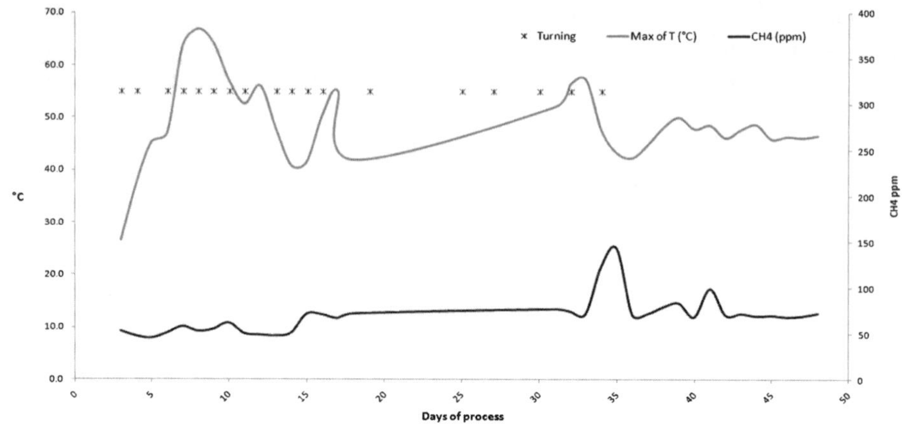

Fig. 1.1 Daily peaks of Temperature and Methane produced during the active phase of a turning windrow on-farm composting process

1.3.3 Keys for Successful Composting Processes

Controlling the composting process consists of two main operations: turning and wetting, which could be combined, too. The goal of the turning and wetting processes is not only to control the temperature of the composting process but also the biomass hygienization. The aerobic process takes place in the core of the pile while at the bottom of the heap the anaerobic process may start quickly after each turning with an area around the core part, like a cylinder inside the pile, that is cold and considered a buffer area. The turing operation exchanges cold air with the hot core area in a way that the cold area is actively involved in the process while the microbial activity in the core area, that is relatively cooler, slows down. During the turning process a tyndallization-like process (Gould 2006) may occur, which can destroy pathogens, spores, and weeds' seeds without affecting composting microbial communities. Hygienization during composting is not done only through high temperature but also through the chemical compounds that are capable of pathogen inactivation. However, hygienization might not be evenly effective throughout the compost pile, the outer zones is usually cooler and dried than the center of the pile preventing it from reaching the high temperature required for hygienization. The turning process is seriously important as it turns the outer parts into inner ones; mixing the outer zones with the hygiene materials favors beneficial microorganisms to outcompete the harmful ones (Fuchs 2010).

The composting temperature trend is also an indication of the process itself beside chemical and biological changes, as a matter of fact, the succession of microbial communities is affected by composting temperature. Furthermore, the maximum activity recorded for some enzymes such as cellulase, xylanase and protease were manifested mainly during the active phase of the composting process (Goyal et al. 2005).

1.4 Organic Farming: Feed the Soil to Feed the Plant

Intensive cropping systems resulted in decline of SOM and lead to loss of soil fertility. Large inputs of chemically synthesized fertilizers were used to sustain crop production year after year. Conventional agriculture, by land use intensification, decreased soil microbial activity and microbial functional diversity (Shen et al. 2008). Furthermore, the application of N-fertilizers with rates higher than crops need has altered the abundance and the diversity of soil microbial communities (Enwall et al. 2005; Chu et al. 2007). The negative consequences of the increased use of chemical fertilizers resulted in the implementation of governmental environmental policies that underline the role and importance of SOM in crop production (Rosenani et al. 2003). Organic farming is considered an alternative farming system that overcomes the environmental, economic, and ethical consequences of the industrialized high external inputs agriculture. The philosophy of organic farming is to feed/fertilize the soil rather than the plant. Synthetic fertilizers directly feed the plant and not the soil, which contradicts the principles of organic farming. Moreover fertilizers in mineral forms are not permitted in organic agriculture due to high solubility in water and potential risk of environmental pollution. Accordingly, only organic fertilizers can be used, which involves an active soil microflora that should be able to mineralize organic forms of nutrients into plant accessible forms, which is a priority in organic farming (Fließbach and Mäder 2000). However, the integration of bacteria with traditional inorganic fertilizers in the field may prove to be effective means to increase the solubility of inorganic phosphorous ions and other nutrients to plants (Kumar et al. 2009). Bacteria take systemic and simultaneous account of environmental aspects, quality of the produce and profitability of agriculture (Maene 2000). Substitution of chemicals with bacterial fertilizers and biopesticides, especially blending of chemical fertilizers with chemical adaptive bacteria is a promising approach to obtain sustainable fertility of the soil and plant growth (Kumar et al. 2011). However, bacteria in the form of biofertilizers with synthetic chemical fertilizers enhance significance crop yield in comparison to individual application of synthetic chemical fertilizer. Sustainable management of natural resources and progressive enhancement of soil quality, biodiversity and productivity are some of the steps to achieve the evergreen revolution (Keshvan and Swaminathan 2006).

Compost application—a core practice in organic farming—is used by farmers to preserve soil fertility for a long-term. Compost contains small portions of plant nutrients in inorganic forms that do not require mineralization to be solubilized in the soil-water solution. This fraction has a direct effect; its nutrients are easily available for plant absorption, while the organic forms of plant nutrient in compost provide the slow release effect for nutrients through the growing season, given that environmental and soil conditions are suitable for the gradual release of plant nutrients. It is important to consider a match between nutrient release from compost and plant demand for nutrients; mismatch might result in nutrient leaching (mainly nitrate in humid areas), toxicity or nutrient deficiency. Nevertheless, sole compost is not enough to cover the whole nutrients requirements on a short-term period to satisfy the farm production. Meeting crop demand for nutrients through organic matter supply does not guarantee sufficient supply for crop yield, as the decomposition of organic matter is a climate-dependent process. On the other hand, organic matter application in soil stimulates the biological activity, which makes the nutrients cycle less predictable (Kirchmann et al. 2008). Therefore, organic farming practices include rotation with legumes, green manuring, returning uncomposted agricultural wastes to soil, and crop residues incorporation in the fertilization schemes to provide, as well, soluble fractions of the essential nutrients. Incorporation of green manure crops into soil provides considerable amount of soluble nutrients into the soil system. Organic liquid fertilizers (bloods, extracts, algae and molasses), which provide easily available nutrients, are largely used to sustain plant needs during the growth. It is worth mentioning that synthetic mineral fertilizers are not fully water-soluble (Kirchmann et al. 2008).

1.4.1 Off Farm Input Reduction by Nutrient Recycling

Agricultural production systems loss nutrients through leaching, runoff, gaseous emissions, export of plant and animal materials out of the production system. On the long run, if these lost nutrients are not replaced or put back into the agricultural system, the system will be depleted (Kirchmann et al. 2008). Organic farming relies on practices that enhance nutrient and energy use and minimize environmental pollution, such as crop rotations and crop diversity, integration of livestock, symbiotic N fixation with legumes, application of organic manure and biological pest control. These practices aim to optimize the use of local on-farm resources; organic systems are inherently adapted to site-specific conditions (Müller-Lindenlauf 2009).

A balanced agroecosystem is able to minimize the use of off-farm inputs increasing its self-sufficiency, improving both its resilience and sustainability, which underline the importance of on-farm resources recycling (Altieri 2007). Organic agriculture claims in recycling principle: nutrients needed for plant growth should be sustained through internal input recycling, biological process as N fixation and functional crop rotations. Organic agriculture aims to minimize the use of off-farm inputs and to establishing as possible a closed farm nutrients cycle (Lampkin 1990). Composting practice is fundamental to help the farmers in the nutrients recycling, improving the resources use efficiency of the internal input matching both goals of high yield and minimum adverse impact on the environment.

Microorganisms are the driving forces for the organic residues recycling processes both during composting and decomposition in soil; they can speed up or slow down the mineralization process in a response to raw materials characteristics and controlled conditions, in the composting process. In soil, decomposing organism are mainly bacteria, fungi and larger organisms such as worms, bugs, nematodes, etc. In organic farming nutrients required to feed the crops are released through soil biological activities during the growing season as a function of mineralization process. Environment, climate and agricultural practices control the background conditions that affect soil microbial activities. Under specific conditions, this complex interaction could result in nutrients sequestration (robbing) or availability or leaching. The soluble fractions of nutrients are released by microorganisms in the soil-water solution resulting in easily available forms ready for plant uptake. Nutrient concentration in soil solution is a very dynamic process over time; it can be higher or lower than plant demand affecting plant growth in terms of quality and quantity. From an environmental perspective, higher concentration of plant nutrients in soil solution might result in nutrients leaching below the root zone leaving less-soluble nutrients forms in the soil surface layers and leading to groundwater contamination on the long run. Soil incorporation of unbalanced, in terms of C/N ratio, organic residues or amendments facilitates nutrients leaching (residues C/N much lower than soil C/N) or nutrients sequestration (residues C/N much higher than soil C/N) from soil pools. Removal of crop residues and appropriately timed application of soil amendment are the most used practices to reduce nutrients 'robbing' and/or leaching.

1.4.2 Crop Residues: Incorporated or Composted?

A remarkable question for farmers and composters is whether to incorporate crop/farm residues directly into soil or use them as a compost substrate. To answer this question many aspects should be considered: (i) raw materials availability in quantity, place and season; (ii) the easiness of the processes; (iii) the cost of the operations; and (iv) the potential benefits and constraints on the yield.

The total amount of crop residues produced globally was estimated in 2001 at $\sim 4 \times 10^9$ mg year^{-1} in which about 75% came from cereals (e.g., corn, rice, wheat, sorghum, barley, rye), those crop residues are mainly constituted of cellulose, hemicellulose and lignin (Barreveld 1989). Crop residues application into soil provides many benefits to the soil ecosystem such as enhanced soil physical (e.g., structure, plant-available water capacity), chemical (e.g., nutrient cycling, cation exchange capacity and soil pH) and biological properties (e.g., microbial biomass C and soil biotic diversity). Crop residues promote sustainable land as a result of its positive impact on the environment and ecosystem services (Lal 2005). Recycling of crop residues through leaving them on soil, incorporating them into soil and/or composts has been practiced in different agricultural systems. Benefits of crop residues recycling include soil protection against wind and water erosion, enrichment with organic matter, enhanced water retention and the principle benefits of nutrient recycling (Smil 1999).

Decomposition of crop residues depends mainly on their chemical and physical characteristic together with the surrounding environmental conditions. The quickness of the decomposition of a crop residue is based on the simplicity of the residues chemical composition. Simple compounds, e.g., sugars, amino acids and low molecular weight organic acids are quickly decomposed while complex polymers and high molecular weight compounds will take longer time to be decomposed (Berg and McClaugherty 2003). Usually, the decomposition of the crop residues in open field is faster at the beginning as the simple compounds decomposed first by large number of microorganisms (Duong et al. 2009), the microbial degradability of plant constituents is well documented in the literature (Insam and de Bertoldi 2007). The relative size of crop residue is an important issue in crop residues decomposition; crop residues that are broken through tillage processes will have smaller size than crop residues that is left un-disturbed on soil. The residues with smaller fraction will have larger surface area and consequently larger microbial activities, which makes the decomposition process faster (Vigil and Kissel 1995). Under field condition, it is important that crop residues get in intimate touch with the decomposing organisms and that the optimum environmental conditions are present in terms of moisture, dissolved oxygen required for decomposition, and balanced C/N of the residues.

1.4.3 C/N Ratio: The Fact of the Case

The C/N ratio has a great influence on the decomposition rate of organic materials in soil. A material with narrow C/N ratio e.g., fresh manure compared to material with a wider C/N e.g., olive pruning, would be much easier for microorganisms to decompose given that it sustain proper supply of C and N.

The benefit of crop residues incorporation depends on the degree of suitability of field conditions and residues characteristics. If soil is not provided with proper moisture content then decomposition of the crop residues will be slow. Moreover, the C/N of the crop residues might necessitate addition of external N to balance the ratio for proper decomposition and controversially increase the cost of the operation. It is potentially possible to avoid the cost of the external N input in the case where different crop residues are provided on the field in a planned manner in order to have a suitable C/N. This is highly recommendable in organic farming as it favors crop polyculture, crop diversity and agroeco systems complexation. Crop residues incorporation into soil might require less energy and machinery than composting process; plant residues or green manure will be incorporated or buried in subsoil all at once while the composting process require frequent turning during the active stages of the process.

However, the cost of the composting process varies depending on the composting method; if composting is to be carried out in a compost facility then cost will include transportation of the crop residues from the field to the composting facility while this cost will be ignored in on-farm composting. The cost in the composting process should include the cost of substrate transport, storage, preparation, e.g., grinding and mixing, turning and operational cost for soil application. A good composter might not need to add external N to adjust the C/N of the starting mixture but can use different types of residues mixed together to provide the proper starting C/N.

Crop residues applied directly to soil might help pathogens and weeds seeds continue their lifecycles and establish in soil while during the composting process high temperature, antibiotics released and antagonism between beneficial and harmful microorganisms provide a hygienized product that helps protecting the soil and crops against soil-borne diseases and weeds. An advantage of the composting process is the suppressive effect of compost on plant disease (Pascual et al. 2002; Bernal-Vicente et al. 2008). Crops growing with compost amendments seem to be less susceptible to disease than those grow without compost application (Suarez-Estrella et al. 2007).

Crop residues incorporation or their compost might have the same impact on the environment; they have to be applied in right timing and under the right conditions. If crop residues/compost is rich in soluble nutrients then in humid environment a risk of leaching these nutrients, especially nitrate, into the groundwater is feasibly possible. Also, if their C/N ratio were not in the optimum range then soil N might be depleted due to microbial growth, which might negatively affect crop growth.

1.5 The Challenge of Ecological Intensification

Food security and safety, environmental pollution, biodiversity loss, climate changes, water shortages, and non-renewable resources depletion are the global challenges for the future agriculture. As a response, inefficient use of land, water and nutrients resources must be terminated or minimized maintaining increased crop production (Foley et al. 2011). Agroecological intensification approaches are under investigation as suitable tools to face these challenges. In this framework, Agroecological practices might play a leading role as relevant part of the solution (Tscharntke et al. 2012). Organic agriculture has to improve and to renew its tools from both scientific and technological point of view to attain economic, social and environmental sustainability, increasing productivity using efficiently natural resource (Altieri 2002). The theoretical maximum yield is based on the crop's genotype and on structural environmental parameters such as available photosynthetically active radiation and temperature. Biotic and abiotic constraints such as nematodes, pest and disease pressure, weeds competition, nutrients deficiency, water stress, lack of pollination, soil quality and cropping history are the cause of obtaining lower yield than the theoretical one (Bennett et al. 2012). Although one sole factor may play a relevant role, it is more likely that combinations of different factors mark the yield decline. Conventional intensification aims to close this yield gap by conventional methods with known negative externalities on the surrounding ecosystems and the long-term depletion of the natural resources of the agroecosystem itself. On the other hand, ecological intensification aims to improve the ecological services without increasing the anthropogenic input. It could be achieved by two approaches, which are not mutually exclusive: service replacement and service enhancement (Bommarco et al. 2012). In the case of on-farm composting and internal nutrients and resources recycling, the replacement effect due to compost application is the reduction of other off-farm sources of nutrients while, the service enhancement is a replacement plus a boost effect of the soil nutrients supply service both increasing the yield and reducing the adverse environmental impact of the agroecosystem.

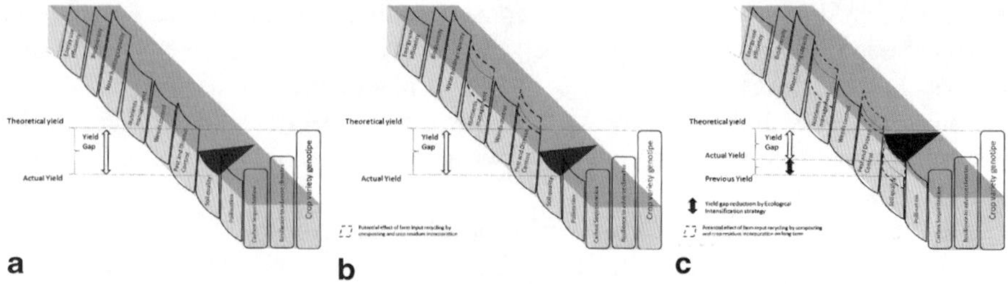

Fig. 1.2 Graphical representation of minimum's law applied to ecosystem services. In the example, 'soil quality' service is considered the limiting factor (**a**). The farm residues recycling and composting practices improve several ecosystem services (Kremen and Miles 2012). The case presented in (**b**) assume that those services are 'nutrients management' and 'pest control'. It results in a partial reduction of off-farm input needs (Ecological Services Replacement) however it doesn't increase the yield that it is limited always by 'soil quality'. On the medium/long-term the same practices ameliorate also soil quality (**c**) leading to yield increasing (Ecological Services Enhancement) by ecological intensification strategy

The Liebig's law (Liebig 1840) could be applied to the ecological services considering each service level a possible limiting factor for the yield (Fig. 1.2). Kremen and Miles (2012) presented a relationship between agroecological practices and the provisioning of ecosystem services showing that compost amendment and cover crops practices enhance, in terms of ecosystem services, biodiversity, soil quality, nutrient management, water holding capacity, pest and disease control, C sequestration and energy use efficiency. It is unsatisfactory to implement agroecological practices to intensify the crop production shrinking the environmental impact but also how appropriately these practices are used (Francis 2003). Effective application of the eco-efficiency concept requires an understanding of the production functions that relate agricultural outputs to the level of resource and other inputs (Dillon and Anderson 1990). Accordingly, the microbial metabolism of decomposer should be well understood to make the most of composting and farm residues recycling.

1.5.1 Feed Microorganism to Feed the Soil

Many microbial processes are essential for the long-term sustainability of agricultural systems (Wardle and Ghani 1995). Microbes grow on plant residues and utilize plant-derived C to build their biomass, then after cell death, part of microbial derived C is transformed into non-living SOM. The bacterial residues after cell death were recorded about 40 times as high as the living microbial biomass (Miltner et al. 2012). Assuming microbial C in soil is 2 % of the SOM, the 80 % of the SOC might have a microbial origin.

In the organic farming systems soil microorganism are: (i) taking benefit of the organic matter management based on compost amendment and crop residues incorporation; (ii) utilizing the available C more efficiently as indicated by a lower metabolic quotient qCO_2 (Fließbach and Mäder 2000); (iii) contributing to nutrients mineralization and temporary storage of potentially leachable elements (Araújo et al. 2008). By time, compost application decrease the required amount of mineral fertilizer needed to obtain a comparable yield. However, amendments and raw materials incorporation with wider C/N affect microbial growth as more C substrates has to be turned over to release enough N for microbial biosynthesis (Ros et al. 2006) which underlines that organic amendments including compost has to be balanced in terms of C/N ratio. The effectiveness of the conversion of organic amendment into microbial biomass is defined by C use efficiency which controls: (i) the conversion of organic residues C into microbial products; (ii) the rate of the agroecosystem C storage; and (iii) energy and material flow to plant biosynthesis among the higher trophic levels (Miltner et al. 2012). The C use efficiency is affected by environmental parameters such as temperature and moisture but it could be decreased by agricultural practices which inhibit microbial growth, like soil incorporation of inappropriate or unbalanced biomasses, or by nutrients limiting conditions (Sinsabaugh et al. 2013). However, microbial C use efficiency increases with the increase in organic forms of soil nutrients (Manzoni et al. 2008), which is explained by the effect of inorganic nutrient availability on decomposer metabolism. Further modeling studies showed a positive correlation between litter decomposer efficiency and soil N content (d'Annunzio et al. 2008). Moreover, C use efficiency changes during the decomposition process (Moorhead et al. 2012). The quality of C degradability could also affect the decomposer C use efficiency, which decreases with the degree of C substrate recalcitrance that is probably due to the lower enzymes biosynthesis requirements to degrade the simple soluble sugars than the more recalcitrant cellulose, hemicellulose, tannins, waxes, and lignin, progressively (Berg and McClaugherty 2003).

1.5.2 Compost in the Turn of the 3rd Millennium

Organic farming agroecological practices aim to preserve, over long-term, soil quality and improve fertility levels through sustainable SOM management, which might result from a synergic effect of livestock integration, on-farm nutrient production and recycling through cover crops and plant residues management and composting. Composting has the great potential to be a natural process that has certain regulatory parameters under human control. Noticeably, it is possible to supply sole compost, or compost and crop residues or compost and green manure at different rates. They could be used to achieve a target C/N ratio for the amendment before its soil incorporation. The C/N ratio goal of the endogenous organic matter to be incorporated should be calculated to have the highest C use efficiency for soil microbial communities. The threshold N ratio required for optimum microbial growth is equal to the C/N ratio of the biomass divided by the C use efficiency and then multiplied by the N assimilation efficiency (Bosatta and Staaf 1982; Frost et al. 2006), which has been evaluated as constant of 0.5 (Sinsabaugh et al. 2013). It means that an increase in the C use efficiency reduces the critical ratio of N for an optimal microbial growth. Cleveland and Liptzin (2007) elaborated the overall soil

C/N ratio of 14.3±0.5 and the overall soil microbial biomass C/N ratio of 8.6±0.3. Accordingly, it is possible to have estimation, from both ecological and biochemical viewpoints, of a C/N range for the biomass to be incorporated in soil feasible to reduce nutrients loss and to maximize microbial growth. This optimum C/N ratio should be higher than soil C/N and lower than 20–25, which has been empirically considered as the transition point in organic matter decomposition from N immobilization to mineralization (Berg and McClaugherty 2003). Values under soil C/N ratio might provide easily available nutrients without keeping SOC depletion. On the other hand, it must be considered that a fertility management strategy based on the organic matter implicates the supply of an amount of nutrients, over time, much higher than the crops demand. For instance, 10 t ha^{-1} of production can be obtained providing a supply of 100 unit of N ha^{-1} of easily soluble fertilizer applied in the appropriate way according to crop requirements. To provide a similar amount of N by compost, assuming 25 % of N mineralization occurs during the cropping period, a compost dosage of about 400 unit of N ha^{-1} has to be provided. It means that the N use efficiency (NUE=Yield/N supply) decreases from 100 kg of production per each kilogram of fertilizers to 25 kg of yield per each kilogram of compost (Dobermann 2005). Indeed, NUE should be evaluated in field condition to have precise values, especially in case of simultaneous incorporation of green manure and/or of residues of the previous crop, as herewith suggested. Moreover, on a medium term the residual effect of mineralization should be accounted to the successive years. Nonetheless, it seems to be in contrast with the ecological intensification assumption of higher resources efficiency, as reported above. A similar example could be done for other nutrients such as P and K (Goulding et al. 2008) and for the crop recovery efficiency of applied nutrients (Dobermann 2005). A reasonable alternative calculation might exclude from the efficiency index calculation the percentage of nutrients which have been provided from an on-farm origin, as follow: NUE=Yield/(N supply–N on-farm). Consequently, more suitable nutrients use efficiency indices must be considered and measured in field conditions to quantify, in terms of nutrients supply, the ecological services provided by on-farm composting and internal resources recycling (Cassman et al. 2002).

1.6 Conclusion

Crop residues recycling and composting has been deeply described to highlight the linkage among starting mixture, process factors and potential use and benefit for the soil, and so for crop production. These practices are fundamental to recycle resources at farm level, improve nutrients use efficiency and to decrease the off-farm input needs. In the organic farming system, a balanced combination of compost application and crop residues incorporation increases the microbial C use efficiency, which regulates the SOM decomposition and the nutrients mineralization resulting in better crops productions. Agroecological fertility management practices contribute both to yield increasing and to reduction of the adverse environmental impacts.

These practices have to be used appropriately under the ecological intensification framework.

References

Abawi GS, Widmer TL (2000) Impact of soil health management practices on soilborne pathogens, nematodes and root diseases of vegetable crops. Appl Soil Ecol 15:37–47
Adani F, Tambone F, Genevini P (2009) Effect of compost application rate on C degradation and retention in soils. Waste Manage 29:174–179
Albiach R, Canet R, Pomares F, Ingelmo F (2000) Microbial biomass content and enzymatic activities after the application of organic amendments to a horticultural soil. Bioresour Technol 75:43–48
Altieri MA (2002) Agroecology: the science of natural resource management for poor farmers in marginal environments. Agric Ecosyst Environ 93(1):1–24
Altieri MA (2007) Fatal harvest: old and new dimensions of the ecological tragedy of modern agriculture. In: Nemetz PN (ed) Sustainable resource management: reality or illuison? Edward Elgar Publishing Ltd., Cheltenham, pp 189–213
Araújo ASF, Santos VB, Monteiro RTR (2008) Responses of soil microbial biomass and activity for practices of organic and conventional farming systems in Piauí state, Brazil. Euro J Soil Biol 44(2):225–230
Atkinson CF, Jones DD, Gauthier JJ (1996) Putative anaerobic activity in aerated composts. J Indian Microbiol 16(3):182–188
Barreveld WH (1989) Rural use of lignocellulosic residues, vol. 75. Food and Agricultural Organization, Rome. ISBN 92–5-102792-7
Beffa T, Blanc M, Marilley L, Fisher JL, Lyon PF (1996) Taxonomic and metabolic microbial diversity during composting. In: De Bertoldi M, Sequi P, Lemmes B, Papi T (eds) The sciences of composting. Blackie Academic and Professional, Glasgow, pp 149–161
Bennett AJ, Bending GD, Chandler D, Hilton S, Mills P (2012) Meeting the demand for crop production: the challenge of yield decline in crops grown in short rotations. Biol Rev 87:52–71
Berg B, McClaugherty C (2003) Plant litter: decomposition, humus formation, carbon sequestration. Springer, Berlin, p 286. (ISBN 978–3-642-38820-0)
Bernal-Vicente A, Ros M, Tittarelli F, Intrigliolo F, Pascual JA (2008) Citrus compost as an organic substrate in growing media for cultivation of melon plants in greenhouse nurseries. Evaluation of nutriactive and biocontrol effects. Bioresour Technol 99:8722–8728
Bommarco R, Kleijn D, Potts SG (2012) Ecological intensification: harnessing ecosystem services for food security. Trend Ecol Evol 28(4):230–238
Bosatta E, Staaf H (1982) The control of N turn-over in forest litter. Oikos 39:143–151
Breitenbeck GA, Schellinger D (2004) Calculating the reduction in material mass and volume during composting. Compost Sci Util 12(4):365–371
Campitelli P, Ceppi S (2008) Chemical, physical and biological compost and vermicompost characterization: a chemometric study. Chemomet Intell Lab Syst 90:64–71
Cassman KG, Dobermann A, Walters DT (2002) Agroecosystems, nitrogen-use efficiency, and nitrogen management. AMBIO J Human Environ 31(2):132–140
Chandler JA, Jewell WJ, Gossett JM, Soest PJ Van, Robertson JB (1979) Predicting methane fermentation biodegradability. Biotechnol Bioeng Symp (16th edn.), 10:93–107
Chen Y, Inbar Y, Hadar Y (1992) Compost residues reduce peat and pesticide use. Biocycle J Compos Organ recycl 33:48–51
Chu H, Fujii T, Morimoto S, Lin X, Yagi K, Hu J, Zhang J (2007) Community structure of ammonia-oxidizing bacteria under long-term application of mineral fertilizer and organic manure in a sandy loam soil. Appl Environ Microbiol 73(2):485–491

Cleveland CC, Liptzin D (2007) C: N: P stoichiometry in soil: is there a "Redfield ratio" for the microbial biomass? Biogeochemistry 85(3):235–252
Cooperband L (2002) The art and science of composting: a resource for farmers and compost producers. Center for integrated agricultural systems. http://www.cias.wisc.edu/wpcontent/uploads/2008/07/artofcompost.pdf. Accessed 4 Aug 2013
d'Annunzio R, Zeller B, Nicolas M, Dhôte JF, Saint-André L (2008) Decomposition of European beech (*Fagus sylvatica*) litter: combining quality theory and ^{15}N labelling experiments. Soil Biol Biochem 40(2):322–333
Debosz K, Rasmussen PH, Pedersen AR (1999) Temporaral variations in microbial biomass C and cellulolytic enzyme activity in arable soils: effects of organic matter input. Appl Soil Ecol 13:209–218
Debosz K, Petersen SO, Kure LK, Ambus P (2002) Evaluating effects of sewage sludge and household compost on soil physical, chemical and microbiological properties. Appl Soil Ecol 19:237–248
Dillon JL, Anderson JR (1990) The analysis of response in crop and livestock production. Pergamon, New York
Dobermann AR (2005) Nitrogen use efficiency—state of the art. Agronomy and horticulture—faculty publications. http://digitalcommons.unl.edu/agronomyfacpub/316. Accessed 4 Aug 2013
Duong TTT, Baumann K, Marschner P (2009) Frequent addition of wheat straw residues to soil enhances carbon mineralization rate. Soil Biol Biochem 41(7):1475–1482
Enwall K, Philippot L, Hallin S (2005) Activity and composition of the denitrifying bacterial community respond differently to long-term fertilization. Appl Environ Microbiol 71(12):8335–8343
Epstein E (1997) The science of composting. CRC, Boca Raton
Feinstein MS, Morris ML (1975) Microbiology of municipal solid waste composting. Adv Appl Microbiol 19:113–151
Fernández JM, Plaza C, García-Gil JC, Polo A (2009) Biochemical properties and barley yield in a semiarid Mediterranean soil amended with two kinds of sewage sludge. Appl Soil Ecol 42:18–24
Fließbach A, Mäder P (2000) Microbial biomass and size-density fractions differ between soils of organic and conventional agricultural systems. Soil Biol Biochem 32:757–768
Foley JA, Ramankutty N, Brauman KA, Cassidy ES, Gerber JS, Johnston M, Mueller ND, O'Connell C, Ray DK, West PC, Balzer C, Bennett EM, Carpenter SR, Hill J, Monfreda C, Polasky S, Rockstro J, Sheehan J, Siebert S, Tilman D, Zaks PM (2011) Solutions for a cultivated planet. Nature 478:337–342
Francis C, Lieblein G, Gliessman S, Breland TA, Creamer N, Harwood R, Poincelot R (2003) Agroecology: the ecology of food systems. J Sust Agric 22(3):99–118
Frost PC, Benstead JP, Cross WF, Hillebrand H, Larson JH, Xenopoulos MA, Yoshida T (2006) Threshold elemental ratios of C and phosphorus in aquatic consumers. Ecol Lett 9(7):774–779
Fuchs JG (2010) Interactions between beneficial and harmful microorganisms: From the composting process to compost application. In: Insam H, Franke-Whittle I, Goberna, M (eds) Microbes at work. Springer, Berlin
Garcia-Prendes R (2001) Evaluation of dairy manure compost as a peat substitute in potting media for container grown plants. Ph. D. Thesis, University of Florida
Gould GW (2006) History of science–spores. J Appl Microbiol 101(3):507–513
Goulding K, Jarvis S, Whitmore A (2008) Optimizing nutrient management for farm systems. Philos Trans Royal Soc B Biol Sci 363(1491):667–680
Goyal S, Dhull SK, Kapoor KK (2005) Chemical and biological changes during composting of different organic wastes and assessment of compost maturity. Bioresour Technol 96:1584–1591
Haug RT (1993) The practical handbook of compost engineering. Lewis Publishers, Boca Raton
Hermann RF, Shann JF (1997) Microbial community changes during the composting of municipal solid waste. Microb Ecol 33:78–85
Höper H, Alabouvette C (1996) Importance of physical and chemical soil properties in the suppressiveness of soils to plant diseases. Euro J Soil Biol 32(1):41–58

Huber B, Schmid O, Kilcher L (2009) Standards and regulations. In: Willer H, Yussefi-Menzler M, Sorensen N (eds) The world of organic agriculture, statistics and emerging trends: IFOAM. Bonn, FiBL, Frick, ITC, Genf
Insam H, de Bertoldi M (2007) Microbiology of the composting process. In: Diaz LF, de Bertoldi M, Bidlingmaier W, Stentiford E (eds) compost science and technology. Waste management series 8. Elsevier, Amsterdam
International Panel on Climate Change (IPCC) (2000) Land use, land use change and forestry. A special report of the IPCC. Cambridge University Press, Cambridge
International Panel on Climate Change (IPCC) (2001) Climate change: the scientific basis. Cambridge University Press, Cambridge
Keshvan PC, Swaminathan MS (2006) From green revolution to evergreen revolution: pathways and terminologies. Curr Sci 91(2):145–146
Kirchmann H, Kätterer T, Bergström L (2008) Nutrient supply in organic agriculture—plant availability, sources and recycling. In: Kirchmann H, Bergström L (eds) Organic crop production—ambitions and limitations. Springer, Dordrecht, pp 89–116
Kremen C, Miles A (2012) Ecosystem services in biologically diversified versus conventional farming systems: benefits, externalities, and trade-offs. Ecol Soc 17(4):40
Kumar S, Pandey P, Maheshwari DK (2009) Reduction in dose of chemical fertilizers and growth enhancement of Sesame (*Sesamum indicum* L.) with application of rhizospheric competent *Pseudomonas aeruginosa* LES4. Euro J Soil Biol 45:334–340
Kumar S, Aeron A, Pandey P, Maheshwari DK (2011) Ecofriendly management of charcoal rot and Fusarium wilt disease in Sesame (*Sesamum indicum* L). In: Maheshwari DK (ed) Bacteria in agrobiology: crop ecosystem. Springer, Heidelberg, pp 387–406
Kutzner HJ (2001) Microbiology of composting. In: Rehm HJ, Reed G, Pühler A (eds) Biotechnology Vol. 11c: environmental processes III. Wiley-VCH, Weinheim
Lal R (2005) World crop residues production and implication of its use as a biofuel. Environ Int 31:575–586
Lampkin N (1990) Organic farming. Farming Press Books, Ipswich
Larsen KL, McCartney DM (2000) Effect of C: N ratio on microbial activity and N retention: bench-scale study using pulp and paper biosolids. Compost Sci Util 8(2):147–159
Lhadi EK, Tazi H, Aylaj M, Genevini PL, Adani F (2006) Organic matter evolution during co-composting of the organic fraction of municipal solid waste and poultry manure. Bioresour Technol 97:2117–2123
Liebig J (1840) Chemistry in its application to agriculture and physiology. Taylor and Walton, London
Luttikholt LW (2007) Principles of organic agriculture as formulated by the international federation of organic agriculture movements. NJAS-Wageningen J Life Sci 54(4):347–360
Maene LM (2000) The 10th world food prize congress, Washington DC, pp 169–171
Manzoni S, Jackson RB, Trofymow JA, Porporato A (2008) The global stoichiometry of litter N mineralization. Science 321(5889):684–686
Miltner A, Bombach P, Schmidt-Brücken B, Kästner M (2012) Soil organic matter genesis: microbial biomass as a significant source. Biogeochemistry 111(1–3):41–55
Moorhead DL, Lashermes G, Sinsabaugh RL (2012) A theoretical model of C-and N-acquiring exoenzyme activities, which balances microbial demands during decomposition. Soil Biol Biochem 53:133–141
Müller-Lindenlauf M (2009) Organic agriculture and carbon sequestration. Possibilities and constrains for the consideration of organic agriculture within carbon accounting systems. Natural Resources Management and Environment Department, Food and Agriculture Organization of the United Nations, Rome. ftp://ftp.fao.org/docrep/fao/012/ak998e/ak998e00.pdf. Accessed 4 Aug 2013
Oades JM (1984) Soil Oraganic matter and structural stability mechanisms and implications for management. Plant Soil 76:319–337
Pascual JA, Hernández MT, García C, Lerma S, Lynch JM (2002) Effectiveness of municipal solid waste compost and its humic fraction in suppressing *Pythium ultimum*. Microb Ecol 44:59–68

Perucci P, Dumontet S, Bufo SA, Mazzatura A, Casucci C (2000) Effects of organic amendment and herbicide treatment on soil microbial biomass. Biol Fertil Soils 32:17–23

Rebollido R, Martinez J, Aguilera Y, Melchor K, Koerner I, Stegmann R (2008) Microbial populations during composting process of organic fraction of municipal solid waste. Appl Ecol Environ Res 6(3):61–67

Richard TL, Woodbury PB (1992) The impact of separation on heavy metal contaminants in municipal solid waste composts. Biomass Bioenergy 3:195–211

Richard TL, Hamelers HVM (Bert), Veeken A, Silva T (2002) Moisture relationship in composting processes. Compost Sci Util 10:286–302

Rivero C, Chirenje T, Ma LQ, Martinez G (2004) Influence of compost on soil Organic matter quality under tropical conditions. Geoderma 123:355–361

Ros M, Klammer S, Knapp B, Aichnerger K, Insam H (2006) Long-term effects of compost amendment of soil on functional and structural diversity and microbial activity. Soil Use Manage 22:209–218

Rosenani AB, Mubarak AR, Zauyah S (2003) Recycling of crop residues for sustainable crop production in a maize-groundnut rotation system. IAEA Book: Management of Crop Residues for Sustainable Crop Production, pp 3–22. http://www-pub.iaea.org/MTCD/publications/PDF/te_1354_web.pdf. Accessed 4 Aug

Ryckeboer J, Mergaert J, Vaes K, Klammer S, De Clercq D, Coosemans J, Insam H, Swings J (2003) A survey of bacteria and fungi occurring during composting and self-heating processes. Ann Microbiol 53:349–410

Shen W, Lin X, Gao N, Zhang H, Yin R, Shi W, Duan Z (2008) Land use intensification affects soil microbial populations, functional diversity and related suppressiveness of cucumber Fusarium wilt in China's Yangtze River Delta. Plant Soil 306(1–2):117–127

Sinsabaugh RL, Manzoni S, Moorhead DL, Richter A (2013) Carbon use efficiency of microbial communities: stoichiometry, methodology and modeling. Ecol Lett 16(7):930–939

Smil V (1999) Crop residues: Agriculture's largest harvest crop residues incorporate more than half of the world's agricultural phytomass. Bioscience 49:299–308

Suarez-Estrella F, Vargas-Garcia C, Lopez MJ, Capel C, Moreno J (2007) Antagonsitic activity of bacteria and fungi from horticultural compost against *Fusarium oxysporum* f. sp. melonis. Crop Protect 26:46–53

Tscharntke T, Clough Y, Wanger TC, Jackson L, Motzke I, Perfecto I, Vandermeer J, Whitbread A (2012) Global food security, biodiversity conservation and the future of agricultural intensification. Biol Conserv 151:53–59

Tuomela M, Vikman M, Hatakka A, Itävaara M (2000) Biodegradation of lignin in a compost environment: a review. Bioresource Technol 72(2):169–183

Vigil MF, Kissel DE (1995) Rate of nitrogen mineralized from incorporated crop residues as influenced by temperature. Soil Sci Soc Am J 59(6):1636–1644

Wardle DA, Ghani AA (1995) A critique of the microbial metabolic quotient (qCO_2) as a bioindicator of disturbance and ecosystem development. Soil Biol Biochem 27(12):1601–1610

Wiegel J, Tanner R, Rainey FA (2006) An introduction to the family Clostridiaceae. In: Dworkin M, Falkow S (eds) The prokaryotes. Springer, US, 4:654–678

Yamamoto N, Otawa K, Nakai Y (2009) Bacterial communities developing during composting processes in animal manure treatment facilities. Asian Australian J Anim Sci 22:900–905

Chapter 2
Intensification of Aerobic Processing of the Organic Wastes into Compost

A. N. Ivankin, Urja Pandya and Meenu Saraf

Abstract The review is devoted to modern achievement in area of the aerobic processing of the organic wastes in compost for improvement of an ecological condition of the environment and purposeful use at intensification of the agriculture. The methods of reception and characteristic of products of microbiological transformation household, agricultural and industrial wastes are considered.

Keywords Aerobic · Bioprocess · Organic matter · Pathogens · Waste treatment · Kinetics · Animal manure

2.1 Introduction

As a result of ability to live of the person the significant amount of waste of different structure is formed. These wastes demand the subsequent destruction or reception from them the products, capable to find of the further application in various areas. The simplest way of destruction of waste is their burning. However in some cases these wastes have a liquid consistence and are not capable to ignition. During their burning such quantity of toxic substances which neutralization demands the expenses which are not compensated by the offered kind of processing can be allocated (Neklyudov and Ivankin 2003). A way of waste processing having popularity increase from years of composting. As a rule, it consists in natural biological decomposition of organic substance of waste by means of various kinds of microorganisms. The end-product of decomposition is humus substance which with success

A. N. Ivankin (✉)
Moscow State University of Forest, Mytishchi,
Moscow oblast, 141005 Russia
e-mail: aivankin@mgul.ac.ru

U. Pandya · M. Saraf
Department of Microbiology and Biotechnology,
University School of Sciences, Gujarat University,
Ahmedabad 380009, Gujarat, India

D. K. Maheshwari (ed.), *Composting for Sustainable Agriculture,* Sustainable Development and Biodiversity 3, DOI 10.1007/978-3-319-08004-8_2

can be used, first of all, as the means stimulating restoration soil properties and in the second—as organic fertilizer.

Many pathogenic microorganisms do not maintain a competition to other microorganisms collecting during composting. Besides warmly, allocated at reception compost, promotes final destruction pathogens. Inorganic waste, glass for example, metals and plastic, separate more often before biochemical processing organic waste, and then if necessary recycle (Weber et al. 2007; Bustamante et al. 2008; Kuba et al. 2008; Hort et al. 2009; Ben et al. 2010). It is known also, that now there are the special enterprises which are engaged both composting, and release of the equipment necessary for these purposes. Such manufactures as has shown an expert, are profitable enough, as compost, received at these enterprises as already it was marked earlier, is widely used at reclamation arable lands in rural, garden and park facilities (Neklyudov and Ivankin 2003).

Process of composting is known since deep times. Still ancient Egyptians have found out that agricultural crops considerably increased after animal manures in fields. They revealed that fresh manure renders essentially smaller influence on ground, and even can sometimes harm in comparison with the manure preliminary mixed with silt or the vegetative rests. In many cases the increase in efficiency of such mix needed long enough time, its maturing sometimes reached year and more (Lima et al. 2004; Zmora-Nahum et al. 2005; Smith 2009; Raj and Antil 2011; Gabhane et al. 2012). In the further it has appeared, that at correctly picked up parity of components reception of fertilizers for ground can be carried out for shorter terms (Neklyudov and Ivankin 2006).

However only after opening the fact of existence of microorganisms and creations of new scientific discipline—of microbiology became clear features of reception of similar fertilizers. It has appeared, that efficiency of a method is defined by participation of various microorganisms and entirely depends on their activity (Marr and Facey 1995; Neklyudov and Ivankin 2006). Process proceeded essentially more quickly increase in aeration of processed components and consequently it began to name aerobic decomposition of organic substances, or composting (Marr and Facey 1995; Smith 2009). With growth of animal industries, increase in an intensification of agricultural crops, development of cities and occurrence of the increasing quantity of waste composting became one of effective means of their processing and improvement of structure not only the ground, but also a condition of an environment (Miller 1996; Tognetti et al. 2011).

The purpose of the review—consideration of publications for last years, the organic waste devoted to aerobic decomposition for reception of composts, the analysis of methods of reception and properties received compost.

2.2 Raw-material Base of Compost Reception

As raw material for reception of composts any waste containing organic substances can practically serve: manure of agricultural animals and birds, dirtiness, silt, firm and liquid organic waste from sewage, city waste, a grass, leaves, hay, straw, waste

Table 2.1 Raw sources of reception of composts

Household organic wastes	Agricultural and a garden-kitchen garden organic wastes	Commercial and industrial organic wastes
Waste of fruit and vegetables	Leaves	Agro industrial waste of beer, tobacco, cotton
Shell of eggs and nuts	Scraps of a grass	Waste from industrial reservoirs (seaweed, water hyacinths), deposits from sewage and drainage systems
Waste of tea and coffee	Scraps of trees and bushes	Various kinds of city dust
Wood ashes, Flowers, garden plants, the vegetative rests, Toilet paper, packing paper, brown, paper and so forth	The fallen down fruit, Waste from the markets, The rests of agricultural crops (roots, stalks, shanks, Straw, Vegetative waste, from lawns and parks	Waste of paper and timber manufacture
Food waste Manure and laying for pets contents of toilets	Various kinds of manure of agricultural animals and birds	Waste from agricultures and farms
Various kinds of house dust	Corpses of the dead animals and birds	Waste of the food- processing industry

of a sugar beet, a cotton, fat, paper, wood and the brewing industries. This raw-material base can be presented in the form of a Table 2.1 (Benito et al. 2006). It is possible to add other kinds of waste which quantity on literary data annually increases in 2–3 times (Aviani et al. 2010; Hartley et al. 2010).

Waste contained up to 60–80 % of the most various organic substances, the main things from which are such high-molecular connections, as fibers, polysaccharides and lignine. If fibers rather easily are exposed biochemical decomposition and the subsequent mineralization cellulose and especially lignine are split essentially worse and at them composting there are the greatest difficulties (Marr and Facey 1995; Tuomela et al. 2000). Many of these waste have high microbic infection, promoting after activation of microorganisms occupying them to synthesis of corresponding enzymes which split and partially saline the organic connections containing in them necessary as for development on them corresponding microflora, and for biosynthesis of biologically active substances. Essential lack composting waste of all kinds is duration of the process composting, sometimes proceeding during many months. It is no wonder therefore, that the basic purpose of many publications—consideration of ways of acceleration of process composting on the basis of deeper understanding of process, identification of microorganisms participating in it, allocation and studying of the enzymes promoting splitting and a mineralization of organic connections.

2.3 Key Parameters of Aerobic Composting

The main key parameters of aerobic composting include such as temperature, a degree of aeration, moisture content and value pH in a punched mix.

2.3.1 Temperature

One of the major parameter provides efficiency composting. It essentially varies during composting due to thermal effect which appears as a result of oxidizing degradation of covalent bond at decomposed substances. It is known, that any process of composting has 3 basic temperature stages—mesophilic, thermophile and a stage of cooling, or final maturing compost. Duration of all of 3 stages depends on structure of a mix of the organic waste intended for composting, and a degree of aeration of this mix. Usually mesophilic the stage begins in the beginning composting or after some period necessary for microorganisms for their adaptation and growth (Walsh et al. 1991). The temperature mesophilic can vary phases from 40 up to 55 °C. In opinion of many researchers, at this stage there is the most active oxidizing degradation of organic waste (Mahimairaja et al. 1995; Tuomela et al. 2000). This temperature is characteristic and for thirds of stage composting—stages of cooling of a punched mix when it is observed final destruction organic connections. Thermophile the stage can proceed from 5 up to 25 day and more, the temperature at this stage composting can reach 70 °C and above. As a rule, at this stage there is an active destruction of pathogenic microorganisms and the highest losses of flying organic substances (Mahimairaja et al. 1995). Last stage can proceed within many weeks and even months and comes to the end, when the temperature of a punched mix will achieve an ambient temperature (Boldrin et al. 2010; Paradelo et al. 2011; Dilek and Yasemin 2013; Serramib et al. 2013).

2.3.2 Aeration

Aeration of a punched mix is closely connected with temperature. For example, with increase of temperature the output of harmful connections and a poisoning of the air environment increase. Such connections formed during composting, sulfur, odorous components are CH_4, methanol. Usually in this case methanol and 2-butanol is the cores admixture of air. However issue of odorous substances of type dimethyl sulfide amplifies at increase in the contents of sulfur-containing substances in raw material, thus high concentration of methane, ammonia and sulfurs in air usually testify to bad aeration during process composting (Skajaa and Hannibal 1991). There are three kinds of aeration of punched mixes: the compulsory aeration which is carried out by forcing of air in clamps, compost heaps or corresponding devices for aerobic process; the passive aeration which is carried out by a special arrangement horizontal clamps in a direction of a wind rose and a lining inside of punched clamps of special pipes with punching, providing passive receipt of air inside of a processed mix, and at last, a method composting with natural aeration, without use of any adaptations, but in view of a wind rose (Imbeah 1998; Aviani et al. 2010).

At compulsory aeration speed of receipt of air usually makes 0.20–1.33 l/min kg of dry flying substances for composting city dust and deposits of sewage and 0.87–

1.9 l/min kg of dry flying substances at processing mixes which structure includes manure. At passive aeration speed of receipt of air in compost heaps is insignificant and makes 0.04–0.08 l/min kg of dry flying substances (Bernal et al. 1993; Haines 1995). Passive aeration can essentially reduce the price of process in comparison with compulsory or active aeration. Passive aeration demands for the realization special aeration channels and appropriate calculation of speeds of the air streams created due to a difference in temperature between the punched mix and air. To establish the similar interrelation, modes of temperature have been studied and convection a stream of air in is punched mixes with 3 kinds loading: wooden shavings, hay and straw and also 3 levels of the contents of humidity: 60, 65 and 70 %. All experiments were spent in a passive and active mode of aeration. For experiments laboratory reactors in volume 105 l have been used.

2.3.3 Humidity

Passive humidity of 60–70 % had the temperature peak arising between 2 and 6 day of composting. After 6 day of composting the moisture content did not render any influence on a temperature mode of process because of loss of moisture in a mix. The interrelation between number Grasholff (GR—the relation of buoyancy to force of internal friction) and speed convection an air stream has been established. Generally speed convection an air stream varied from 1.5 up to 0.7 mg of dry air/kg s of dry material compost from 0 up to 20 day accordingly for all samples compost. This speed of a stream of air was provided aeration with the channels made inside compost of heaps for passive aeration. In comparison with straw where speed of an air stream fell below a level of number GR, wood shavings and hay have appeared are more effective as loading. At use of these substances speed of a stream of air increased constantly together with GR (Altieri and Esposito 2010; Sato et al. 2010; Yu et al. 2011; Pane et al. 2013). On the other hand, advantage of compulsory aeration is higher speed composting and an opportunity of partial regulation of temperature of process; lack—significant losses of flying organic substances at carrying out composting (Brinton and Brinton 1994).

Influence of compulsory aeration on a degree of formation of ammonia was studied at composting waste of housekeeping on pilot installation. It is shown, that significant clearing NH_3 occurred at the first stages of degradation of fibers and amplified with increase in speed of aeration during this period. Good correlation between change of temperature, clearing CO_2 and allocation NH_3 was observed (Atkinson et al. 1996). In this connection separate researchers recommend to spend composting at passive or natural aeration which though proceeds with slower speed, but provides fuller maturing compost and does not demand high monetary expenses (Hemmat et al. 2010; Jolanun and Towprayoon 2010; Bustamante et al. 2011; Turrión et al. 2012). Parallel experiments on influence of various kinds of aeration on quality compost have shown, that any processes of artificial aeration reduce time composting, however on occassion, can worsen quality compost. It is most effective, in comparison with other kinds, aeration of section type lengthways horizontally laid composting ledge by means of the compressor if speed of aeration is supervised in conformity with speed of process of decomposition of an organic material. Besides the gases departing at controllable section aeration compost ledge, contain smaller quantity bed smelling and flying substances.

Various methods aerobic composting are discussed in works (Mahimairaja et al. 1995; Bustamante et al. 2011). From the considered methods it becomes clear, that the most suitable are methods of physical and microbiological processing in aerobic thermophile conditions. The basic problem at such kinds of processing—foaming during process. Methods which can solve this problem are offered. Aerobic thermophile composting is a simple and inexpensive method which can resolve problem decontamination on farms where the given way is constantly used for waste disposal. The content of water in punched mixes has essential value for reception high-quality of compost. Usually process proceeds effectively enough at the maintenance of water in a punched mix of 35–65 % (Walsh et al. 1991). To support these values especially important at compulsory aeration when water leaves together with departing air (Doublet et al. 2010; Yu et al. 2011; Cellier et al. 2012; Mostafid et al. 2012; Angin et al. 2013). For understanding of interrelation between humidity, temperature, and microbial activity have been carried out researches at composting mixes of firm oozy deposits from city sewage. Microbial activity was defined, how speed of consumption of oxygen, mg/kg s. Moisture load was the dominating factor influencing microbial activity of the punched mix. Humidity of 50 %, apparently, is that minimal limit below which it should not fall. In this case activity of microorganisms above, than 1.0 mg/kg s. The temperature also was the important factor at composting mixes of waste. However it rendered smaller influence, than moisture content. In particular, increase of activity of the microorganisms, caused by increase in temperature; it was possible to reach one increase in humidity.

2.3.4 pH Value

Value pH in a mix—one of parameters of efficiency of process composting. Usually sizes pH vary from sub-acidic up to neutral and alkalescent values due to formation of ammonium, i.e., in an interval pH 4.5–8.1. As a rule, these values are closely connected with activity of microorganisms which participate during composting (Bustamante et al. 2011; Carbonell et al. 2011; Mostafid et al. 2012). Composting in a mode pH-controlled it has been shown on various kinds of city dust. In pH-controlled process to compost added lime to neutralize pathogenic microorganisms and to prevent downturn pH below 7.0, in particular at early stages composting. Speed of decomposition of organic materials was above in pH-controlled a reactor in comparison with a reactor without the control pH. The quantity of losses of nitrogen in pH-controlled rose, but the degree of transformation of nitrogenous components remained low. Also pH-dependence of the micro-sorbents participating in composting components. Optimum value pH for activity of the microorganisms

Table 2.2 Nutrient contents of animal manure composition

Name of Animals	Nutrient contents (%)			
	H_2O	Organic substances	Nitrogen	P_20
Horned Livestocks	81	16	0.4	0.2
Hens	57	29	1.5	1.3
Ducks	54	25	1.0	1.4
Horses	73	22	0.5	0.4
Pigs	78	17	0.5	0.4
Rabbits	75	23	1.1	1.2
Sheeps	64	31	0.7	0.4

participating in reception compost, in particular in degradation of fibers, was value pH 7.0–8.0 whereas the greatest speed of degradation of the carbohydrates, observed at early stages of maturing compost, happened at pH 6.0–9.0 (Glenn 1990).

Change of time composting for processing house waste due to initiation of a phase of low value pH at the control mesophilic temperatures has been investigated at an initial stage composting house waste until value pH did not reach some constant value. For this purpose pH value was measured in a condensate received after cooling compost of gas. The assumption has been put forward, that activity of microorganisms at low value pH suppresses with fast growth of temperature. The sharp increase in value pH due to decomposition of fat acids, apparently, is a good marker of the termination of the temperature control (Ohins 1994).

2.4 Types of Various Substrates for Compost and their Importance

Composting with immense importance in agriculture is concern under uses of various substrates. These substrates are following:

2.4.1 Composting Manure of Agricultural Animals

As already it was marked earlier, reception of fertilizers from manure of agricultural animals is known from an extreme antiquity. Some concepts about structure of manure of agricultural animals and birds can be gathered in Table 2.2 (Walsh et al. 1991; Neklyudov and Ivankin 2003), and also from works of other authors (Logsdon 1993). Apparently from Table 2.2, manure contains a lot of necessary nutrients for plants: significant amounts of carbon and nitrogen, and 80–85 % are necessary on organic connections of these substances (Walsh et al. 1991; Logsdon 1993) and also significant amounts of phosphorus. Besides phosphorus (Table 2.2), manure contains significant amounts of potassium and magnesium and practically all the microcells necessary for growth and development of plants (Logsdon 1993).

Apparently from Table 2.2, the relation of carbon to nitrogen which is the most important parameter of quality compost, has values 19–25 for manure of various kinds of poultry and rabbits, for pigs and ruminants higher parameters are characteristic: 34–44, i.e., the smaller contents of amount of nitrogen. However these parameters below parameters of relation C:N at other kinds of organic waste which can achieve the values 50–60 and even 200–300 at lignine cellulose materials.

The amount of organic substances (OS), containing in manure of a horse, the cow, a pig, a sheep, a goat, the rabbit, chickens and an ostrich, the general contents of carbon (CC), total of nitrogen (CN), ratio C_{CC}:N_{CN}, amount of soluble organic carbon (SOC), amount of organic nitrogen (AON), a ratio of carbohydrates to AON, carbon of ulmification acids (CUA), carbon fulvo acids (CFA), an index of ulmification ($C_{CUA}/C_{CC}100$), relation of C_{CUA}/C_{CFA} and a ratio of nitrogen NH_4^+/NO_3^- is certain (Logsdon 1993). An unduly high content of water in manure in some cases prevents effective composting firm dry substances (Lopez-Aranda et al. 2002; Guanzon and Holmer 2003). However, on occassion, this lack can be eliminated, adding to manure various loading (Guanzon and Holmer 2003), that promotes allocation of hotbed gases (Wallace et al. 1992). Practical examples similar composting are resulted below.

Allocation of hotbed gas (CO_2, CH_4, N_2O) in containing manure a laying for pigs (a storage time 113 day during a winter season) defined quantitatively, using an awning with which covered all clamp during all period of composting. Liberation of gas paid off in the form of equivalent CO_2/g of dry substance. In addition the defined time of keeping (using gas SF_6) and concentration of gases in various parts of clamp. Average time of keeping of gases in clamp was less than 2 h. Formation and allocation of methane was observed only in the central part of clamp whereas accumulation CO_2 occurred on all volume. The highest allocation of CH_4, CO_2 and N_2O was observed in the beginning of storage manure lying in a clamp when the temperature rose and by that conditions for development thermophile microorganisms were created. It is possible to make the conclusion of these data, that thermophile microorganisms first of all influence formation of hotbed gas. It has been established, that by the most important gas determining global warming in environment, is N_2O (Wallace et al. 1992). In this connection to punched manure as already it was marked earlier, began to add loading, providing partial adsorption of allocated gases and increase in process of aeration, for example straw, sawdust, shavings, household dust and even deposits of sewage (Imbeah 1998; Walsh et al. 1991).

Intensive allocation NH_3 during composting pork manure in compost heaps without compulsory aeration was observed at an initial stage when there was a heating a punched mix. The sharp increase in amount of CH_4 happened immediately after pork manure has been collected in compost heaps, but active temperature of compost started to decrease. Speeds of allocation of each gas in compos heaps small and great volume were 112.8 and 127.4 g nitrogen of NH_3/kg of the general contents of nitrogen, 37.2 and 46.5 g nitrogen of N_2O/kg of the general contents of nitrogen and 1.0 and 1.9 g CH_4/kg of organic substances accordingly. It is shown, that the sizes of compost heaps are one of the major factors influencing for speed of allocation of gases (Maeda et al. 2011). Composting manure and waste of cattle-breeding farms.

In clamp or compost heaps, volume 0.25 m^3 and height 1 m with natural aeration, it was possible to receive compost with satisfactory properties during 10–12 week carrying out of process at the contents of a dry material in the beginning of composting 17–21 % (Wetterauer and Killorn 1996). It is offered to punch also manure of cattle together with waste of a house and domestic economy as loading compost at their ratio with compost 1:1. At such method of reception compost it has appeared it is effective for restoration of the disturbed structure of the ground which is taking place at a crop rotation of agricultural crops (Guerra-Rodrigues et al. 2001).

Composting dry manure of large horned livestock with fallen down leaves of trees it was spent in a reactor for composting which have been executed from polyurethane by thickness of 3 mm. The volume of a reactor was 108.8 l. The mass ratio of components was 3:1. Composting proceeded during 103 day. The thermophile stage of composting came through 14 day and had extent during 10 day then during 20 day the temperature fell up to 20 °C. Final value pH of compost is equal 8.89. The ratio C:N in the beginning of composting was 12.86, in the end composting 13.27. Indexes of germination of seeds were the following: for ryegras—55.35, wheats—56.56, an oats of 100 % (Schmitt et al. 1996). Considering, that pig-breeding farm last two decades especially effectively develops in many countries, processing of manure of these animals becomes not only actual, but also a necessary problem (Imbeah 1998; Antil et al. 2013). Composting manure of pigs application-dependent with its necessity preliminary of separation as initial fractions contain up to 97 % of water and only 2–3 % of firm substances (Walsh et al. 1991) and the subsequent neutralization of liquid fraction (Imbeah 1998) which can be added to a punched mix of firm waste for maintenance in them of a necessary water level (Yu et al. 2011; Bernal et al. 1993; Haines 1995; Antil et al. 2013).

In clamp or compost heaps, volume 0.25 m^3 and height 1 m with natural aeration, it was possible to receive compost with satisfactory properties during 10–12 week carrying out of process at the contents of a dry material in the beginning of composting 17–21 % (Wetterauer and Killorn 1996). It is offered to punch also manure of cattle together with waste of a house and domestic economy as loading compost at their ratio with compost 1:1. At such method of reception compost it has appeared it is effective for restoration of the disturbed structure of the ground which is taking place at a crop rotation of agricultural crops (Guerra-Rodrigues et al. 2001).

Composting dry manure of large horned livestock with fallen down leaves of trees it was spent in a reactor for composting which have been executed from polyurethane by thickness of 3 mm. The volume of a reactor was 108.8 l. The mass ratio of components was 3:1. Composting proceeded during 103 day. The thermophile stage of composting came through 14 day and had extent during 10 day then during 20 day the temperature fell up to 20 °C. Final value pH of compost is equal 8.89. The ratio C:N in the beginning of composting was 12.86, in the end composting 13.27. Indexes of germination of seeds were the following: for ryegras—55.35, wheats—56.56, an oats of 100 % (Schmitt et al. 1996). Considering, that pig-breeding farm last two decades especially effectively develops in many countries, processing of manure of these animals becomes not only actual, but also a necessary problem (Imbeah 1998; Antil et al. 2013). Composting manure of pigs application-dependent with its necessity preliminary of separation as initial fractions contain up to 97 % of water and only 2–3 % of firm substances (Walsh et al. 1991) and the subsequent neutralization of liquid fraction (Imbeah 1998) which can be added to a punched mix of firm waste for maintenance in them of a necessary water level (Yu et al. 2011; Bernal et al. 1993; Haines 1995; Antil et al. 2013).

Composting pork manure with sawdust it is reached at mixing pork manure and sawdust in compost heaps with usual and compulsory aeration. It is revealed, that time of maturing compost is identical to usual and compulsory aeration (60 day), but quality compost at compulsory aeration above. In this case there is also essentially greater destruction of pathogenic microorganisms, such as *Salmonella* sp. (Balakrishnan and Batra 2011). Composting one pork manure and other waste of animal industries with waste of wheat and food waste it was spent in an isothermal reactor in volume 10 m^3. The temperature increased during processing waste up to 70 °C and above, and could remain such during 19 day even if process spent in winter conditions when the temperature of air was below 0 °C. The amount which has formed during processing of thermal energy was more, than necessary amount of the electric energy which is required for heating. Hygienic parameters of the formed product completely corresponded to sanitary requirements. Process proceeded well except for small losses of nitrogen during aerobic processing. The end-product can be used as fertilizer and for improvement of structure and quality of ground. The best results turned out at aerobic processing a mix of waste of animal industries, waste of wheat and food waste (Nandy et al. 2002).

It is investigated also composting the pork manure mixed with rice straw to determine characteristics of 3 kind's aeration systems: compulsory aeration, passive aeration and natural aeration. Results have shown that warming up of a mix during composting is enough to satisfy requirements of standards of sanitary and to provide maturing pork manure. Indexes of composting, including physical changes, such as value pH, size of the general organic carbon, the contents of organic substances in compost, size of the common nitrogen, moisture content, ratio C:N in firm particles, ratio C:N in a water-soluble part, the contents of NH_4^+ nitrogen, the contents of anion nitrogen ($NO_3^- + NO_2^-$) vary slightly at all 3 ways of aeration, except for a temperature structure. The economic analysis has shown, that passive aeration is most convenient for small pig-breeding farms, compulsory aeration should be spent on pig-breeding farm average and big productivity with a high degree of industrialization. For avoidance of rise in temperature during composting up to high values, it is necessary to spend the control of compulsory aeration over process (Atkinson et al. 1996). Instead of pork manure with success can be used bird's excrement which as it is visible from Table 2.2, possesses essentially higher parameters.

2.4.2 Composting of Vegetative Waste

The chemical compound and structure of the various vegetative wastes containing in the structure lignine cellulose materials, and also the microorganisms participating in their transformation, are in details stated in recently left monographies

(Rabinovich et al. 2001). However during composting in some cases action of these microorganisms and the enzymes produced by last is essentially distorted due to various factors which usually do not manage to be considered at drawing up of a punched mix. Therefore those laws which are observed during composting though partially and coincide with the data published in monographs (Bolobova et al. 2002), each time demand additional specification. As is known, lignine cellulose the waste containing cellulose, hemicellulose and lignine, are the mixed sources of carbon and the nitrogen, necessary for composting. However in the majority of them the contents of nitrogen is not enough for reception effective compost and consequently at them composting to them add manure, urea, waste of the meat, dairy, fish, brewing industry and other sources of carbon (Imbeah 1998). One of widely widespread kinds are waste of the paper and timber industry. So, for example, manufacture 10 t can bring papers, on one data, in occurrence 10–20 t waste, including lignine, sewage, scraps of the paper and a pulp (Ekinci et al. 2006), on other data, these waste make more than 20% for a small paper-mill (Arner 1995). Such big divergence in data speaks first of all that in the second case waste only one paper whereas in the first case practically all waste of paper manufacture is considered. Not the smaller amount of waste is formed and as a result processing of wood and preparation of it various products. The part of these waste, especially in carpentry the industries, goes on manufacturing of furniture whereas the most part at the best is burnt, and in the worst sends on dumps, being one of many contaminant environments.

2.4.3 Composting of Paper Waste

One of the most simple and fast methods of processing of waste of a paper is their mixing with manure of cattle as an additional source of nitrogen and microorganisms. Depending on a ratio of paper waste and manure can be received compost with a various ratio of carbon and nitrogen (Buchanan and Fulford 1992). Instead of manure additional sources of nitrogen, for example urea or waste of the food-processing industry can be added (Bannick and Joergensen 1993; Oshins and Fiorina 1993; Sun et al. 2011).

2.4.4 Decomposition of Lignin Cellulose Waste

Effective decomposition paper and others lignine cellulose waste during composting is closely connected and with decomposition of lignine. Degradation of lignine under action basidiomycets—white decay is intensively investigated last years (Rabinovich et al. 2001; Bolobova et al. 2002; Zwart 1990). As is known, the organic substance of similar waste turns in CO_2, humus and heat at composting at presence of microorganisms. It is come out with the assumption, that humus it is formed mainly from lignine. Thus, lignine is not completely salinity during composting a component. The increased temperature observed during thermophile of a phase, is important for splitting lignine and cellulose. The complex of the organic components similar lignine, is split mainly under action thermophile mushroom cultures and actinomycete. The optimum temperature for thermophile mushroom cultures 40–50 °C also is optimum for splitting lignine in compost (Tuomela et al. 2000). Challenge is composting wood waste which, as is known, contain 22–27 % of lignine (Rabinovich et al. 2001; Bolobova et al. 2002). Reception of composts from waste of wood can be carried out even in regions of the Far North at ambient temperature −30 to −40 °C as during composting the temperature inside compost achieves 60 °C and above. Duration composting 90–27 °C day, but compost can be used as fertilizer and after 3 month of composting. Use of such compost essentially increased productivity of a potato, and tubers had low the contents of nitrates (Bannick and Joergensen 1993).

Especially successful compost it is possible to receive at biodegradation of shavings, sawdust and so forth after addition to them of superphosphate, nitrate of ammonium, urea, sulfate potassium, and also salts of microcells: copper, cobalt, molybdenum and a pine forest. In this case composting it is possible to spend at 23 °C under a layer of the punched polythene. As a result occurs the bio destruction on 50–53 % readily available carbohydrates and on 65 % remote carbohydrates. Time composting made 90 day at periodic humidifying compost. The deduction of compost nutrients and nitrogen took place also. Final ratio C:N varied within the limits of 20–35 depending on structure of punched mixes (Mondini et al. 1996). Close results receive and after 4-month's composting hydrolyzates of the wood cellulose containing lignine (McEvoy 1993).

Influence of relation C:N and humidity on efficiency of composting was investigated on an example of poplar wood the crushed wood (the size of particles of 1,7 mm) was punched on pilot installation. Ratio C:N were: 10:1, 30:1 and 50:1 at use of urea as a source of nitrogen. For each mix the humidity (30, 50 and 70 %) got out. Ratio C:N equal 50:1 or 30:1, and humidity of 70 % were most effective for composting. As the indicator of a level of a mineralization of wood from a poplar used amount allocated CO_2 (Bernal et al. 1993). The bulked up chips remaining after cutting down of trees, punched in a mix with various organic and inorganic connections in an aerobic reactor for composting during 5 months and then in clamps too within 5 months. As organic substances used bird's excrement, urea, connections of nitrogen with $Ca(OH)_2$, and also compost, received by fast processing of food waste and dust during 24 h in bacterial fermenter. As inorganic components ashes of a various origin has been used. On the basis of chemical properties received compost the conclusion is made, that it corresponds to necessary requirements (Andersen et al. 2011). Examples of hydrolysis of waste of wood materials are described in works of other researchers (Atkinson et al. 1996; Bolobova et al. 2002; Sellami et al. 2008).

2.4.5 Composting the Hydrolytic Lignine

Lignine on the structure it is close to structure of humus (Tuomela et al. 2000) and consequently represents especially interesting object for composting. It is known, that hydrolytic lignine—a product of acid processing of wood at reception of a paper, consists from actually lignine (88%), the rests poly- and mono- carbohydrates, organic acids, pitches, wax, nitrogenous substances and cindery elements (Tuomela et al. 2000; Rabinovich et al. 2001). At composting such lignine under influence of association of microorganisms there is a change of molecule-mass structure, the acid-cores of properties, molecular structures. So, by the end of the third month composting the increase the contents phenolic OH^- and $COOH^-$ groups, and also amount of negative charges on macromolecules was observed. Cultivation of barley with entering into ground composting lignine as fertilizer has shown, that these changes increased biological activity hydrolytic lignine (Kanotra and Mathur 1994). The ulmification substances of compost from lignine and their influence on properties of podsolic ground were investigated on an example of compost from manure with a lignine material and a lignine hydrolysate. Spectral and other data have shown, that hydrolized lignine as from compost, and received to straight lines by, differs on the structure from ulmification acids, and in a greater degree for the connections containing carbon, than for the connections containing nitrogen. However, introduction in ground ulmification substances renders positive influence on the status ulmification components in ground, it cationic capacity and buffer ability (Neklyudov and Ivankin 2006; Holikulov and Ortikov 2007).

2.5 Laws of Composting

Unfortunately, among the big abundance of works on various composting organic wastes seldom there are the works, devoted to studying of kinetic laws composting such waste. In this connection special interest represents one their works in which spent studying kinetic of composting mixes of some kinds of agricultural waste: a grass (54%), tomatoes (10,6%), pepper (20%) and eggplants (15,4%). For carrying out of reaction vertical reactors with compulsory aeration and one usual a reactor of convection type have been used. Allocation CO_2 and change of temperature recorded speed in 3 various places of reactors. Every day supervised moisture content, pH and speed of decomposition of organic substances in a punched mix. Kinetic models of process have been investigated some. According to the received data the optimal model which mathematical description appeared as follows has been chosen:

$$\boldsymbol{K_T} = \frac{\boldsymbol{a}}{\dfrac{\boldsymbol{M_c}}{\boldsymbol{T}-(\boldsymbol{C}.\boldsymbol{b})}} \mathbf{exp}\left[(\boldsymbol{Tc}) - \left(\boldsymbol{d}\frac{\boldsymbol{M_c}}{\boldsymbol{T}}\right)\right]$$

Where, K_T—speed of decomposition (total of flying substances, g/g total of flying substances a day); T—Temperature of process, °C; M_C—Daily moisture content in a mix, %; C—daily speed of allocation CO_2 during composting an organic mix in a reactor for composting, %; a, b, c and d—constants of reaction. According to the received results the highest speed of splitting of organic substances and optimum value of temperature composting were observed at speed of aeration 0.41 l/min kg of organic substances (Thambirajah et al. 1995).

2.6 Case Study of Compost with Different Substrates

Food processed materials and sewage waste substrate provide significant approach in composting process. Both case studies are following:

2.6.1 Food Processed Materials

After extraction of sugar and a filtration of treacle as a withdrawal there is a sugar flat cake and an oil cake which can be substrata for composting. Aerobic composting this flat cake during 6 week at its shovelling through everyone 4 day for providing of access of air, led to reception compost. In the beginning of fermentation the relation C:N made 30.6. After 6-week fermentation it became equal 17.0–18.8. For composting, it was not required any other additional substances and materials (Chao et al. 1996; Entry et al. 1996). With the purpose of reduction of the ratio C:N and by that decrease in losses of nitrogen during carrying out of process it was spent composting a sugar flat cake together with an oil cake, taken in the ratio 2:1. The components mixed with each other, analyzed with an interval in 3–5 both days compost were analyzed every week for definition of temperature, pH, NH_4^+, NO_3^-, the general contents of nitrogen, loss of carbon and an index of germination of seeds. Both mixes had thermophile a stage during 15–20 day and the further temperature composting above an ambient temperature during 80 days fast splitting organic substances in both mixes approximately during 40 day then there came the period of stabilization was observed. Both mixes reached a maturity approximately through 90 day, on what specified stable ratio C:N, low ratio NH_4^+/NO_3^-, absence of allocation of heat and an index of germination of seeds which was above 80 %. Mixing of a filtration flat cake 12–15 %. Addition of an oil cake can lead to even greater reduction of loss of nitrogen during composting. Both composts can be used for agriculture (Jones et al. 1995; Chao et al. 1996).

Composting bards of a sugar beet and on used loading investigated influence of the nature of organic substance on two composts received composting the concentrated solutions bards of a sugar beet and 2 kinds of agricultural waste with various amount of organic connections in waste of a grapevine and waste of cotton. Composting it was carried out in the aerated heaps with mechanical hashing under controllable conditions within 4 months. After 71 day of composting, to heaps bards similar initial added the new portion. Change of temperature, value pH and contents of inorganic nitrogen happened in both heaps. However the structure of organic fractions depends on a kind of the added organic components to the bard. Lower decomposition of organic connections was observed in a punched mix to which waste of the grapevine containing a lot of lignine have been added. It was not revealed any phytotoxicity in both kinds of composts. Chemical and physical properties of composts testified to an opportunity of their further use as fertilizers (Bujang and Lopez 1993; Jones et al. 1995). Similar results earlier have been received also by other researchers (Bujang and Lopez 1993; Grobe 1995; Oshins and Kelvin 1992; Verville 1996).

The waste of housekeeping containing not less of 80% of organic substances, have been collected in compost heaps. One of heaps consisted of one food organic waste of housekeeping, another from a mix of food organic waste of housekeeping lignine and cellulose waste in a mass ratio 50:50. Value pH during of composting increased from 5.0 up to 8.3 and the contents of ashes from 30% up to 61% after composting both mixes within 6 months the contents of fat acids in investigated of compost made 8.9% up to and 2.95% after 1 month of composting in recalculation on dry weight. The contents of sodium in compost, consisting of one organic waste, was rather high and made 0.4–0.6%, including on dry weight. Most the heat during composting rose up to 68–70 °C. Later 3 month after the beginning of composting the temperature inside of compost decreased to an ambient temperature. The contents of heavy metals was much more below in compost, received of the sorted food organic waste of housekeeping, than in compost from a mix of waste of housekeeping and scraps of trees. Nevertheless, this content in compost was essential below maximum permissible concentration, except for lead which contained about 100 mg/kg in compost, received of a mix of organic waste of a house and park economy. The general losses of gaseous nitrogen made at composting waste of housekeeping of 50–52 and 26.2% at composting mixes of organic waste of house and park economy (Liao et al. 1994).

2.6.2 Sewage Waste

For composting firm organic waste and the silt, the sewage which have remained after a filtration, usually use anaerobic fermentation with the purpose of water treating and reception of biogas, however, in opinion of separate researchers, for processing such waste can be used with success aerobic composting. In this case waste of sewage usually mixes with other organic sources, such, as manure, waste of the timber industry and inorganic sorbents (Logsdon 1991, 1993; Kjolhede 1994; Bye 1991; Lufkin et al. 1995; Lopez and Baptista 1996). A mix of manure, dirtiness and silt from waste of sewage in the beginning process highly active $Ca(OH)_2$ for destruction of pathogens and then subject fermentations bacteria, for example such, as *Bacillus badius* or *Cellulomonas,* at a controllable level pH. The received fermented product can be used as fertilizer for ground (Lufkin et al. 1995; Lopez and Baptista 1996). The simple and cheap method of partial sewage treatment and the subsequent reception of compost by passing sewage through a layer of wood shavings or large sawdust is offered. Received after that the product is punched, and water after additional clearing can be used for various needs (Lufkin et al. 1995). Dry deposits from city sewage mixed with a lignocellulose material in the ratio 2:1 for the subsequent composting at 28 °C and humidity of 70 %. A source for reception lignocellulose a material can be scraps from city trees and (or) a grapevine. Compost it has appeared it is suitable for improvement of properties of ground (Logsdon 1991).

Impurity in city sewage in the form of the dissolved toxic organic connections, emulsions or dispersions delete by passing it through the columns filled by sand, glass, etc. at contact to warm air for increase in viscosity of liquids. The received viscous liquid mixes up with a punched substratum adsorbing it, for example hay and is punched by usual methods (Logsdon 1991, 1993). Organic impurity from sewage were crushed and homogenated by the minerals containing potassium, such, as feldspar then were punched for reception of fertilizers. The mix can contain also $CaSO_4$ (for example, waste from neutralization SO_2), bentonite and siliceous materials, for example ashes, a dust, ore, lava material, etc.) ≥ 1 %, a material containing phosphate, for example ≥ 1 % of apatite and microcells, for example ore, slags, alloys, sea seaweed or extracts from them. Process is suitable as well for reception of composts from manure, waste of an agriculture, paper waste, waste of leather, textiles, waste of food and the food-processing industry (Logsdon 1993; Kjolhede 1994).

2.7 Conclusion

In recent years, human activities have reached such a point of progress that the recycling capacity of nature has been exceeded, and the accumulation of waste has become a serious environmental and economic problem. To manage this, there is a marked drift towards the use of new technologies, mainly based on biological processes like composting for recycling and resourceful utilization of different organic agro and industrial waste. It is obvious from the research that plant response varies with the amount of compost in the growing substrate as well as with the type of compost being used. Inoculants may be a useful tool in composting processes when the capabilities of microorganisms are suitable for the characteristics of the waste to be composted. Knowledge of different phases of composting and substrate composition is important for specific inoculation in composting and thereby maximum biotransformation of organic matter can be achieved.

Compost products can vary. Strategy to further improve compost quality will be continuously needed. The impacts of long-term use of composts as part of growing medium on soil and water quality have yet to be fully explored. An area where knowledge needs to be strengthened is associated with the leaching of soluble organic carbon on ground and surface water quality, availability, and translocation of compost metal, and the long-term availability of compost metal in soil. It without a doubt is a valuable product as compost improves soil organic matter content, nutrient availability soil aeration, and water holding capacity, and reduces soil bulk density.

References

Altieri R, Esposito A (2010) Evaluation of the fertilizing effect of olive mill waste compost in short-term crops. Int Biodeterior Biodegrad 64:124–128

Andersen JK, Boldrin A, Christensen TH, Scheutz C (2011) Mass balances and life cycle inventory of home composting of organic waste. Waste Manage 31(9):1934–1942

Angin I, Aksakal EL, Oztas T, Hanay A (2013) Effects of municipal solid waste compost (MSWC) application on certain physical properties of soils subjected to freeze-thaw. Soil Tillage Res 130:58–61

Antil RS, Raj D, Narwal RP, Singh JP (2013) Evaluation of maturity and stability parameters of composts prepared from organic wastes and their response to wheat. Was Biomass Valoris. 4(1):95–104

Arner R (1995) By George! Composting at mount Vernon. Biocycle 36:78–80

Atkinson CF, Jones DD, Gauthier JJ (1996) Biodegradability and microbial activities during composting of poultry litter. poult Sci 75:608–617

Aviani I, Laor Y, Medina S, Krassnovsky A, Raviv M (2010) Co-composting of solid and liquid olive mill wastes: management aspects and the horticultural value of the resulting composts. Bioresour Technol 101:6699–6706

Balakrishnan M, Batra VS (2011) Valorization of solid waste in sugar factories with possible application in India. J Environ Manage 92:2886–6706

Bannick CG, Joergensen RG (1993) Change in N fractions during composting of wheat straw. Biol Fertil Soil 16:269–274

Ben AW Achiba WB, Lakhdar A, Gabteni N, Du Laing G, Verloo M, Boeckx P, Van Cleemput O, Jedidi N, Gallali T (2010) Accumulation and fractionation of trace metals in a Tunisian calcareous soil amended with farmyard manure and municipal solid waste compost. J Hazard Mater 176:99–108

Benito M, Masaguer A, Moliner A, De Antonio R (2006) Chemical and physical properties of pruning waste compost and their seasonal variability. Bioresour Technol 97:2071–2076

Bernal MP, Lopez-Real JM, Scott KM (1993) Application of natural zeolites for the reduction of ammonia emissions during the composting of organic wastes in a laboratory composting simulator. Bioresour Technol 43:35–39

Boldrin A, Hartling KR, Laugen M, Christensen TH (2010) Environmental inventory modelling of the use of compost and peat in growth media preparation. Resour Conserv Recy 54:1250–1260

Bolobova AV, Askadskij AA, Kondrashchenko VI, Rabinovich ML (2002) The theoretical basis of biotechnology of wood composites. Enzyme Model Process Science, Moscow

Brinton RB, Brinton WF (1994) Bavarian on-farm composting. Biocycle 35:47–49

Buchanan M, Fulford B (1992) California egg ranch solves manure problem. Biocycle 33:56–57

Bujang KB, Lopez RJM (1993) On-farm composting in the U.K. Biocycle 34:72–73

Bustamante MA, Paredes C, Marhuenda-Egea FC, Purez-Espinosa A, Bernal MP, Moral R (2008) Co-composting of distillery wastes with animal manures: carbon and nitrogen transformations in the evaluation of compost stability. Chemosphere 72:551–557

Bustamante MA, Said-Pullicino D, AgullÑ E, Andreu J, Paredes C, Moral R (2011) Application of winery and distillery waste composts to a Jumilla (SE Spain) vineyard: Effects on the characteristics of a calcareous sandy-loam soil. Agric Ecosyst Environ 140:80–87

Bye J (1991) Privatization composting in a farm setting. Biocycle 32:60–61

Carbonell G, de Imperial RM, Torrijos M, Delgado M, Rodriguez JA (2011) Effects of municipal solid waste compost and mineral fertilizer amendments on soil properties and heavy metals distribution in maize plants. Chemosphere 85:1614:1623

Cellier A, Francou C, Houot S, Ballini C Gauquelin T, Virginie B (2012) Use of urban composts for the regeneration of a burnt Mediterranean soil: a laboratory approach. J Environ Manage 95:S238–S244

Chao WL, Tu HJ, Chao CC (1996) Nitrogen transformation in tropical soils under conventional and sustainable farming systems. Biol Fertil Soils 21:252–256

Dilek K, Yasemin K (2013) Effects of olive solid waste and olive solid waste compost application on soil properties and growth of *Solanum lycopersicum*. Int Biodeterior Biodegrad 82:157–165
Doublet J, Francou C, Putraud JP, Dignac MF, Poitrenaud M, Houot S (2010) Distribution of C and N mineralization of a sludge compost within particle-size fractions. Bioresour Technol 101:1254–1262
Ekinci K, Keener HM, Akbolat D (2006) Effect of feedstock, airflow rate and recirculation ratio on performance of composting systems with air circulation. Biores Technol 97:922–932
Entry JA, Wood BH, Edwards JH, Wood CW (1996) Municipal waste becomes asset to farm land: proper carbon:nitrogen ratio is key to success. Highlight Agr Res 43:8–9
Gabhane J, William SPM, Bidyadhar R, Bhilawe P, Anand D, Vaidya AN, Wate SR (2012) Additives aided composting of green waste: effects inorganic matter degradation, compost maturity, and quality of the finished compost. Bioresour Technol 114:382–388
Glenn J (1990) Yard waste composting enters a new dimension. Biocycle 31:30–36
Grobe K (1995) Organic farms and urban yard trimming. Biocycle 36:63–65
Guanzon YB, Holmer RJ (2003) Basic cultural management practices for vegetable production in urban areas of the Philippines. Urban Agric Mag 10:14–15
Guerra-Rodrigues E, Vazquez M, Dias-Ravina M (2001) Co-composting of chestnut burr and lead litter with solid poultry manure. Bioresour Technol 78:107–109
Haines J (1995) Aspergillus in compost: straw man or fatal flaw. Biocycle 36:32–35
Hartley W, Dickinson NM, Riby P, Leese E, Morton J, Lepp NW (2010) Arsenic mobility and speciation in a contaminated urban soil are affected by different methods of green waste compost application. Environ Pollut 158:3560–3570
Hemmat A, Aghilinategh N, Rezainejad Y, Sadeghi M (2010) Long-term impacts of municipal solid waste compost, sewage sludge and farmyard manure application inorganic carbon, bulk density and consistency limits of a calcareous soil in central Iran. Soil Tillage Res 108:43–50
Holikulov ST, Ortikov TK (2007) Technological options to enhance humus content and conserve water in soils of the Zarafshan valley, Uzbekistan. Climate Change and Terrestrial Carbon Sequestration in Central Asia, 257
Hort C, Gracy S, Platel V, Moynault L (2009) Evaluation of sewage sludge and yard waste compost as a biofilter media for the removal of ammonia and volatile organic sulfur compounds (VOSCs). Chem Eng J 152:44–53
Imbeah M (1998) Composting piggery waste: a review. Bioresour Technol 63:197–203
Jolanun B, Towprayoon S (2010) Novel bulking agent from clay residue for food waste composting. Bioresour Technol 101:4484–4490
Jones OR, Willis WM, Smith SJ, Stewart BA (1995) Nutrient cycling from cattle feedlot manure and composted manure applied to southern high plains drylands. Animal waste and the land-water interface. Lewis Publishers, Boca Raton, pp 265–272
Kanotra S, Mathur RS (1994). Biodegradation of paddy straw with cellulolytic fungi and its application on wheat crop. Bioresour Technol 47(2):185–188
Kjolhede J (1994) Regional source separated composting. Biocycle 35:38–39
Kuba T, Tschwll A, Partl C, Meyer K, Insam H (2008) Wood ash admixture to organic wastes improvescompost and its performance. Agric Ecosyst Environ 127:43–49
Liao PH, Vizcarra AT, Chen A, Lo KV (1994) Aeration experiments for swine wastecomposting. Bioresour Technol 41:145–152
Lima JS, de Queiroz JEG, Freitas HB (2004) Effect of selected and non-selected urban waste compost on the initial growth of corn. Resour Conserv Recy 42:309–315
Logsdon G (1991) Producing and marketing manure compost. Biocycle 32:44–47
Logsdon G (1993) A profit center grows on composted manure. Biocycle 34:66–67
Lopez RJ, Baptista M (1996) A preliminary comparative study of three manure composting systems and their influence on process parameters and methane emissions. Compost Sci Util 4:71–82
Lopez-Aranda JM, Medina JJ, Miranda L (2002). Demonstration stage on MB alternatives for strawberry production in Huelva (Spain). Annual international research conference on methyl bromide alternatives and emissions reduction, pp 6–8

Lufkin C, Loudon T, Kenny M, Scott J (1995) Practical applications of on-farm composting technology. Biocycle 36:76–78

Maeda K, Hanajima D, Toyoda S, Yoshida N, Morioka R, Osada T (2011) Microbiology of nitrogen cycle in animal manure compost. Microb Biotechnol 4:700–709

Mahimairaja S, Bolan NS, Hedley MJ (1995) Agronomic effectiveness of poultry manure composts. Commun Soil Sci Plant Anal 26:1843–1861

Marr JB, Facey RM (1995) Agricultural waste. Water Environ Res 67:503–507

McEvoy M (1993) Analyzing contaminants in the organic waste stream. Biocycle 34:71–72

Miller FC (1996) Composting of municipal solid waste and its components. In: Palmisano AC, Barlaz MA (eds) Microbiology of solid waste. CRC Press, Boca Raton, Florida, pp 115–154

Mondini C, Chiumenti R, Da Borso F, Leita L, De Nobili M (1996) Changes during processing in the organic matter of composted and air-dried poultry manure. Bioresour Technol 55:243–249

Mostafid ME, Shank C, Imhoff PT, Yazdani R (2012) Gas transport properties of compost-woodchip and green waste for landfill biocovers and biofilters. Chem Eng J 191:314–325

Nandy T, Shastry S, Kaul SN (2002) Wastewater management in a cane molasses distillery involving bioresource recovery. J Environ Manage 65:25–38

Neklyudov AD, Ivankin AN (2003) Ecological basis of manufactures: interrelation of ecology, chemistry and biotechnology. MGUL Publishing House, Đoscow

Neklyudov AD, Ivankin AN (2006) Processing of the organic wastes. MGUL Publishing House, Đoscow

Ohins C (1994) Turn leaves into loam: is co-composting right for you. New Farm 16:33–40

Oshins C, Fiorina L (1993) Challenges of on-farm composting. Biocycle 34:72–73

Oshins C, Kelvin R (1992) On-farm composting of municipal organics. Biocycle 33:50–51

Pane C, Piccolo A, Spaccini R, Celano G, Villecco D, Zaccardelli M (2013) Agricultural waste-based composts exhibiting suppressivity to diseases caused by the phytopathogenic soil-borne fungi *Rhizoctonia solani* and *Sclerotinia minor*. Appl Soil Ecol 65:43–51

Paradelo R, Villada A, Barral MT (2011) Reduction of the short-term availability of copper, lead and zinc in a contaminated soil amended with municipal solid waste compost. J Hazard Mater 188:98–104

Rabinovich ML, Bolobova AV, Kondrashchenko VI (2001) The theoretical basis of biotechnology of wood composites. Wood and destroying mushrooms. Science, Đoscow

Raj D, Antil RS (2011) Evaluation of maturity and stability parameters of composts prepared from agro-industrial wastes. Bioresour Technol 102:2868–2873

Sato A, Takeda H, Oyanagi W, Nishihara E, Murakami M (2010) Reduction of cadmium uptake in spinach (*Spinacia oleracea* L.) by soil amendment with animal waste compost. J Hazard Mater 181:298–304

Schmitt MA, Schmidt DR, Jacobson LD (1996) A manure management survey of minnesota swine producers: Effect of farm size on manure application. Appl Eng Agric 12:595–599

Sellami F, Jarboui R, Hachicha S, Medhioub K, Ammar E (2008) Co-composting of oil exhausted olive-cake, poultry manure and industrial residues of agro-food activity for soil amendment. Bioresour Technol 99:1177–1188

Serramib N, Snchez-Monedero MA, Roig A, Contin M, De Nobili M (2013) Changes in soil humic pools after soil application of two-phase olive mill waste compost. Geoderma 192:21–30

Skajaa J, Hannibal E (1991) Anaerobic gasification advances. Biocycle 32:74–77

Smith SR (2009) A critical review of the bioavailability and impacts of heavy metals in municipal solid waste composts compared to sewage sludge. Environ Int 35:142–156

Sun W, Huang GH, Zeng G, Qin X, Yu H (2011) Quantitative effects of composting state variables on C/N ratio through GA-aided multivariate analysis. Sci Total Env 409:1243–1254

Thambirajah JJ, Zulkali MD, Hashim MA (1995) Microbiological and biochemical changes during the composting of oil palm empty-fruit-bunches. Effect of nitrogen supplementation on the substrates. Bioresour Technol 52:133–144

Tognetti C, Mazzarino MJ, Laos F (2011) Comprehensive quality assessment of municipal organic waste composts produced by different preparation methods. Waste Manage 31:1146–1152

Tuomela M, Vikman M, Hatakka A, Itavaara M (2000) Biodegradation of lignin in a compost environment: a review. Bioresour Technol 72:169–183
Turrión MB, Lafuente F, Mulas R, López O, Ruipérez C, Pando V (2012) Effects on soil organic matter mineralization and microbiological properties of applying compost to burned and unburned soils. J Environ Manage 95:S245–S249
Verville RR (1996) Organic feedstock generators team up with local farmers. Biocycle 37:58–61
Wallace R, Silva FF, Chen Y (1992) Unsaturated hydraulic characteristics of composted agricultural wastes, tuff, and their mixtures. Soil Sci 153:434–441
Walsh JL, Valentine GE, Ross CC (1991) Agricultural wastes. Res J Water Pollut Control Fed 63(4):452–458
Weber J, Karczewska A, Drozd J, Licznar M, Licznar S, Jamroz E, Kocowicz A (2007) Agricultural and ecological aspects of a sandy soil as affected by the application of municipal solid waste composts. Soil Biol Biochem 39:1294–1302
Wetterauer D, Killorn R (1996) Composting animal manure with municipal yard trimmings. Biocycle 10:54–57
Yu GH, Wu MJ, Luo YH, Yang XM, Ran W, Shen QR (2011) Fluorescence excitation-emission spectroscopy with regional integration analysis for assessment of compost maturity. Waste Manage 31:1729–1736
Zmora-Nahum SO, Markovitch J, Tarchitzky J, Chen Y (2005) Dissolved organic carbon (DOC) as a parameter of compost maturity. Soil Biol Biochem 37:2109–2116
Zwart K (1990) Biotechnological aspects of manure and manure cake composting. In: Dekkers JJ, van der Plas HC, Vuijk DH (eds) Agricultural biotechnology in focus in the Netherlands. Pudoc, Wageningen, pp 240–242

Chapter 3
Lignocellulose Biodegradation in Composting

Martin A. Hubbe

Abstract Plant-derived material, i.e., lignocellulosic biomass, makes up a major proportion of the initial mass in a typical composting operation. Such biomass plays some key roles as the mixture is being converted to prepare a useful soil amendment. For instance, the lignocellulosic component can provide bulking, can help to balance the C:N elemental composition, and serves as the main source of energy for the bacterial processes that go on during composting. This chapter reviews recent research helping to clarify these roles and to explain the underlying mechanisms. Recent studies have highlighted the importance of bacterial communities, as well as the succession in the composition of those communities during the different thermal phases of composting. Progress also has been made in understanding the flows of heat resulting from metabolism, aeration, and chemical changes in the compost mixture. Advances have been reported in the chemical analysis of compost, revealing details of chemical transformations occurring during the decomposition and stabilization of compost. The lignin component in a compostable mixture provides chemical building blocks that give rise to humic acids and other substances that resist further biodegradation and allow mature compost to retain water and bind minerals. Based on the literature one can conclude that composting, especially when lignocellulosic materials are employed under suitable conditions, is an environmentally responsible, relatively mature technology that can be expected to receive increasing research attention in the future.

Keywords Bacterial communities · Aeration · C:N ratio · Compost pile · Detoxification

---

M. A. Hubbe (✉)
Dept. of Forest Biomaterials, North Carolina State University,
Campus Box 8005, Raleigh, NC 27695-8005, USA
e-mail: hubbe@ncsu.edu

D. K. Maheshwari (ed.), *Composting for Sustainable Agriculture,* Sustainable Development and Biodiversity 3, DOI 10.1007/978-3-319-08004-8_3

3.1 Introduction

Composting depends intimately on the use of plant materials, i.e., lignocellulosic biomass. The goal of this chapter is to review the results of recent research that help to clarify the various roles of lignocellulosic biomass when mixtures of biodegradable materials are composted. Four key roles will be emphasized. First, lignocellulosic materials typically make up a large proportion of the starting mixture in a compost pile. Second, the cellulose and hemicellulose (polysaccharide) components of biomass can be seen as a main source of energy that fuels the composting process. Third, the lignin component of the biomass can be regarded as a main source of what later becomes chemically transformed into humus—the component of mature compost that holds onto moisture and minerals, enhancing the quality of soils. Fourth, the bulky nature of lignocellulosic materials provides bulking and access to aeration not only during composting, but also after the compost has been applied to soil.

Building upon an extensive review article published three years earlier (Hubbe et al. 2010), the present article is mainly concerned with research articles published more recently. Indeed, based on the pace and quality of such publications, it appears that research related to composting has been accelerating. Also, some other review articles have appeared that deal with different aspects of the topic (Singh et al. 2010; Kumar 2011; Maeda et al. 2011; Wichuk et al. 2011; Yeoh et al. 2011; Franke-Whittle and Islam 2013).

A few definitions will be provided here on behalf of readers who may not have studied the composition of plant materials. The term "lignocellulose" denotes material derived from the photosynthetic growth of plants, leading to the buildup of cell walls. The cell walls of plants are comprised mainly of two kinds of chemicals—lignin, which is a phenolic polymer material, and the polysaccharides cellulose and hemicelluloses, which are polymers made up of sugar units. Cellulose, which has a tendency to form somewhat crystalline, highly fibrous assemblies, is one of the main polysaccharides in all lignocellulosic materials. Hemicellulose does not readily form crystalline domains. Rather, side-groups in its linear structure inhibit crystallization and allow it to act as a more compliant component in the structure of plants. Some of the hemicellulose is covalently bound to lignin (Tunc et al. 2010), and therefore it helps to achieve a better integration of the inherently hydrophilic cellulose with the much less hydrophilic lignin component. Lignocellulosic materials include not only wood, which is a prime example, but also grasses, sugarcane residues, and various other byproducts from agriculture.

Though the term "composting" already will have been defined in preceding sections of this monograph, a few points can be emphasized here. Composting can be viewed as a process whereby a mixture of organic materials is subjected to enzymatic action leading to the heating up of the mixed material, some loss of dry mass, changes in the chemical nature of the solids, and the ultimate stabilization of the material so that it becomes more suitable as an additive for fertile soil. The enzymes promoting the chemical changes are mainly secreted by naturally present bacteria and fungi, though, as will be discussed, there has been recent research related to supplementary treatments either with micro-organisms (Kausar et al. 2010; Amira et al. 2011) or with enzymes (Feng et al. 2011).

3.2 Roles of Lignocellulose Biodegradation in Composting

As implied by the prefix "bio" in biodegradation, one of the key roles of lignocellulosic materials in a compost pile is to help support life. Since most composting processes take place in the dark, the energy needed to support life cannot come directly from photosynthesis. Instead, the bacteria and fungi within an active compost pile mainly consume the energy from sugars stored within the cellulose and hemicellulose components. Meanwhile, and especially in the later stages of composting, the lignin component is chemically transformed by further enzymatic action. The subsections that follow will emphasize recent findings regarding the biological and biochemical aspects.

3.2.1 Conditions Affecting Composting

As noted by Karadag et al. (2013), the outcome and speed of a composting operation can be profoundly affected by the starting composition of a pile. Differences in ingredients of a mixture intended for composting govern such factors as the ratio of carbon to nitrogen, the pH, and ultimately the distribution and abundance of biological organisms. Thus, the starting composition is a key determinant that affects the goals of composting. In addition to the four main goals of composting that were listed in the Introduction, Karadag et al. (2013) also list stabilization of organic materials, minimization of landfilling, destruction of pathogens, minimization of greenhouse gases, and reducing the cost of obtaining fertile soil.

3.2.1.1 C:N Ratio

It is well known that efficient composting requires a suitable proportionality between the elements carbon and nitrogen within the organic materials. Most lignocellulosic materials, such as wood and dry grasses, are rich in carbon but contain little if any bound nitrogen. Thus, it is a common practice to add plant-based biomass at an optimized ratio when one wants to compost nitrogen-rich waste products, such as manure. Recent research results have provided additional support for such practices (Doublet et al. 2011; Hachicha et al. 2012; Luz Cayuela et al. 2012; Shan et al. 2013; Thomas et al. 2013). Several researchers have provided further details of how the C:N ratio typically decreases during the course of composting, such that the ratio often has been used as an indication of the relative maturity of compost

(Kausar et al. 2010; Hachicha et al. 2012; Razali et al. 2012; Cheng et al. 2011; Paradelo et al. 2013; Thomas et al. 2013). Zhao et al. (2011b) carried out the analysis in yet finer detail by determining the C:N ratio for different soluble fractions, when using either straw or sawdust as the cellulosic component for composting. Losses in carbon content in the course of composting were mainly traceable to degradation of the cellulose and hemicellulose components.

3.2.1.2 Bulking and Air Access

In addition to allowing some control over the C:N ratio, another key purpose of adding plant-based materials to a compost mixture can be to provide structure and channels for aeration (Hubbe et al. 2010; Zhao et al. 2011a, b). Access to oxygen, throughout a composting pile, is essential in order to minimize the possibility of anaerobic micro-zones (Luz Cayuela et al. 2012). Conventional composting is essentially an aerobic process, and the presence of at least some oxygen is necessary to avoid the formation of greenhouse gases, foul odors, and toxic conditions. The particle size of the lignocellulosic component needs to be optimized. For instance, Yanez et al. (2010) found that reduction of the particle size to about 1 cm allowed more rapid chemical changes during composting and a higher relative content of humic substances after a fixed period of composting. Doublet et al. (2011) found that the choice of bulking agent affected mainly the rate of composting, rather than the final results evaluated after a long period. Serramia et al. (2010) found that the nature of the bulking agents mainly affected the rate of changes during early phases of composting, while not having much effect on the time required to reach full maturity of compost.

3.2.1.3 Moisture

As confirmed by the work of Yanez et al. (2010), the moisture content of a compost pile can be expected to affect the outcome of the process. In the cited work relatively high moisture content tended to promote the progress of composting. As documented by Karadag et al. (2013), the moisture content can be expected to decrease during the course of composting; such an effect is consistent with the elevation of temperatures during composting, as well as the convection of air through a compost pile (Hubbe et al. 2010). Zambra et al. (2011) showed, by means of a dynamic model, that the initial moisture content can affect the maximum temperatures that are reached during composting; control of such temperatures can be important to avoid cases of spontaneous combustion within compost piles.

3.2.1.4 Inoculation

The term "inoculation" implies the addition of living bacteria or fungi (including spores) to a mixture of materials to be composted. Because bacteria and fungi are

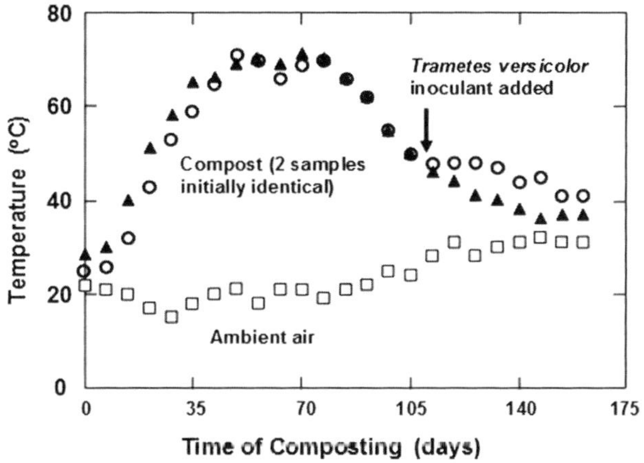

Fig. 3.1 Example showing effect on temperature when compost is inoculated during the cool-down phase. White-rot fungus was added to the compost only in the case represented by unfilled circles. The unfilled squares, lower in the figure, represent the ambient air temperature. (Data replotted from Hachicha et al. 2012)

present naturally in the environment, and due to rapid rates of reproduction under suitable conditions, it is not always clear whether significant benefits can be achieved by inoculation (Hubbe et al. 2010). Thus it is notable that several recent studies have shown significant increases in the rate of changes within compost files following inoculation (Kausar et al. 2010; Amira et al. 2011; Parveen and Padmaja 2011; Wang et al. 2011; Yang et al. 2011; Hachicha et al. 2012; Saha et al. 2012). Saha et al. (2012) demonstrated an innovative strategy in which straw biomass was precomposted with addition of white-rot fungus, followed by combination with poultry droppings in a second stage of composting. Superior overall results were obtained.

Some interesting evidence, supporting the ability of inoculation to promote metabolic activity as well as self-heating during composting (Hachicha et al. 2012) (Fig. 3.1). Olive mill wastewater sludge was mixed with spent coffee grounds and poultry manure. Two matching piles were prepared. On day 106 of composting one of the two piles was treated with white-rot fungus. As shown, the inoculation resulted in significantly higher temperatures following treatment. Such an increase in temperature can be tentatively attributed to increased metabolism due to the enzymatic processes.

3.2.1.5 Enzyme Addition

Enzymes can be regarded as being the tools by which bacteria and fungi bring about chemical changes during composting, changes that can account for the improvements noted in the previous section. It follows that, as a possible alternative to inoculating a compostable mixture with bacteria or fungi, it is possible to just add enzymes directly. Thus, Feng et al. (2011) showed that addition of ligninolytic enzymes hastened the degradation of lignin, in comparison to hemicellulose. The enzymatic treatment not only decreased the time required for composting, but it also gave rise to higher temperatures during composting and enhanced the populations of biota taking part in the decomposition processes. Innovative work by Zhao et al. (2012)

showed that treatment with ligninolytic enzymes also affected the rates of transfer of proteins during the process. The ability of the enzymes to adsorb and desorb from surfaces within the compost pile were affected by the ratio of manganese peroxidase to lignin peroxidase. More work is needed in such areas to clarify how such interactions can be expected to affect composting processes.

3.2.1.6 Compost as a Source of Enzymes

Enzymes capable of breaking down different components of lignocellulosic biomass have been the focus of intense research efforts directed toward the preparation of cellulose-based liquid fuels and related chemicals (Taherzadeh and Karimi 2007; Viikari et al. 2012). As might be expected, compost has served as a convenient source from which researchers have isolated enzyme-producing biota for use in such efforts. Recent work of this type was reported by Liu et al. (2011) and Harun et al. (2013). In many cases, the obtained organisms have been further engineered by recombinant technology to enhance the expression and activity of the resulting enzymes (Liu et al. 2011).

3.2.2 Bacterial Communities

There is increasing realization within the published literature that different species of bacteria or fungi often co-exist in a seemingly synergistic manner within compost mixtures. As noted by Lü et al. (2012), given the complexity of materials present in a compost pile, it is not reasonable to expect that a single organism would be able to supply all of the different kinds of enzymes needed for efficient and timely biodegradation. Rather, biodegradation is typically brought about by communities of organisms. For example, ligninolytic enzymes can be expected to promote access to polysaccharides within a compostable mixture, allowing other microorganisms to flourish and take care of other aspects of biodegradation (Feng et al. 2011).

The word "community" suggests the possibility that certain species may ordinarily tend to share the same environmental niche. Indeed, several recent studies have suggested that relatively stable, synergistic groupings of biota tend to occur together at different phases of composting (Huang et al. 2010; Kausar et al. 2010; Fontenelle et al. 2011; Reddy et al. 2011; Xiao et al. 2011; Eida et al. 2012; Gladden et al. 2012; Lü et al. 2012; Wei et al. 2012; de Gannes et al. 2013; Karadag et al. 2013). In general, the cited studies showed that the communities of bacteria and fungi dominating each successive phase of composting tended to be distinct from each other.

Figure 3.2 gives an example of how microbial populations often rise and fall substantially during the course of composting (Xiao et al. 2011). A compost mixture comprising kitchen waste, leaves, grass, and small branches was shredded. Due to the small batch sizes employed (20 kg) the temperature was able to be controlled primarily by incubation settings. As shown, the higher temperature of incubation resulted in

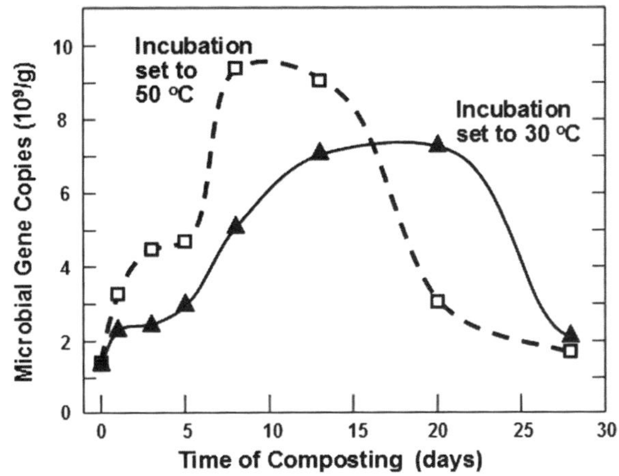

Fig. 3.2 Numbers of copies of acetomycetes genes detected in compost mixtures vs. time, depending on the temperature of incubation. (Data replotted from Xiao et al. 2011)

higher microbial populations earlier. By contrast, the lower temperature resulted in slower increases in the number of genes detected, and about six more days had passed before a major decline in gene count was observed.

3.2.2.1 Quantification of Biota

An ongoing need to quantify populations of bacterial or fungal types in a given compost mixture has been addressed in various ways in recent studies. Perhaps the most straightforward approach entails weighing of the microbial biomass produced under different conditions (Shan et al. 2013). Reddy et al. (2011) and de Gannes et al. (2013) employed pyrosequencing to analyze distributions of microbial communities. The method makes it possible to distinguish the DNA extracted from different samples in which microbial communities are present. Li et al. (2013) employed denaturing gradient gel electrophoresis to separate the genes responsible for enzyme production. As a part of the analysis, the respective genes were first amplified. Primer molecules having specific DNA sequences were used to amplify the results of polymerase chain reaction analysis. Wei et al. (2012) employed DNA sequence detection and were able to document a shift from bacteria-dominated communities to fungi-dominated communities during maturation of compost. Huang et al. (2010) compared the levels of different quinones as a means of judging which micro-organisms were dominant during different phases of composting. The authors found that certain quinones were indicative of certain fungi.

Another practical way to estimate the relative abundance of different microorganisms is by comparing the activities of different identifiable enzymes present in compost as a function of time (Adams and Umapathay 2011; Amira et al. 2011; Eida et al. 2012; Gladden et al. 2012). Adams and Umapathay (2011) showed that extracellular enzyme analysis can be combined with physiological profiling of the communities of microbes.

3.2.2.2 Factors Affecting Microbial Populations

The passage of time, i.e., the "phase" of composting, is well known to play a dominant role with respect to what kind of organisms are dominant in a given sample of organic matter subjected to composting (de Gannes et al. 2013; Karadag et al. 2013; Li et al. 2013). De Gannes et al. (2013) found that the dominant populations of microbes tended to become more diverse as composting progressed through the mesophilic and mature phases. In addition, there was no evidence found to support a hypothesis that certain microbes entered a dormant phase, facilitating their return to prominence at a later stage of composting. Karadag et al. (2013) also concluded that the phase of composting had the most dominant influence on population levels. One may hypothesize that such a succession of populations facilitates a relatively rapid initial decomposition of the more readily biodegradable materials, such as hemicellulose and cellulose, leaving the more recalcitrant lignin moieties to be degraded during later phases of composting, when other micro-organisms are dominant.

When seeking to answer the question, "Why should time seem to play such a dominant role relative to microbial populations?" one of the most frequent answers has been "temperature". Many studies have shown that during composting the temperature first rises over a few days from ambient conditions to about 50 to 65 °C (thermophilic phase), then more gradually falls back towards ambient temperature as the compost approaches maturity (Hubbe et al. 2010; Caricasole et al. 2011; Hu et al. 2011). In other studies the expected "rise and more gradual fall" temperature trend has been shown to be highly punctuated by temporary severe drops in temperature associated with turning of compost piles (Wang et al. 2011; Luz Cayuela et al. 2012). The following authors have recently reported evidence that microbial populations are strongly affected by the prevailing temperatures within a composting mixture (Xiao et al. 2011; Lü et al. 2012; Karadag et al. 2013). Lü et al. (2012) showed that by imposing a certain temperature during composting it was possible to obtain different stable communities of biota. Other researchers, however, have reported cases in which temperature differences did not seem to play a major role in determining microbial populations (Adams and Umapathy 2011). In summary, it seems that the variables of time and temperature can never be completely separated due to complex interactions between those two variables.

Though it would make logical sense to control the relative dominance of different microbial groups by control of the pH during composting, there is little evidence that such an approach has been tried. Lin et al. (2011) showed that such an approach can be used effectively during anaerobic digestion of lignocellulosic materials. The pH was observed to decline during the course of anaerobic processing of waste materials. By contrast, under the aerated conditions of typical composting, the pH has been found to rise to a weakly alkaline range as the compost matures (Hu et al. 2011; Karadag et al. 2013).

An example of pH changes during composting is given in Fig. 3.3 (Lü et al. 2012). As shown within the re-plotted figure, the observed changes are consistent with expected chemical changes that are typical of composting. The initial downward trend (see first large arrow) is consistent with the saponification of acetate

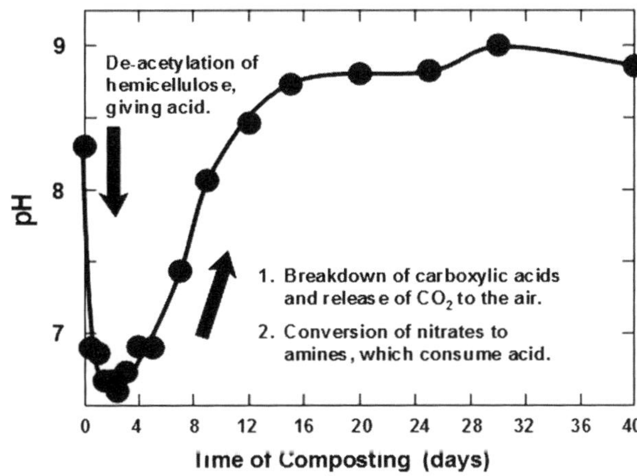

Fig. 3.3 Changes in pH with the passage of time during the composting of wheat straw, cattle manure, and chicken manure. (Data replotted from Lü et al. 2012)

esters within the easily-degraded hemicellulose component of biomass (Hubbe et al. 2010). However, the carboxylic acids created by saponification are subject to breakdown, resulting in the release of CO_2, which is an acidic gas, during the further progress of composting. Such loss of an acidic component renders the remaining material less acidic. An additional contribution to pH rise is provided by accumulation of amines or of the ammonium ion in the compost (see section 5.2).

3.3 Thermal Effects: Cellulose as a Fuel for Heating during Composting

The overall chemical reaction taking place during composting is essentially the same as that of combustion. As in combustion, two of the main products are carbon dioxide and heat. Water vapor is released as well. All of these facts are consistent with a view that the polysaccharide components in the mixture somehow serve as a fuel to drive the process forward (Hubbe et al. 2010). Unlike burning, however, the objective of composting is to only partially consume the organic material. Substantial organic content, i.e., the humus component of matured compost, is needed in order to retain moisture and minerals in soil. Overheating, which sometimes leads to self-ignition of compost piles, is usually taken as an indication that the starting conditions in the pile were somehow inappropriate, e.g., too large a pile (Zambra et al. 2011, 2012).

3.3.1 Distribution of Temperature within a Compost Pile

One of the most challenging aspects of composting is the strong contrast between the temperature at the outside of a compost pile and the simultaneously much higher temperatures that can be measured in the interior (Zambra et al. 2011). The cited study showed that the evolution of moisture content, oxygen levels, as well as temperatures within a compost pile are strongly dependent on both time and location. Surely, one might conclude, if the temperature at the core of a compost pile can be considered to be at an ideal level, then the temperature at the outside of such a pile, especially near to the base of the pile, could be regarded as being in a substandard condition. Support for such a hypothesis can be found in the widespread practice of periodically "turning" compost piles (Wang et al. 2011; Zhao et al. 2011a; Luz Cayuela et al. 2012). The outer layers of a compost pile provide some insulation, allowing the pile to heat up internally as a consequence of metabolic activity.

3.3.2 Research Strategies Related to Energy Flows during Composting

Progress has been reported recently with respect to experimental systems and theoretical models, both having the aim of better understanding the dynamics of thermal interactions during composting.

3.3.2.1 Experimental Systems

On the experimental side, one of the most notable advances has entailed the design of small-scale reactors that can faithfully replicate the thermal effects that are routinely observed in a full-scale compost pile (Lashermes et al. 2012). This was achieved by use of a set of matching reactors (smaller than 10 L), the wall temperatures of which could be precisely controlled. During the initial six days of composting the wall temperatures were controlled to be within the range of 1–2 °C below the measured core temperature of each mini-batch. In this way, "self-heating" within each batch was allowed to proceed despite the fact that the composted mass was insufficient to provide self-insulation against excessive cooling. Then, during later phases of composting, the temperature was programmed to mimic the prevailing temperatures measured for full-scale operations. The cited authors reported high reproducibility when carrying out side-by-side replications. By contrast, inadequate heating is sometimes observed when composting is carried out in insufficiently large batches in the absence of temperature control (Paradelo et al. 2013). As has been shown by Hermann et al. (2011), attempts to biodegrade materials under near-ambient temperature conditions can be expected to yield substandard results in terms of such outcomes as the rates of chemical changes, the killing of pathogens, and the ability to degrade certain materials such as poly-lactic acid (PLA).

Research undertaken at full-scale composting facilities offers another highly promising and practical approach. For example, the following recent studies were carried out at industrial-scale facilities (Hachicha et al. 2012; He et al. 2013; Karadag et al. 2013). An advantage of such systems is that they are inherently realistic and commercially relevant. A disadvantage is that the experimental conditions are seldom allowed to deviate from within a comfort zone prescribed by those in charge of the operation.

3.3.2.2 Modeling and Simulation

Various research teams have achieved important advances in the mathematical modeling and simulation of composting, usually with an emphasis on thermal effects (Barneto et al. 2010; Fontenelle et al. 2011; Zambra et al. 2011, 2012; Zhang et al. 2012). Vidal-Beaudet et al. (2012) carried out related modeling work to predict long term effects after compost has been added to soil.

Barneto et al. (2010) employed an autocatalytic kinetic model to represent a composting process. Model predictions were compared with results obtained by thermo-gravimetric analysis and calorimetry. The main components, cellulose, hemicellulose, and lignin, were assigned different parameters. Water and carbon di-oxide flows, as well has hydrogen production, were modeled. The positive effect of the lignin component on the production of hydrogen during composting was successfully predicted.

Fontenelle et al. (2011) developed a mathematical model incorporating empirically-determined rate expressions. The effects of yeast, bacteria, and fungi were separately modeled for a mixture including sugars, starches, hemicelluloses, and cellulose with the uptake of oxygen and production of heat. The dynamics of composting were faithfully captured by the model. The model was able to predict trends of temperature with time and also oxygen levels at various locations in the pile.

The principle contribution of Zambra et al. (2011, 2012) was to capture, in their model, the three-dimensional aspects of a compost pile, such as generation of heat in the core and dissipation of heat at the surfaces. A nonlinear model based on diffusion rates was used to predict the effects of aerobic bacteria and their metabolism during decomposition of biomass. As noted earlier, their model was unique in its ability to predict conditions leading to combustion due to excessive temperature and insufficient moisture in the core of a large compost pile.

The model of Zhang et al. (2012) was based on "biochemical fractionation", in which different rate constants were applied to different components, effectively dividing the sample into fractions. Observed versus predicted proportions of different fractions were compared for a set of contrasting experiments. Variables such as temperature and oxygen content were inserted as constants. As noted by the authors, in future work it might be possible to achieve better calibration between such a model and experimental outcomes by including the evolution of temperature in the simulation.

3.3.3 Aeration during Composting

The cycling of air through a compost pile has a complex relationship to temperatures and moisture. On the one hand, introduction of fresh air, which is usually cooler, is necessary to support the ongoing metabolic processes. But on the other hand, excess fresh air, beyond what is needed for respiration, can cool the pile.

3.3.3.1 Convective Aeration

Due to self-heating, along with the porous nature of a compost pile, air tends to rise from the core of a compost pile and become filtered by cooler matter near the top surface, then be released to the environment (Hubbe et al. 2010; Zambra et al. 2011). Fresh air tends to be drawn in through the sides of the pile, completing the convective cycle. Lignocellulosic materials, which are often referred to as the "bulking agents" of composting, play an essential role in maintaining suitable porosity (Luz Cayuela et al. 2012). Zambra et al. 2011 were able to model the turbulent flows of air associated with such convection. However, in view of the widespread use of passive convection, especially in household composting efforts, there is a need for more study in this area.

3.3.3.2 Forced Aeration

The use of pressurized air, usually injected at the base of a compost pile, has become a common practice, especially in large-scale composting operations. Several recent studies have considered the optimization of such systems (Yanez et al. 2010; Arslan et al. 2011; Fontenelle et al. 2011; Lashermes et al. 2012; Wang et al. 2012b). Arslan et al. (2011) reported finding an optimum level of aeration. Studies considered in an earlier review article (Hubbe et al. 2010) have indicated that aeration rate can be used as a practical control tool to more reliably achieve optimal thermal conditions during composting.

3.3.3.3 Anaerobic Processing

Though anaerobic processing does not fall within the usual definition of composting, various recent researches have revealed some common features. For instance, even though composting is supposed to entail aerobic conditions, oxygen levels often become depressed within a compost pile (Zambra et al. 2012) and localized anaerobic zones can result (Luz Cayuela et al. 2012). One of the key differences between the two kinds of processes is that whereas a well-managed composting process can be carried out in the open air, anaerobic processes need to be covered in order to collect foul-smelling, toxic, and sometimes flammable gases (Monlau et al. 2013). However, if a system has been set up to collect and utilize the gases, then a positive net impact on the environment can be expected from anaerobic digestion of organic wastes (Hermann et al. 2011; Chairattanamanokorn et al. 2012). Gases such as hydrogen and methane produced by anaerobic digestion of organic wastes have the potential to be used in place of fossil fuels (Chairattanamanokorn et al. 2012). Hermann et al. (2011) regarded anaerobic digestion as preferable to composting in certain cases, since it allows for the recovering of bio-based fuels, while also providing a residue of solids that are suitable for use as soil conditioners.

One of the most interesting findings, from the standpoint of conventional composting, is the fact that anaerobic treatment at least sometimes can be effective in breaking down aromatic structures (Tambone et al. 2013). These are the very structures that tend to be most resistant to breakdown during conventional composting. It follows that anaerobic digestion could be considered as a pretreatment stage, either before conventional composting or before saccharification and fermentation for the purpose of making ethanol (Taherzadeh and Karimi 2007).

3.3.4 The Energy Footprint of Composting

From the standpoint of environmental sustainability, it is important to compare different processing schemes in terms of such factors as their ability to store carbon, helping to curb increases in gaseous carbon dioxide. Composting can be regarded as usually having a favorable net impact, since it can minimize the release of methane, which has a much greater global warming effect than CO_2. Composting also tends to stabilize humic carbonaceous substances so that they can persist in the soil for many years.

To go one step further, recent research has shown that composting can be used as a pretreatment step in preparation for pyrolysis to produce bio-based fuel products (Barneto et al. 2010). This approach can make sense because composting tends to reduce cellulose and hemicellulose-related materials, while leaving mainly lignin-related materials behind. Per unit mass, pyrolysis of the lignin component yields more hydrogen, which then can be used as a fuel in place of fossil fuels. Lemee et al. (2012) likewise used composting as a pretreatment in preparation for catalytic hydroliquefaction to produce biofuels.

When assessing the value of composting as a preliminary step for various applications, as just noted, it can be important to gain some knowledge about the energy content and susceptibility to thermal decomposition. Such information has been obtained, in the case of compost samples, by thermogravimetric analysis (Bernabé et al. 2011, 2013).

3.4 Chemical Changes During Composting

The humus component of mature compost has chemical structures indicating its probable origin mainly from the decomposition of lignin (Hubbe et al. 2010; Hachicha et al. 2012). Meanwhile, the polysaccharides and various minor chemical constituents of biomass are more fully decomposed and depleted during conventional composting. This section considers recent research progress in understanding such chemical changes.

3.4.1 Biodegradation Processes

Recent studies have continued to increase our understanding of what happens to cellulose, hemicellulose, and lignin as a consequence of biological activity during composting (Caricasole et al. 2011; Zhao et al. 2011a, b; Bikovens et al. 2012; Hachicha et al. 2012; Luz Cayuela et al. 2012; Wang et al. 2012a; Bernabé et al. 2013; He et al. 2013; Iwai et al. 2013; Paradelo et al. 2013). In general, the cited studies confirm the extensive loss of cellulose and hemicellulose, while confirming the increasing proportion of humus. He et al. (2013) found that in addition to a build-up of fulvic- and humic-acid components in the maturing compost, there was also production of water-extractable aromatic compounds. Such compounds might be subject to leaching in some situations. Iwai et al. (2013) showed that some such compounds can be extracted from compost by seawater. On the other hand, work by Zhao et al. (2011a) revealed that the water-soluble fractions of compost tend to be susceptible to biodegradation, at least in some cases. It follows that the water-soluble components are likely to be chemically transformed or consumed by the time the compost has been judged to be sufficiently matured, depending on the intended usage. Bikovens et al. (2012) showed the presence of lignin-protein complexes in compost derived from meat processing grease wastes.

3.4.2 The Recalcitrance of Lignin and Its Decomposition

The importance of lignin biodegradation in composting is highlighted by studies showing increased rates of compost maturation following inoculation with fungal species that produce ligninolytic enzymes (Hachicha et al. 2012). Likewise, Feng et al. (2011) demonstrated an increased rate of degradation of lignin and other materials in compost following direct treatment with ligninolytic enzymes. Also, as shown by Zhao et al. (2011a), biomass that has higher lignin content typically is more resistant to change during composting. Another important piece of evidence is that the main changes to the lignin component within a compost pile generally occur only towards the end of a composting process (Huang et al. 2010; He et al. 2013). Nakhshiniev et al. (2012) found evidence that following hydrothermal treatment of palm biomass, a layer of lignin tended to cover and protect some of the more easily degradable components, thus slowing the rate of biodegradation of those materials. Though Wang et al. (2012a) suggested a different mechanism, their results from fluorescent labeling studies of composted mixtures are consistent with a mechanism in which lignin blocks access of enzymes to some of the hemicellulose, thus slowing its degradation. It is possible that naturally occurring covalent bonds between lignin and hemicellulose (Tunc et al. 2010) would tend to promote co-location of lignin and hemicellulose within micro-zones within the decomposed biomass.

Humic acids, such as those produced during the course of composting, are noted for their relatively high resistance to further biodegradation (Hubbe et al. 2010).

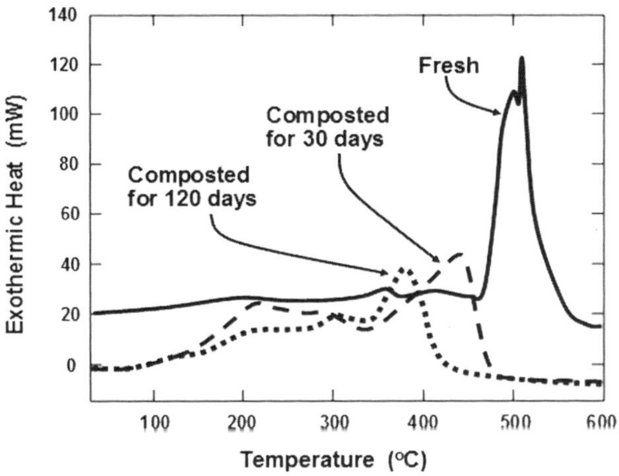

Fig. 3.4 Effect of composting time of lignin samples on the results of thermogravimetric analysis carried out in an inert atmosphere. (Data replotted from Bernabé et al. 2013)

However, as shown by Bernabé et al. (2013), that kind of stability does not imply greater thermal stability. Typically thermodegradation of the lignin component of biomass requires higher temperatures to be reached in an inert atmosphere, compared to hemicellulose and cellulose. The thermogravimetric analysis results in Fig. 3.4 reveal a significant decrease in the decomposition temperature of lignin samples as a function of their time of composting.

It is important to keep in mind that components of lignocellulosic materials other than lignin also can play a role in resistance to biodegradation. For instance, the substantially crystalline nature of cellulose in its natural form can be expected to slow down its biodegradation relative to the amorphous zones within the cellulose component (Hubbe et al. 2010). Thus, the inclusion of microcrystalline cellulose in certain bio-composites can be expected to slow down their biodegradation (Maiti et al. 2011). Caricasole et al. (2011) showed that pine needles are particularly difficult to degrade, presumably because of a high content of terpene-related compounds.

3.4.3 Detoxification in the Course of Composting

As shown by Hu et al. (2011), composting has potential to be used as a tool for decomposition of toxic waste products. Addition of the toxic substance tetracycline to a compost pile was shown to have little if any effect on such parameters as temperature or the main chemical transformations. Meanwhile, most of the tetracycline was decomposed, presumably through enzymatic action. The decomposition of toxic components during ordinary composting has been noted in recent studies (Hachicha et al. 2012; Karadag et al. 2013).

3.4.4 Analysis of Chemical Transformations

Significant progress has been reported recently in various aspects of chemical analysis related to composting. Methods employed have included ^{13}C NMR spectroscopy (Caricasole et al. 2011), X-ray diffraction (Hu et al. 2011; Razali et al. 2012), pyrolysis followed by gas chromatography and mass spectroscopy (Iwai et al. 2013), denaturing gradient gel electrophoresis (Lü et al. 2012; Li et al. 2013), DNA sequence detection (Xiao et al. 2011; Wei et al. 2012), and two-dimensional Fourier transform infrared spectroscopy with fluorescent labeling of specific components in compost (Wang et al. 2012a). He et al. (2013) employed multivariate statistical analysis to look for correlations among different chemical changes accompanying composting. Spectral information was considered in two groups, i.e., the progress of decomposition and the stabilization or maturation of components in the compost.

3.5 Composting of Lignocellulosic for the Stewardship of Soils

The word "stewardship" implies taking responsibility in a caring, active manner. The word has special applicability in the case of composting, because natural, generally eco-friendly processes can be used to restore the fertility to soils. Several recent studies have emphasized the importance of the lignocellulosic component in compost mixtures relative to soil quality. Indeed, many studies have focused on indices such as the C:N ratio to assess the degree or maturity and suitability of composts for different kinds of applications in soil, e.g., lawn care, vs. flowers, vs. vegetables, etc. (Hubbe et al. 2010; Cheng et al. 2013). According to Vidal-Beaudet et al. (2012) it is not uncommon for reconditioned urban soils to contain as much as 50% compost by volume. Consequently, it is very important that such material be chemically stable and has suitable structure and density.

3.5.1 Lignocellulosic Materials to Provide Bulk and Aeration in Soils

The ability of lignocellulosic materials to provide bulking and permeability to composts and/or to the soils to which compost is added has been documented in recent studies (Serramia et al. 2010; Doublet et al. 2011; Zhao et al. 2011a, b; Luz Cayuela et al. 2012). Such studies have shown, for instance, that the use of "bulking agents" can decrease the overall time needed to reach fixed levels of maturation of compost (Doublet et al. 2011). One needs to be careful, however, in attributing all such effects to bulking. As shown earlier, lignocellulosic materials are often essential to achieving an optimum C:N ratio such that a compost pile is able to self-heat effectively and to support a vibrant microbial community.

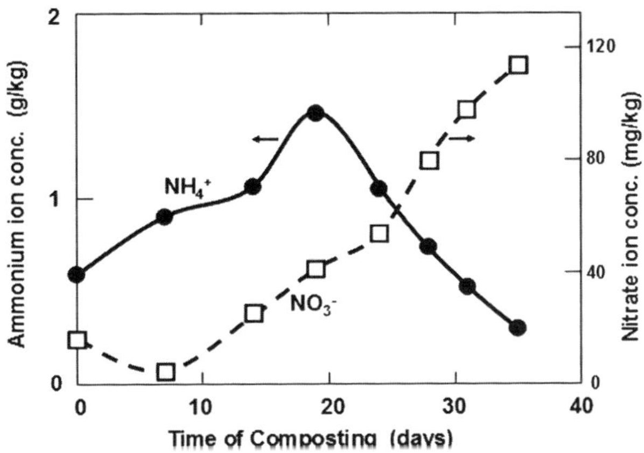

Fig. 3.5 Changes in the concentrations of ammonium ion (filled circles and left-hand axis) and nitrate ions (open squares and right-hand axis) during the course of composting. (Data replotted from Wang et al. 2011)

3.5.2 Lignocellulosic Material and the Processing of Nitrogen

Relative to the fertility of soils, conditions affecting the processing of nitrogen during composting are especially critical (Hubbe et al. 2010). Under favorable circumstances, the nitrogen content of compost can displace the need to use synthetic fertilizers (Hermann et al. 2011). Recent studies have shown that, when composting conditions are well chosen, the content of fixed nitrogen tends to rise during the course of composting (Amira et al. 2011; Bikovens et al. 2012). Work by Doublet et al. (2011) showed that lignocellulosic materials had little effect on nitrogen processing under the studied conditions; rather they mainly affected the rates of chemical changes. However, the same authors noted that a high C:N ratio, achieved by the addition of lignocellulosic materials, tends to favor the beneficial transformation of nitrogen compounds and to limit the production and volatilization of ammonia gas. Likewise, Orrico et al. (2012) found that addition of various lignocellulosic materials tended to reduce losses of nitrogen during composting. Leconte et al. (2011) found that addition of the more easily degraded rice hulls was better able to inhibit the release of ammonia, in comparison to sawdust. Particle sizes also could be optimized to help maximize levels of nitrogen and phosphorous in the final compost. Serramia et al. (2010) and Luz Cayuela et al. (2012) showed that nitrogen processing in compost also depends on the nature of the N-containing organic material employed. For instance, urea was found to be a good promoter for composting (Serramia et al. 2010). Wang et al. (2011) found that higher ammonium contents in compost were achieved following inoculation by white-rot fungus. Yang et al. (2011) reported related results for compost piles inoculated with a pair of thermophilic bacteria.

Figure 3.5 shows how the levels of different nitrogen-containing species can change during the course of compositing (Wang 2011). Wheat straw was composted with manure from cattle and chickens. The observed increase in the content of ammonium ions is consistent with the breakdown and transformation of proteins in the mixture. The later rise in nitrate ion concentration was attributed to the action of ammonium- and nitrite-oxidizing bacteria.

The two-step composting procedure pioneered by Saha et al. (2012) yielded promising results in terms of nitrogen retention. As noted earlier, these researchers first treated the lignocellulosic component with white rot fungus, which is capable of breaking down the lignin portion. Then, in a second step, the pretreated lignocellulosic component was combined with poultry droppings. Losses of nitrogen were lower than in the case of conventional processing, in which all components are combined at the beginning of the process.

3.5.3 Soil Fertility and Plant Growth

The ultimate test of compost, in many cases, is whether it will support vibrant plant growth. Only a few recent articles have included fertility tests as part of their analysis (Wang et al. 2011; Thomas et al. 2013). Rather, the recent studies have more often considered other factors that are expected to be beneficial for plants. For instance, Saha et al. (2012) used ion exchange capacity and humic substance content of the compost as measures of quality. Alternatively, some studies have been concerned with the suitability of the compost as a medium for the growth of edible mushrooms (Farnet et al. 2013).

3.5.4 Restoration of Soil under Challenging Circumstances

As reported by Bonoli and Dall'Ara (2012), composting can be considered as a contributing strategy for restoration of soil in difficult sites. The cited work provides an example of what can happen when organic matter is employed in such efforts in the absence of composting. Sludge obtained from paper mill operations was applied to cover an area that previously had been used as a quarry. It became evident that decomposition of the sludge under anaerobic conditions was giving rise to substantial amounts of biogases, which were being released to the atmosphere. The immediate problem was solved by constructing "chimneys" to direct such gases to a centralized area, where the gases were required to pass through a 1 m thick "hat" of compost material. Presumably, biochemical interactions within the compost were able to consume much of the problematic gases.

Iwai et al. (2013) described a case in which a compost mixture was evaluated for restoring of barren coastal ground in contact with seawater. Due to the contamination of the site with iron, the strong chelating ability of composts was regarded as a positive attribute. Factors contributing to solubility of compost components in the salt water were studied.

3.5.5 Composting and Sustainability in Life Cycle Assessment

Life cycle assessment can be defined as an attempt to numerically account for the various contributing impacts of a certain process or product on the environment. Typically the goal is to quantify impacts all the way from the acquisition of raw materials ("cradle") up to the point of final disposal ("grave"). Questions can be raised such as, "What else could have been done, instead of the proposed process or product, and what would be the net effect on the environment?" Criteria can include the amounts of greenhouse gases produced, the amounts of fossil fuel used, or the amounts of carbon converted into a form that is likely to remain stable for many decades.

Important recent work relative to the life cycle assessment of composting has been reported by Hermann et al. (2011). These authors set out to compare conventional composting (which the authors call "industrial composting"), versus biodegradation at ambient temperature (which was misleadingly called "home composting"), anaerobic digestion, and incineration. Based on data from the literature, the four processes were compared relative to the decomposition and processing of seven commonly composted materials. When emphasis was placed on minimization of carbon dioxide (with allowance made for the displacement of fossil fuel by produced biofuel), anaerobic digestion was judged to be the most eco-friendly technology in cases where it could be applied. Surprisingly, composting under ambient temperature conditions was judged to produce less CO_2 than conventional composting, for which temperatures are allowed to rise during the process. Incineration was also judged to be more eco-friendly than composting, when the main attention is paid to just CO_2 and energy. Although these findings may help to curb an overzealous view of composting, two points need to be considered: First, none of the alternative systems considered in the study—with the possible exception of composting under ambient temperatures—was able to provide an amendment for rich topsoil. And second, the study seems to have neglected problems due to very slow biodegradation and an increased likelihood of anaerobic conditions if the temperature of composting is not allowed to rise enough to stimulate convection of air, as well as proliferation of the most effective microbial communities.

3.5.6 Composting and Sustainability in Greenhouse Gas Emissions

Arguably, one of the most important aspects of the addition of lignocellulosic materials during composting of organic waste materials consists of the minimization of greenhouse gas emissions (Yang et al. 2011). In general, generation of gases such as methane, hydrogen, and volatile organic compounds can be minimized by establishing an initial C:N ratio greater than about 25, or preferably above 60 (Hubbe et al. 2010). Application of sludge and/or inadequately matured compost to soil is to be avoided, since anaerobic conditions in the soil can lead to uncontrolled production of biogases (Huang et al. 2010; Bonoli and Dall-Ara 2012).

As shown by Luz Cayuela et al. (2012), both methane and nitrous oxide (N_2O) can be generated during ordinary composting. The challenge is to maintain conditions such that such releases are minimized, and/or the generated gases are subsequently changed to benign or beneficial forms before they can be released from the pile. In the cited study methane emissions were detected during the whole composting cycle, whereas the nitrous oxide was observed only during the first day of composting, before thermal conditions had been properly established. Ideally methane would tend to be oxidized to CO_2 during its filtration through over-lying layers of the compost pile.

3.6 Concluding Statement

The recent publications reviewed in this article make it clear that interest in the science and technology of composting is very strong. Lignocellulosic materials clearly play a major role in composting systems, and this fact is reflected in the topics of recent publications. A general picture that emerges from the literature is that composting, if done with care, can be considered to be a relatively mature, reliable, and environmentally sound technology. Composting is an appropriate means of handling otherwise unwanted wastes from yards, agriculture, meat productions, sewage, and the like and converting well-balanced mixtures into a valuable component for addition to soils.

References

Adams JDW, Umapathy D (2011) Investigating microbial activities during a starch-amended co-composting process at mesophilic and thermophilic temperatures. Environ Technol 32:1817–1823

Amira RD, Roshanida AR, Rosli MI, Zahrah MFSF, Anuar JM, Adha CMN (2011) Bioconversion of empty fruit bunches (EFB) and palm oil mill effluent (POME) into compost using *Trichoderma virens*. African J Biotechnol 10:18775–18780

Arslan EI, Unlu A, Topal M (2011) Determination of the effect of aeration rate on composting of vegetable-fruit wastes. Clean-Soil Air Water 39:1014–1021

Barneto AG, Carmona JA, Blanco MJD (2010) Effect of the previous composting on volatiles production during biomass pyrolysis. J Phys Chem A 114:3756–3763

Bernabé GA, Almeida S, Ribeiro CA, Crespi MS (2011) Evaluation of organic molecules originated during composting process. J Thermal Anal Calorim 106:773–778

Bernabé GA, Kobelnik M, Almeida S, Ribeiro CA, Crespi MS (2013) Thermal behavior of lignin and cellulose from waste composting process. J Thermal Anal Calorim 111:589–595

Bikovens O, Dizhbite, T, Telysheva G (2012) Characterisation of humic substances formed during co-composting of grass and wood wastes with animal grease. Environ Technol 33:1427–1433

Bonoli A, Dall'Ara A (2012) A bioremediation case of an ex-quarry area restored by paper sludge. J Biotech 157:499–504

Caricasole P, Provenzano MR, Hatcher PG, Senesi N (2011) Evolution of organic matter during composting of different organic wastes assessed by CPMAS C-13 NMR spectroscopy. Waste Manage 31:411–415

Chairattanamanokorn P, Tapananont S, Detjaroen S, Sangkhatim J, Anurakpongsatorn P, Sirirote P (2012) Additional paper waste in pulping sludge for biohydrogen production by heat-shocked sludge. Appl Biochem Biotechnol 166:389–401

Cheng QY, Guo SS, Ai WD, Tang YK, Qin LF (2013) A straw-soil co-composting and evaluation for plant substrate in BLSS. Advan Space Res 51:483–491

de Gannes V, Eudoxie G, Hickey WJ (2013) Prokaryotic successions and diversity in composts as revealed by 454-pyrosequencing. Bioresour Technol 133:573–580

Doublet J, Francou C, Poitrenaud M, Houot S (2011) Influence of bulking agents on organic matter evolution during sewage sludge composting; Consequences on compost organic matter stability and N availability. Bioresour Technol 102:1298–1307

Eida MF, Nagaoka T, Wasaki J, Kouno K (2012) Isolation and characterization of cellulose-decomposing bacteria inhabiting sawdust and coffee residue composts. Microbes Environ 27:226–233

Farnet AM, Qasemian L, Peter-Valence F, Ruaudel F, Savoie JM, Ferre E (2013) Capacity for colonization and degradation of horse manure and wheat-straw-based compost by different strains of *Agaricus subrufescens* during the first two weeks of cultivation. Bioresour Technol 131:266–273

Feng CL, Zeng GM, Huang DL, Hu S, Zhao MH, Lai C, Huang C, Wei Z, Li NJ (2011) Effect of ligninolytic enzymes on lignin degradation and carbon utilization during lignocellulosic waste composting. Process Biochem 46:1515–1520

Fontenelle LT, Corgie SC, Walker LP (2011) Integrating mixed microbial population dynamics into modeling energy transport during the initial stages of the aerobic composting of a switchgrass mixture. Bioresour Technol 102:5162–5168

Franke-Whittle IH, Islam H (2013) Treatment alternatives of slaughterhouse wastes, and their effect on the inactivation of different pathogens: a review. Crit Revs Microbiol 39:139–151

Gladden JM, Eichorst SA, Hazen TC, Simmons BA, Singer SW (2012) Substrate perturbation alters the glycoside hydrolase activities and community composition of switchgrass-adapted bacterial consortia. Biotech Bioeng 109:1140–1145

Hachicha R, Rekik O, Hachicha S, Ferchichi M, Woodward S, Moncef N, Cegarra J, Mechichi T (2012) Co-composting of spent coffee ground with olive mill wastewater sludge and poultry manure and effect of *Trametes versicolor* inoculation on the compost maturity. Chemosphere 88:677–682

Harun NAF, Baharuddin AS, Zainudin MHM, Bahrin EK, Naim MN, Zakaria R (2013) Cellulase production from treated oil palm empty fruit bunch degradation by locally isolated *Thermobifida fusca*. BioResources 8:676–687

He XS, Xi BD, Jiang YH, He LS, Li D, Pan HW, Bai SG (2013) Structural transformation study of water-extractable organic matter during the industrial composting of cattle manure. Microchem J 106:160–166

Hermann BG, Debeer L, De Wilde B, Blok K, Patel MK (2011) To compost or not to compost: carbon and energy footprints of biodegradable materials' waste treatment. Polym Degrad Stabil 96:1159–1171

Hu ZH, Liu YL, Chen GW, Gui XY, Chen TH, Zhan XM (2011) Characterization of organic matter degradation during composting of manure-straw mixtures spiked with tetracyclines. Bioresour Technol 102:7329–7334

Huang DL, Zeng GM, Feng CL, Hu S, Lai C, Zhao MH, Su FF, Tang L, Liu, HL (2010) Changes of microbial population structure related to lignin degradation during lignocellulosic waste composting. Bioresour Technol 101:4062–4067

Hubbe MA, Nazhad M, Sanchez C (2010) Composting as a way to convert cellulosic biomass and organic waste into high-value soil amendments: a review. BioResources 5:2808–2854

Iwai H, Fukushima M, Yamamoto M, Komai T, Kawabe Y (2013) Characterization of seawater extractable organic matter from bark compost by TMAH-py-GC/MS. J Anal Appl Pyrol 99:9–15
Karadag D, Ozkaya B, Olmez E, Nissila ME, Cakmakci M, Yildiz S, Puhakka JA (2013) Profiling of bacterial community in a full-scale aerobic composting plant. Intl Biodet Biodeg 77:85–90
Kausar H, Sariah M, Saud HM, Alam MZ, Ismail MR (2010) Development of compatible lignocellulolytic fungal consortium for rapid composting of rice straw. Intl Biodeter Biodegrad 64:594–600
Kumar S (2011) Composting of municipal solid waste. Crit Revs Biotechnol 31:112–136
Lashermes G, Barriuso E, Le Villio-Poitrenaud M, Houot S (2012) Composting in small laboratory pilots: performance and reproducibility. Waste Manage 32:271–277
Leconte MC, Mazzarino MJ, Satti P, Crego MP (2011) Nitrogen and phosphorus release from poultry manure composts: the role of carbonaceous bulking agents and compost particle sizes. Biol Fertil Soils 47:897–906
Lemee L, Kpogbemabou D, Pinard L, Beauchet R, Laduranty J (2012) Biological pretreatment for production of lignocellulosic biofuel. Bioresour Technol 117:234–241
Li HT, Xu XH, Chen HH, Zhang Y, Xu J, Wang J, Lu XY (2013) Molecular analyses of the functional microbial community in composting by PCR-DGGE targeting the genes of the beta-glucosidase. Bioresour Technol 134:51–58
Lin CW, Wu CH, Tran DT, Shih MC, Li WH, Wu CF (2011). Mixed culture fermentation from lignocellulosic materials using thermophilic lignocellulose-degrading anaerobes. Proc Biochem 46:489–493
Liu DY, Zhang RF, Yang XM, Xu YC, Tang Z, Tian W, Shen QR (2011) Expression, purification and characterization of two thermostable endoglucanases cloned from a lignocellulosic decomposing fungi *Aspergillus fumigatus* Z5 isolated from compost. Protein Express Purif 79:176–186
Lü YC, Li N, Gong DC, Wang XF, Cui ZJ (2012) The effect of temperature on the structure and function of a cellulose-degrading microbial community. Appl Biochem Biotech 168:219–233
Luz Cayuela M, Sanchez-Monedero MA, Roig A, Sinicco T, Mondini, C (2012) Biochemical changes and GHG emissions during composting of lignocellulosic residues with different N-rich by-products. Chemosphere 88:196–203
Maeda K, Hanajima D, Toyoda S, Yoshida N, Morioka R, Osada T (2011) Microbiology of nitrogen cycle in animal manure compost. Microbial Biotech 4:700–709
Maiti S, Ray D, Mitra D, Sengupta S, Kar T (2011) Structural changes of starch/polyvinyl alcohol biocomposite films reinforced with microcrystalline cellulose due to biodegradation in simulated aerobic compost environment. J Appl Polymer Sci 122:2503–2511
Monlau F, Barakat, A, Trably E, Dumas C, Steyer JP, Carrère H (2013) Lignocellulosic materials into biohydrogen and biomethane: Impact of structural features and pretreatment. Crit Reviews Environ Sci Technol 43:260–322
Nakhshiniev B, Gonzales HB, Yoshikawa K (2012) Hydrothermal treatment of date palm lignocellulose residue for organic fertilizer conversion: effect on cell wall and aerobic degradation rate. Compost Sci Util 20:245–253
Orrico ACA, Centurion SR, de Farias RM, Orrico MAP, Garcia RG (2012) Effect of different substrates on composting of poultry litter. Revista Brasil Zootecn Braz J Animal Sci 41:1764–1768
Paradelo R, Moldes AB, Barral MT (2013) Evolution of organic matter during the mesophilic composting of lignocellulosic winery wastes. J Environ Manag 116:18–26
Parveen AA, Padmaja CK (2011) Efficacy of fungi and actinomycetes in converting municipal solid waste (MSW) and water hyacinth (WH) into organic manure. Res Crops 12:167–172
Razali WAW, Baharuddin AS, TarmezeeTalib A, Sulaiman A, Naim MN, Hassan MA, Shirai Y (2012) Degradation of oil palm empty fruit bunches (OPEFB) fibre during composting process using in-vessel composter. BioResources 7:4786–4805
Reddy AP, Allgaier M, Singer SW, Hazen TC, Simmons BA, Hugenholtz P, VanderGheynst JS (2011) Bioenergy feedstock-specific enrichment of microbial populations during high-solids thermophilic deconstruction. Biotechnol Bioeng 108:2088–2098

Saha N, Mukherjee D, Sen S, Sarkar A, Bhattacharaya KK, Mukhopadyay N, Patra PK (2012) Application of highly efficient lignocellulolytic fungi in cocomposting of paddy straw amended poultry droppings for the production of humus rich compost. Compost Sci Util 20:239–244

Serramia N, Sanchez-Monedero MA, Fernandez-Hernandez A, Civantos CGO, Roig A (2010) Contribution of the lignocellulosic fraction of two-phase olive-mill wastes to the degradation and humification of the organic matter during composting. Waste Manage 30:1939–1947

Shan YN, Chen JH, Wang L, Li F, Fu XH, Le YQ (2013) Influences of adding easily degradable organic waste on the minimization and humification of organic matter during straw composting. J Environ Sci Health B—Pesticides Food Contam Ag Wastes 48:384–392

Singh RP, Ibrahim MH, Esa N, Iliyana MS (2010) Composting of waste from palm oil mill: a sustainable waste management practice. Revs Environ Sci Biotech 9:331–344

Taherzadeh MJ, Karimi K (2007) Enzyme-based hydrolysis processes for enthanol from lignocellulosic materials: a review. BioResources 2:707–738

Tambone F, Adani F, Gigliotti G, Volpe D, Fabbri C, Provenzano MR (2013) Organic matter characterization during the anaerobic digestion of different biomasses by means of CPMAS C-13 NMR spectroscopy. Biomass Bioenergy 48:111–120

Thomas GV, Palaniswami C, Prabhu SR, Gopal M, Gupta A (2013) Co-composting of coconut coir pith with solid poultry manure. Current Sci 104:245–250

Tunc MS, Lawoko M, van Heiningen A (2010) Understanding the limitations of removal of hemicelluloses during autohydrolysis of a mixture of southern hardwoods. BioResources 5:356–371

Vidal-Beaudet L, Grosbellet C, Forget-Caubel V, Charpentier S (2012) Modelling long-term carbon dynamics in soils reconstituted with large quantities of organic matter. Eur J Soil Sci 63:787–797

Viikari L, Vehmaanpera J, Koivula A (2012) Lignocellulosic ethanol: from science to industry. Biomass Bioenergy 46:13–24

Wang HY, Fan BQ, Hu QX, Yin ZW (2011) Effect of inoculation with *Penicillium expansum* on the microbial community and maturity of compost. Bioresour Technol 102:11189–11193

Wang LP, Shen QR, Yu GH, Ran W, Xu YC (2012a) Fate of biopolymers during rapeseed meal and wheat bran composting as studied by two-dimensional correlation spectroscopy in combination with multiple fluorescence labeling techniques. Bioresour Technol 105:88–94

Wang ZC, Gao MC, Tian Y, Wang BY, Wang M (2012b) Study on forced aeration composting of digested sludge and sawdust under low ambient temperature. Prog Environ Sci Eng 356–360:2027–2030. (Book ser. Advanced Materials Research)

Wei H, Tucker MP, Baker JO, Harris M, Luo YH, Xu Q, Himmel ME, Ding SY (2012) Tracking dynamics of plant biomass composting by changes in substrate structure, microbial community, and enzyme activity. Biotech Biofuels 5. doi:10.1186/1754-6834-5-20

Wichuk KM, Tewari JP, McCartney D (2011) Plant pathogen eradication during composting: a literature review. Compost Sci Util 19:244–266

Xiao Y, Zeng GM, Yang ZH, Ma YH, Huang C, Xu ZY, Huang J, Fan CZ (2011) Changes in the actinomycetal communities during continuous thermophilic composting as revealed by denaturing gradient gel electrophoresis and quantitative PCR. Bioresour Technol 102:1383–1388

Yanez R, Bueno P, Rivera A, Garcia-Barneto A, Diaz MJ (2010) Selective organic compounds degradation under controlling composting conditions. Waste Manage 30:755–763

Yang MY, Zeng XL, Tianjia, Zhang XQ (2011) Screening of complex thermophilic microbial community and application during municipal solid waste aerobic composting. African J Biotechnol 10:15163–15169

Yeoh CY, Chin NL, Tan CS (2011) Co-composting of palm oil mill wastes. J Food Agric Environ 9:880–885

Zambra CE, Moraga NO, Escudey M (2011) Heat and mass transfer in unsaturated porous media: moisture effects in compost piles self-heating. Intl J Heat Mass Transf 54:2801–2810

Zambra CE, Moraga NO, Rosales C, Lictevout E (2012) Unsteady 3D heat and mass transfer diffusion coupled with turbulent forced convection for compost piles with chemical and biological reactions. Intl J Heat Mass Transf 55:6695–6704

Zhang Y, Lashermes G, Houot S, Doublet J, Steyer JP, Zhu YG, Barriuso E, Garnier P (2012) Modelling of organic matter dynamics during the composting process. Waste Manage 32:19–30
Zhao L, Gu WM, He PJ, Shao LM (2011a) Biodegradation potential of bulking agents used in sludge bio-drying and their contribution to bio-generated heat. Water Res 45:2322–2330
Zhao L, Wang XY, Gu WM, Shao LM, He PJ (2011b) Distribution of C and N in soluble fractionations for characterizing the respective biodegradation of sludge and bulking agents. Bioresour Technol 102:10745–10749
Zhao MH, Zeng ZT, Zeng GM, Huang DL, Feng CL, Lai C, Huang C, Wei Z, Li NJ, Xu P, Zhang C, Liu ZF, Xie GX (2012) Effects of ratio of manganese peroxidase to lignin peroxidase on transfer of ligninolytic enzymes in different composting substrates. Biochem Eng J 67:132–139

Chapter 4
Bio-composting of Aquatic Biomass Residue and its Amendments in Soil Reclamation

Dinesh Kumar Maheshwari and Mohit Agarwal

Abstract The practice of solid waste treatment known from Biblical era to present era, composting is significantly subsidize and manage natural resources defined as renewable, biodegradable organic matter generated through life processes. Composting under variety of applications is quite important for farming with interesting aspect of large cultivation of aquatic plants and it's composting to generate animal feed, soil amendments and other energy sources. Interestingly, microbial population associated with decaying litter of macrophytes utilizes organic matter and facilitate degradation of cell wall polymers of plant detritus. *Eichhornia crassipes* is a troublesome aquatic weed of water body habitant. On after compost, it is found similar to organic compost or bio-dung. Similarly, *Ipomoea aquatica* is favorable compost with composition of moisture, crude oil, crude fiber, carbohydrate, crude protein and mineral elements including K, Fe, Mn, Zn, Na, Ca and Mg. Compost of both plant residues is entirely suitable as organic manure which, off-set the cost of fertilizers in farming system. Addition of compost corroborates nutrients, micronutrients and organic matter availability to soil and favor growth/activity of symbiotic bacteria resulting into improved biomass. The immense importance of organic manures/compost in the form of humus rich with plant nutrients increases the fertility of several kind of soil. Though, chemical fertilizer thought as only way to increase soil fertility, but severally effects soil fertility too. So, in this scenario, composting for agricultural benefits is the need of today to help in reclamation of waste lands. It is an effective way to stepping out agricultural production and extension of area for cultivation through reclamation of waste lands. Millions of farmers in developing countries need adequate resources for augmenting the crop productivity, and ensuring continued maintenance and building up of the soil fertility for greater productivity from agro-waste residues.

Keywords Aquatic biomass · *Eichhornia crassipes* · *Ipomoea aquatica* · Bioconversion · Lignocellulosics · Soil fertility

---

D. K. Maheshwari (✉) · M. Agarwal
Department of Botany and Microbiology,
Gurukul Kangri University, Haridwar, India
e-mail: maheshwaridk@gmail.com

D. K. Maheshwari (ed.), *Composting for Sustainable Agriculture,* Sustainable Development and Biodiversity 3, DOI 10.1007/978-3-319-08004-8_4

4.1 Introduction

Compost is the ancient practice of solid waste treatment known since Biblical era and persists to the present era. Composting was modernized in the 1920s as a tool for organic farming (Heckman 2006) in a broadest means that "artificially accelerated complete or partial decomposition of heterogeneous complex organic matter by microbial populations in aerobic and/or anaerobic environment". The contributions in development of this thrust field of composting have been paid by various scientists. Initially, Howard and Wad (1931) conceptualized composting process. Gaur (1984) reviewed an account on composting for scientific conceptualization. Composting differ from other decomposing processes in respect to its physiochemical processes particularly temperature and rate of organic matter decomposition (Crawford 1983). The microorganisms significantly contribute in process of compost formation as they multiply and decompose the biomass residue under aerobic condition for composting. The community of microorganisms alters the composting process due to variation in temperature, moisture, nutrient levels etc. The initial decomposition accelerates with due heat generation through oxidation and respiratory activity that leads to further increase in decomposition process of compost formation resulting production of uniform fermented dark colored material called humus or compost. This context considers on aquatic biomass residue of two aquatic weed plants *Eichhornia crassipes* and *Ipomoea aquatica* and their composting process under influences of indigenous microbial community.

4.2 Composting of Aquatic Biomass Residue (ABR)

The "biomass" of the natural resources may be defined as renewable, biodegradable organic matter generated through life processes. It is an attractive and easily degradable matter. That is enriched with amorphous cellulose, acts as renewable source of carbon for synthetic fuel and energy. The selection of best technology and desired biomass as raw materials for compost production permits specific applications, complicated by the multitude of organic residue and plant species and the many processing combinations which yield various products for human values (Klass 1981). The potential use of biomass residue of aquatic origin has been worth as animal feed especially in South Asian countries.

The significance of aquatic biomass residue as compost has importance due to characteristically advantageous to soil ecosystem due to easy mixing and potentially beneficial to strengthen soil fertility. However, high percentage of crystalline cellulose presents in aquatic weeds and presence of lignin are always hindrance in bioconversion and composting processes. The lignin is the cementing material that reduces the accessibility of cellulose (Kirk and Harkin 1973). The low amounts of proteins in the aquatic biomass residue also contribute to their low value.

Now-a-days world's increasing attention is focusing on aquatic biomass residue because the aquatic surface on earth is several times larger than the land available and most of the dry land of whole globe and surfaces is in the process of being exploited. The land available for terrestrial biomass growth is becoming limited. In this endeavor, many aquatic plants multiply so fast and termed as weeds (Agrawal and Gopal 2013) can be utilized as important biomass resources for composting process as well as in variety of applications compatible with the ecosystems. Aquatic weeds have high biomass productivity in comparison to that of terrestrial plants because they need not to be water stressed and can optimally supplied with nutrients in solubilized form. Another interesting aspect is that the large cultivation of aquatic plants and their degradation has been suggested to generate animal feed as well as potential energy source (Benemann 1981).

The concept of cultivation of aquatic plants/biomass for energy dates back 25 years when microalgae were suggested as renewable source of methane (Meier 1977). A related study (Ashare 1978) using a similar design concept and analysis concluded that emergent aquatic plants would be favored over microalgae because these do not be limited by availability of an enriched CO_2 source. Thus, aquatic biomass seems to exhibit higher yield than that of land biomass. Hence, water-based biomass considered to be most suitable for their diverse applications. According to Benemann (1981), there are several advantages of aquatic biomass to that of terrestrial plants as the former has continuous hydraulics production system, independent of soil characteristics and short generation time. Besides, the plants do not weather stressed resulting in higher productivity. Further, these can be grown in brackish and waste water, contain low lignin and thus easily accessible for bioconversion of cellulose components by microbial cellulases.

Interestingly, floating and submerged macrophytes supported considerably more number of bacteria and fungi than emergent macrophytes. Bacteria were better equipped than fungi to act under partial anaerobic environment of lakes indicated that most of bacteria and fungi associated with decaying litter of macrophytes that utilizes only leached dissolved organic matter and only a few of them are involved in degradation of cell wall polymers of plant detritus (Gaur 1987).

Thus, free-floating *Eichhornia crassipes* and rooted floating *Ipomoea aquatica* occur throughout year; referred to as macrophytes, constitute an important component of aquatic ecosystem. Free-floating water hyacinth is the aquatic weed originated in the 23.15 % wetland area of North Eastern region of India. At an average annual productivity of dry (ash-free) *E. crassipes* is 50 t ha^{-1} $year^{-1}$. Water hyacinth is one of the most productive plants in the world attributing the weed to cover water surfaces faster than most other plants.

4.2.1 *Eichhornia crassipes*

Eichhornia crassipes (common name—water hyacinth), is a member of family Pontederiaceae and is most widespread aquatic weed (Fig. 4.1). It also withstands, under considerable drought and grows as a land plant. The plant is perennial, floating in water

Fig. 4.1 Vegetation showing morphology of *Eichhornia crassipes*

partly or buried in mud, in shallow water with long fibrous roots. *Echhornia crassipes* is native plant of Amazon basin, and a perennial aquatic plant. Its habitat ranges from tropical desert to subtropical desert to rainforest zones. It is widely introduced in North America, Australia, Africa, India, New Zealand etc. (Duke 1983). *Eichhornia crassipes* is a troublesome aquatic weed and habitant of small and large water bodies, irrigation canals, dams and many other important aquatic or semi-aquatic places.

The biological characteristics that makes this plant as a source of high productivity, makes its interest in aquatic energy farming thus biomass of plant can be used for energy source because the presence of cellulose, hemicelluloses, lignin and pectin substances are accessible for microbial conversion processes. The standing crop has been estimated to produce 100–200 mt ha^{-1} $year^{-1}$ and under ideal conditions each plant can produce 248 offspring in 90 days. While, the aquatic biomass could always be used for burning, the high water content is the limit as the biomass requires drying before consumption. On the other hand, the aquatic biomass (plants) because of their water content provided ideal raw material for anaerobic digestion provided they are suitably harvested and chopped to the required size. There are universally accepted that to control or eradicate this obnoxious weed, it has to be converted to compost using different methods. The chemical analysis of these composts obtained after microbial bioconversion found similar to that of organic compost or bio-dung (Basarkar et al. 2013). Water hyacinth compost or manure was also experimentally applied to water bodies and the periodic biological changes were noted along with the algal succession (Naskar et al. 1986).

4.2.2 Ipomoea aquatica

Similarly *Ipomoea aquatica* (common name—water spinach), is a member of family Convolvulaceae, and mainly found in India, South Asia, Malaysia, Taiwan and

Fig. 4.2 Vegetation showing morphology of *Ipomoea aquatica*

China (Edie and Ho 1969) (Fig. 4.2). Besides, it is also been introduced in Hawaii, Brazil, Cuba, Jamaica, Trimidad and other Islands of West Indies. It is widely distributed throughout tropical and warm climate regions of southeastern Asia. and also been distributed in Arabia, western Africa, Egypt and India (Upendra and Sinha 2001)

It is a tropical trailing, habituating muddy stream banks but it is also found in fresh water ponds and marshy areas (Jain et al. 1987). The components of the plants which are submerged showed some resistance to adverse or unfavorable conditions (Chin and Fong 1978). Vegetation formed by aquatic weeds provides support for the luxurious growth of *I. aquatica*, which also grows on the borders of wet paddy fields. The roots grow rapidly and hardly require any attention. Heavy clay, water lodging and organic soil are ideal for its growth (Morton and Snyder (1976). Unlike water hyacinth, it is suitable to spread and compete with native vegetation. Growing seasonal productivity of *I. aquatica* marked upto 27.48 g m^{-2} day^{-1} and extrapolated productivity upto 100.30 mt ha^{-1} $year^{-1}$ was recorded by Kanungo et al. (2001). It contains rich nutritional composition of moisture, crude oil, crude fibre, carbohydrate, crude protein and mineral elements including K, Fe, Mn, Zn, Na, Ca and Mg (Umar et al. 2007).

4.3 Bio-decomposition of Aquatic Macrophytes

The decomposition of aquatic macrophytes as aquatic biomass residues involves three fundamental steps which may occur simultaneously or after time layers, i.e., physical, autolytic and microbial leaching. The leaching of dissolved inorganic/organic matter through a biotic decomposition facilitates microbial colonization and catabolism of the decaying litter due to depolymerization of structural materials like cellulose, hemicellulose and lignin (Martínez et al. 2010; Bugg et al. 2011). Available literature revealed that, there is little information earned on these aspects of decomposition of aquatic macrophytes.

The decomposition in aquatic macrophytes generally initiates with natural senescence and is caused by abiotic and biotic processes. The abiotic process includes physical breakdown i.e., leaching and fragmentations. On the other hand, the biotic process includes autolysis and microbial breakdown i.e., degradation of structural polymers into smaller units. The different genera of bacteria and fungi and consumption of debris by macro invertebrates perform the process of decomposition. Earlier studies on water hyacinth decomposition revealed that the decay rates may vary with tissue nitrogen and fiber content. About 30 % of initial dry weight is lost solely by abiotic processes and remainder by microbial processes (Gaur 1987).

The relative contributions of the bacteria and fungi to the overall decay of water hyacinth (*E. crassipes* (Mart.) Solms) determined by Gaur (1987). Less number of bacteria (13) and more number of fungi (24) associated with in-situ decaying of litter of water hyacinth (Singhal et al. 1992). Most bacteria were Gram-negative, facultative anaerobes, and able to degrade polysaccharides and proteins. The terrestrial fungi were predominant. Six fungi grew well on water hyacinth leaves and were actively involved in the degradation of lignocellulose. Rest of the fungi grew slowly or failed to grow on water hyacinth leaves. Decomposition experiments have demonstrated that the first 4 days of decay are dominated by non-microbial processes (Singhal et al. 1992). Scientists have observed rates of abiotic and microbial decomposition in all types of hyacinth leaves is dominated by physical leaching in initial phase of 4 days duration and later by microbial processes. The largest part of physical leaching takes place within the first 4 days. Thereafter, the weight loss occurs due to physical leaching that declines exponentially. The weight loss by microbial decomposition is minimal in thc initial phase but increased exponentially in the later phase. The bloom leaves decompose significantly faster than post bloom leaves, and post bloom green leaves decompose faster than post-bloom brown leaves. The rate constants of abiotic decomposition recorded significantly higher in post bloom leaves and microbial decomposition is 15 and 55 %, respectively, on the other hand, in pre-bloom leaves it was 33 and 19 % where as in post bloom green leaves and 24 and 6 % in post bloom leaves (Singhal et al. 1992). The variation in rate of decomposition is due to age of the leaves but role of other factors cannot be ruled out.

The compost maturity is generally observed by color appearance showing yellowish coloration and without precipitation with starch—iodine test indicating the presence of simple sugars. Absence of sulfides can be depicted by no coloration on lead acetate strip. On the other hand, presence of nitrate is possible to detect by red coloration. Change in C:N ratio during biodegradation revealed that the total organic carbon content in mature compost should be declined by 50 % and total N contents must be increased in composed to that of green biomass.

Although initial pH of the compost remain in acidic range as it is found in cell sap of most plant. The production of organic acids during early stages of composting, initial pH turned towards acidic range but with rise in temperature, shift in pH

Fig. 4.3 Composting of *Eichhornia crassipes* with biogas slurry

towards alkalinity was noticed and pH maxima was recorded on 8th day. This obviously is due to the fact that the optimum temperature during aerobic decomposition and active microbial activity did not allow accumulation of the organic acids as the inter-mediatory products and thereby, the pH of the biomass remained alkaline in later period of decomposition. Further, interestingly low pH was observed at mesophilic range of composting while during thermophilic range of temperature, pH increased rapidly because of the evolution of ammonia (Kate 1990), that during facultative anaerobic fermentation, nitrogen in cow-dung in biogas fermenter converted into ammonia. Urea considers as nitrogen fertilizer. It creates too much ammonia for plant to absorb and provide sufficient and rapid nitrogen availability. It is considerably metabolize by majority of urease producing bacteria, facilitate instant nitrogen production. Wei et al. (2012) studied the effect of inorganic fertilizer and composted manure amendments and found that application of composted manure on maize was significantly more effective than that of urea. However, in another study combined application of chemical fertilizer suggests that compost blending increase immobilization of urea-N in soils with high C and N contents and further it increases nitrification of fertilizer-N in soils (Han et al. 2004). Hence, GAB mixed with biogas slurry+urea, release ammonical form of nitrogen resulting in the rise in pH. Moisture percentage in the GAB is another important parameter required to be optimize for aerobic composting. At low moisture content, the decomposition tends to be aerobic. Further, supra-optimal moisture level reduces the microbial activity by hindering the movement of air and thus reduces the O_2 supply (Alexander 1983). Khandelwal (1995) recorded that the optimum decomposition and compost formation occur if initial moisture content is kept at 50–60%. On further increase in moisture content, decomposition of GAB inhibited and also impart unpleasant odor. In-process composting of *E. crassipes* with biogas slurry is shown in Fig. 4.3.

The reduction in total solids (TS) of biomass is directly related to the extent of biodegradation. Variation in C:N ratio is another important parameter to determine the qualitative status of the compost. The C:N ratio of mature compost is found to be low as compared to the initial ratio of 40:1 in GAB mixed with slurry and urea. The assimilation of C in microbial cell and volatilization of CO_2 considered as the possible reasons of decline in C:N ratio. The assimilation is further followed by mineralization of C and N (Alexander 1983).

The temperature is one of the most important physical conditions in determining how rapid natural materials are degraded, assimilated and mobilized. In case of water hyacinth, author observed that maximum biodegradation occurred in the range of 30–40 °C because, mesophilic microorganisms remains more active at this temperature range. On increasing temperature, decline in rapidity of composting was recorded. At high temperature, all reactive compounds such as sugars, fats etc. denatured and microbes are killed and great deal of energy released because of oxidation of carbon to CO_2 (Dalzell 1979). Further, Khandelwal (1995) noticed that composting process absorbed heat energy from the surroundings and its entrapment in the compost bed cannot be ruled out.

In fact, after mulching the compost, the nutrient status of eroded soil showed remarkable increase of total organic C content by six times. The nitrogen contents were 2.5 times and phosphorus exhibited 3 times higher than that of eroded or sub-standard soil. The potassium contents on after mixing compost was observed to be 4 fold higher that leads to a significant increase in its mineral status and correspond to soil fertility. The increase in net primary productivity (NPP) and gross primary productivity (GPP) is thus, directly related to the improved status of soil due to better availability of solubilized nutrients after addition of compost. Several workers (Biswas et al. 1971; Gaur et al. 1972) reported improvement of soil structure and chemical characteristics, texture and appearancc by using agro-compost. Naskar et al. (1986) observed increase in productivity on application of water hyacinth compost in agriculture. Howard and Lee (1993) observed increase yield of sweet potato using EC compost. Water holding capacity is an important parameter and compost also increased water holding capacity of soil.

The chemical composition of both *E. crassipes* and *I. aquatica* showed that along with sufficient quantity of hemicellulose, quite good amount of lignin was also present which makes the residue unsuitable for hydrolysis due to its recalcitrant nature. It is also reported to be toxic to micro-organisms so alters the process of degradation, the greater need of today is to evolve appropriate pretreatment method that may diminish the protective effect of lignin as most of the potential aquatic biomass residues are heavily lignified. The pretreatment is essential to enhance the susceptibility of lignocelluloses to ligno-cellulolytic enzyme action. Over the years, several pretreatments methods including physical, chemical and biological treatments have been proposed by many workers to modify the substrates and render them as more susceptible for bioconversion and enzymatic processing. Physical methods e.g., ball milling (Fan et al. 1980), Hammer milling (Mandels et al. 1974), Vibroenergy milling (Ghose 1969), Extrusion (Han and Callihan 1970), Pyrolysis (Shafizadeh 1977), Irradiation (Kunz and LaRur 1975), Steaming (Bender et al. 1970) etc. Chemical treatment methods are includes acids, alkalies and solvent treatments. However, Alkaline pretreatment is most common and successful method (Mandels et al. 1974). Biological process is better than other methods, because the microbial growth increase protein content. Biological processes facilitate; Lower energy requirement, higher yield of desired product, greater substrate and reaction specificity, lower pollution and non-feasible transformation with chemical reagent. An ideal pretreatment accomplished reduction in crystallinity of cellulosic fibers along with reduction in the lignin content. On these biomasses, alkali alongwith steam treatment is proved the effective methods for complete de-lignification which readily cleaves the ether and carbon bonds leads de-polymerization and solubilization of the lignin polymer. In our studies, pretreatment of aquatic biomass suggested that enzyme cellulase activity increased after the pretreatment of biomass.

Saccharification is another process of hydrolysis of complex carbohydrate (e.g., starch, cellulose, lignin etc.) into its simpler forms or components. In composting this is achieve by microorganisms involved with their enzymatic reaction. However, chemical pretreatment and enzymatic saccharification both are in major consideration (Saha et al. 2005). Wide varieties of cellulosic material are readily saccharified by the cellulases produced by *Trichoderma viride* (Mandels 1975; Ghose 1969), *Coriolus versicolor* (Zafar et al. 1889), *Tremetis versicolor* (Valmaseda et al. 1991), *Aspergillus niger* (Maheshwari et al. 1994) and other higher cellulolytic fungi. Zhu and Pan (2010) and Agbor et al. (2011) used pretreated biomass for cellulose production. The exo-glucanase activity was observed maximum in alkali with steam treated *E. crassipes* following by *I. aquatica*. The high enzyme activity in alkali pretreat substrate promotes swelling of cellulosic fiber and reduction in crystallinity index. The fibres beyond water swollen dimensions allow increased enzyme penetration into the fine structure of cell wall (Tarkow and Feistm 1969). Similar trend was observed in case of endoglucanases and β-glucosidase enzyme actions. The catalase enzyme particularly both the susbstrate *E. crassipes* and *I. aquatica* preferred different pretreatment process for releasing maximum sugar formation during saccharification (Kaur 1993; Kour et al. 1993).

4.4 Efficiency of Aquatic Biomass Residue as Compost

The utilization of plant residues is relatively new technology as a source of carbon compounds among existed resources exhausted in world. ABRs seems to be a good sources of organic carbon and thus used as a compost to meet out the great demand of organic manure. The water hyacinth is valuable in the compost due to their high content of nitrogen. It tops the list of most dreaded aquatic weeds and now spread around the globe. It has successfully resisted all attempts of eradicating even by chemical, biological, mechanical or hybrid means. It is therefore, several successful studies were conducted on pile composting of cattle manure, swine manure, municipal bio-solids, animal mortalities and food residuals. However, limited investigations have been made on high rate windrow/agitated pile composting of water hyacinth. Water hyacinth compost is entirely suitable to be used as organic manure. Further, preparation of such compost will off-set the cost of cleaning the irrigation system of this weed and will prevent the health hazard arising from leaving the plant material on the banks of the river and canals. Water hyacinth compost has been shown to have positive effects on crop growth when using the water hyacinth compost as organic fertilizer.

The compost product already contains a well-working flora of microorganisms will be favorable for the soil. Applying the water hyacinth directly without any other processing than sun-drying seems to be the best alternative in small scale use, due to relatively small loses of nutrients and workload required. This option also does not require any large investments or new technology. If the fresh water hyacinth could be applied as mulch on the fields, the labor need for weeding could be used for handling the water hyacinths instead.

4.5 Compost amended with Nitrogen Fixing Rhizobia

Addition of compost borne nutrients, micronutrients and organic matter to soil may enhance the growth of symbiotic bacteria. Various workers (Bouldin et al. 1985; Madriaga and Angle 1992) observed increase in nodulation in crop legumes. Bhardwaj and Gaur (1970) obtained better growth and activity of rhizobia in soil mulched with compost. The organic acids present in compost helps in release of combined P into free available form (Struthers and Seiling 1950). Acids prevent precipitation of P by Fe or Al (Gaur 1986), for instance increase in carbolic acid group and decrease in chain length showed positive effect on P availability. On the other hand, humic acid released various micro-nutrients like Fe, Mn, Cu and Zn (Elgala and Schlichling 1976). Even, the addition compost detoxifies certain pesticides including DDT (Pareekh and Gaur 1973) and BHC (Gaur et al. 1975).

Interestingly, the shoot weight of the tree legumes increased on application of compost and that the maximum dry matter accumulation was due to that the EC on degradation provided better release of nutrients which favored microbial growth/activity resulting into improved biomass. The influence of compost mulched soil was exhibited more on *Leucaena leucocephala* than that of *Dalbergia sisoo.*

Rhizobia promote plant growth using several mechanistic strategies as well as via fixing nitrogen symbiotically and enhance overall crop protection and production (Maheshwari et al. 2013). Rhizobia symbiosis with plants depends upon several factors. Water hyacinth, *Eichhornia crassipes* is a chief weed to utilize in preparation of compost with its perceived benefits in improvement of crop productivity and limiting its disease. Recent technique of blending *Rhizobium* inoculant with hyacinth compost is compatible option for plant growth promotion and pest suppression. High nutrient content in water hyacinth compost can stimulate *Rhizobium* nodulation and nitrogen fixation, consequently improving plant growth and pest resistance. Naluyange et al. (2014) conducted a field study with two trials to assess the compatibility of commercial *Rhizobium* inoculant, DAP, cattle farmyard manure (FYM), and four formulations of water hyacinth compost i.e., water hyacinth only (H), with molasses (H+Mol), cattle manure culture (H+CMC) or effective microbes (H+EM) and concluded in their study that the commercial *Rhizobium* inoculant is predominantly compatible with water hyacinth compost formulations containing effective microbes and cattle manure culture, which could enhance tolerance of bean plants to aphids and possibly to anthracnose disease. These two water hyacinth compost formulations need further investigation for their potential in enhancing food production and alleviating the water hyacinth problem.

4.6 Significance of Compost

The aquatic weeds are the source of green biomass with low C:N ratio. These aquatic weed in the form of compost not only provides the suitable solution for checking their unsuitability but will also make the soil enriched with nutrients. Due to high percentage of cellulosic fibers, *E. crassipes* is suitable for making compost for agriculture (Naskar et al. 1986). Gaonkar and Kulkarni (1986) compared the chemical composition of *E. crassipes* with that of bagasse, rice hull and paddy straw and observed that presence of low lignin content make this plant suitable as a commercial source of α-cellulose. The mineral status reported in the ranges from 23:1 to 35:1 as also analyzed by Crawford (1983). Thus, the compost from *E. crassipes* can be easily prepared by modified method of Das and Ghatnekar (1979). As stated earlier that green aquatic biomass (GAB) degradation was achieved within 1–2 weeks but the GAB treated with biogas slurry and urea decompose more readily in comparison to other treated combinations due to better microbial action (Joshi 1992). Das and Ghatnekar (1979) reported the degradation of durva grass mixed with cow-dung within 2 weeks duration. Khandelwal (1995) studied on the compost mulching in substandard soil and its influence on two fast growing fuel wood tree legumes viz. *Dalbergia sisoo* and *Leucaena leucocephala.* Application of compost increase the nodule weight and size significantly by 21 and 104% in *D. sisoo* and 44 and 54% in *L. leucocephala* respectively when compared to garden soil and substandard soil respectively. Similarly, fresh weight (FW) of plants cultivated in compost was increased approximately five fold in comparison to *D. sisoo* raised in substandard soil. This value was more than 10 fold higher in case of *L. leucocephala.* The net primary productivity (NPP) and gross primary productivity (GPP) of both the plant species were found to increase significantly when grown in compost mulched soil.

The immense importance of organic manures/compost in the form of humus rich with plant nutrients increases the fertility of several kind of soil. Extensively, chemical fertilizer thought is the only way to increase soil fertility, but fertilizer on soil under long term use leads the several harmful effects, in this scenario, an urgent need for use of composting for agricultural benefits is the need of today as it also helps in reclamation of waste lands. Organic materials are available as by products from the plant (aquatic and terrestrial) and animal origin is considered as the immense source of crop productivity. Composting of aquatic biomass is the best method for converting organic residue in the inputs for agriculture.

The preparation and applications of organic manures or compost in the agricultural fields is a traditional system and widely adopted by the scientific community and farmers around the globe.

4.7 Soil Reclamation and Enrichment

Soil is the important component of land. The degradation and destruction of soils are due to several unforeseen and not-so-unforeseen problems, such as rapid and unscientific spread of agriculture, construction of huge dams, canals, urbanization and industrialization deforestation (Pathak and Rao 1991) and natural calamities. The degraded land does not give enough return rather it is a burden or otherwise useless, hence termed as wasteland. Wasteland is usually uncultivated, un-habitated and bare-land, lacking moisture with stress climate and poor soil quality which is not conducive for the regeneration of any type of vegetation. Loss of the green cover and ability to withhold water and soil, such lands are in critical condition. Mostly, these are skeleton soil lacking in humans, these contain toxic elements with low nutrients. Compost as organic residue includes green manures, animal and plant waste, municipal waste, industrial effluent serves as quite effective sources for plant nutrient and soil reclamations.

Stamp (1948) gave the definition and considered waste land as the land which has been previously used but now abandoned and no further use has been found for such land. These are cultivable waste land such as gullied and or ravenous land; surface water logged and marshy land; salt affected land; shifting cultivation area; strip lands; sands and industrial waste land on the other hand uncultivable waste land includes barren, rocky, stony waste, sheet rock area; steep slopping area and snow covered or glacial area.

Composting is another and effective way to stepping out agricultural production and extension of area for cultivation through reclamation of waste lands. Waste lands formed due to indiscriminate and over utilization of forest produce standing over the area, unscientific land management by putting the area to improper land use and sometime even as an unintended side effect of the very process of development.

Millions of farmers in developing countries need adequate resources for augmenting the crop productivity, and ensuring continued maintenance and building up of the soil fertility for greater productivity from agro-waste residues. The organic matter of the cultivated soil in tropics is low due to high temperature and intense microbial activity with due to increases of need to add humic substance in soil to replenish organic matter in soil. The organic matter provides the soil fertility with both macronutrients as well as micronutrients. Farmyard manure (FYM), compost addition, green manuring are among few options which proved best mode of replenishing the soil humus content during the course of cultivation practices (Gaur 1990). Amendment of soil with plant compost is reported to be the best recycling method used since ancient time. In fact, due to microbial activity, the organic matter in the composting materials decomposes into easily accessible simple chemicals.

4.8 Conclusion

Composting of aquatic biomass residue contribute in management of aquatic weed plants *Eichhornia crassipes* and *Ipomoea aquatica* and their composting process under influences of indigenous microbial community. The "biomass" of the natural resources may be defined as renewable, biodegradable organic matter generated through life processes. The significance of aquatic biomass residue as compost has importance due to characteristically advantageous to soil ecosystem to strengthen soil fertility. The utilization of plant residues as a source of carbon compounds is a relatively new technology in the resource exhausted world. The enormous importance of organic manures/compost is the need of today as it also helps in reclamation of waste lands. Waste lands formed due to indiscriminate and over utilization of forest produce standing over the area, unscientific land management by putting the area to improper land use and sometime even as an unintended side effect of the very process of development. Millions of farmers in developing countries need adequate resources for augmenting the crop productivity, and ensuring continued maintenance and building up of the soil fertility for greater productivity from agro-waste residues.

References

Agbor VB, Cicek N, Sparling R, Berlin A, Levin DB (2011) Biomass pretreatment: fundamentals toward application. Biotechnol Adv 29(6):675–685

Agrawal A, Gopal K (2013) Aquatic weeds: occurrence and distribution. In: Agrawal A, Gopal K (eds) Biomonitoring of water and waste water, Springer, India, pp 15–32

Alexander M (1983) Introduction to soil microbiology. Willey, New York

Ashare E (1978) Cost analysis of aquatic biomass system. Dynatech R/D, Cambridge

Basarkar UG, Kshirsagar GS, Saoji AA (2013) Comparative study of vermicompost using parthenium biodung and usual green biodung. Int J Bioassay 2(5):819–824

Bender F, Heaney DP, Bowden A (1970) Potential of steamed wood as a feed for ruminants. Forest Prod J 20(4):36–41

Benemann JR (1981) Energy from fresh and backish water aquatic plants. In: Donald L Klass (ed) Biomass as a non fossil fuel source. ACS symposium series, ACS

Bhardwaj KKR, Gaur AC (1970) The effect of fulvic acids on the growth and efficiency of nitrogen fixation by *Azotobacter chroococum*. Folia Microbiol 15:364

Biswas TD, Basu SN, Ray RR (1971) Cumulative effect of different levels of manures on the physical properties of soil. J Ind Soc Soil Sci 19:31

Bouldin DR, Gouin FR, Waddington DV, Sopper WE (1985) Benefits of land spreading of sludges in the northeast. In: Frink CR, Huller TL (eds) Criteria and recommendations for land applications of sludges in the northeast. (Pennsylvania Agric Stn 851:23–27)

Bugg TD, Ahmad M, Hardiman EM, Rahmanpour R (2011) Pathways for degradation of lignin in bacteria and fungi. Nat Prod Rep 28(12):1883–1896

Chin LY, Fong FW (1978) Preliminary studies on the productivities of the two most abundant macrophytes in Subang lake. Malaysian Agric J 51(4):422–435
Crawford JH (1983) Composting of agricultural wastes: a review. Process Biochem 18: 14–18
Dalzell HW (1979) Composting in tropical countries principles of composting. The Int Inst Biol Husbandry Ltd. (IIBH), England, pp 1–36
Das CR, Ghatnekar SD (1979) Replacement of cowdung by fermentation of aquatic and terrestrial plants for use as fuel, fertilizer and biogas plant feed. Research planning workshop on energy for rural development. Indian Institute of Management, Ahmedabad, pp 1–19
Duke J (1983) Handbook of energy crops. Center for New Crops and Plants and Products, Purdue University
Edie HH, Ho BWC (1969) *Ipomoea aquatica* as vegetable crop in Hong Kong. Econ Bot 23(1):32–36
Elgala AM, Schlichling E (1976) Der Einfluss von huminsaure und DTAP-Zusatz aut die extrahiernakest von Fe, Mn, Zn and Cu aus geldingten und undedungten. Boden Landw Forschung 29:34–36
Fan LT, Lee YH, Beardmore DH (1980) Major chemical and physical feature of cellulosic materials as substrate for enzymatic hydrolysis. Springer Berlin Heidelberg, pp 101–117
Gaonkar SM, Kulkarni PR (1986) Water hyacinth and some agricultural waste sources as a source of alpha cellulose. Res Ind 31:254–255
Gaur AC (1984) Organic matter and rice. International Rice research Institute, Phillipines, pp 503–514
Gaur AC (1986) Recycling of crop residue—recent trends in India. Izvestiya Akademii Nauk. SSSR Seriya Biologicheskaya 2:210–217
Gaur AC (1987) Recycling of organic waste by improved techniques of composting and other method. Res Conserv 13:157–174
Gaur AC (1990) Phosphate solubilizing microorganism as biofertilizer. Omega Scientific Publications, New Delhi, p 176
Gaur AC, Mathur RS, Bhardwaj KKR (1972) Soil Structure as influenced by organic matter and inorganic fertlilizers. Labdev J Sci Tech India 108:55–56
Gaur YD, Sen AN, Subba Rao NS (1975) Usefulness and limitation of ketolactose test to distinguish *Agrobacterium* from *Rhizobium*. Curr Sci 42(15):545–546
Ghose TK (1969) Continuous enzymatic saccharification of cellulose with culture filtrates of *Trichoderma viridae*. Biotechnol Boengg 6:39–74
Han YW, Callihan CD (1970) Cellulose fermentation, effect of substrate pretreatment of microbial growth. Appl Microbiol 27:159–165
Han KH, Choi WJ, Han GH, Yun SI, Yoo SH, Ro HM (2004) Urea–nitrogen transformation and compost–nitrogen mineralization in three different soils as affected by the interaction between both nitrogen inputs. Biol Fertil Soil 39(3):193–199
Heckman J (2006) A history of organic farming: transitions from Sir Albert Howard's War in the Soil to USDA National Organic Program. Renew Agric Food Syst 21:143–150
Howard AL (1993) Research on use of composted organic waste. Spectrum, British Sci News, pp 5–6
Howard A, Wad YD (1931) The waste product of agriculture, their utilization as humus. Oxford University Press, London
Jain SK, Gujral GS, Vasudevan P (1987) Potential utilization of water spinach. Scient Ind Res 46:77–78
Joshi PB (1992) Studies on decomposition of organic waste with special reference to thermophilic fungi. Ph.D. thesis. Nagpur University, Nagpur
Kanungo VK, Sinha S, Naik ML (2001) Net primary productivity of some aquatic macrophytes in sewage–sullage mixture. J Environ Biol 22(3):219–223
Kate T (1990) Utilization of anaerobic digested biogas effluent slurry or sludge. A theme paper presented at national seminar on biogas technology, Nagpur, p 15
Kaur J (1993) Utilization of lignocellulosic biomass for ethanol production. PhD thesis. Barkatullah University, Bhopal

Khandelwal A (1995) Investigation on growth of some tree legumes in degraded land amended with *Rhizobium* and *Eichhornia* residue. PhD thesis. Gurukul Kangri University, Haridwar
Kirk TK, Harkin JMTL (1973) Lignin biodegradation and the bioconversion of wood. Am Soc Chem Engg Symp 133:124–126
Klass DL (1981) Biomass as a non-fossil fuel source. ACS symposium series. Wiley, New York
Kour J, Maheshwari DK, Mehta P (1993) Microbial degradation of aquatic biomass by *Trichoderma viride* 992 and *Aspergillus wentii* 669 with reference to the physical structure. J Basic Microbiol 33(1):19–25
Kunz WGW, LaRur TA (1975) Nitrogenase activity in rhizobia in absence of plant host. Nature 256:407–408
Madriaga GM, Angle JS (1992) Sludge borne salt effects on survival of *Bradyrhizobium japonicum*. J Environ Qual 16:118–124
Maheshwari DK, Gohade S, Paul J, Verma A (1994) Paper mill sludge as a potential source for cellulose production by *Trichoderma ressei* QM 9123 and *Aspergillus niger* using mixed cultivation. Carbohydrate Polym 23:161–163
Maheshwari DK, Agarwal M, Dheeman S, Saraf M (2013) Potential of rhizobia in productivity enhancement of Macrotyloma uniflorum L. and Phaseolus vulgaris L. Cultivated in the Western Himalaya. In: Maheshwari DK, Saraf M, Aeron A (eds) Bacteria in agrobiology: crop productivity. Springer, Berlin, pp 127–165
Mandels M (1975) Microbial source of cellulase. Biotechnol Bioengg. Symp 5. Wiley, New York, p 81
Mandels M, Hontz L, Nystrom J (1974) Enzymatic hydrolysis of waste cellulose. Biotechnol Bioengg 16:1471–1493
Martínez ÁT, Speranza M, Ruiz-Dueñas FJ, Ferreira P, Camarero S, Guillén F, Martínez MJ, Gutiérrez A, del Río JC (2010) Biodegradation of lignocellulosics: microbial, chemical, and enzymatic aspects of the fungal attack of lignin. Int Microbiol 8(3):195–204
Meier RL (1977) Solar energy research. Wisconsin Press, Madison, pp 179–189
Morton JF, Snyder GH (1976) Aquatic crops vs. organic soil subsidence. In Proc Fla State Hortic Soc 89:125–129
Naluyange V, Ochieno DM, Maingi JM, Ombori O, Mukaminega D, Amoding A, Odendo M, Okoth SA, Shivoga WA, Muoma JVO (2014) Compatibility of *Rhizobium* inoculant and water hyacinth compost formulations in Rosecoco bean and consequences on *Aphis fabae* and *Colletotrichum lindemuthianum* infestations. Appl Soil Ecol 76:68–77
Naskar K, Saha PK, Majuumdar R, Bakshi GDN (1986) Compost value of water hyacinth (*Eichhornia crassipes*) in field of agriculture and aquaculture. Bull Bot Soc Bengal 40:11–17
Pareekh RP, Gaur AC (1973) Effect of DDT on nodulation, growth, yield and nitrogen uptake by *Pisum sativum* inoculated with *Rhizobium leguminosarum*. Indian J Microbiol 9:93–98
Pathak PS, Rao DLN (1991) Wasteland management. Bookwell, New Delhi, pp 1–48
Saha BC, Iten LB, Cotta MA, Wu YV (2005) Dilute acid pretreatment enzymatic saccharification and fermarentation of wheat straw to ethanol. Process Biochem 40:3693–3700
Shafizadeh F (1977) Proc TAPPI Frost Biol Wood Chem Conf, Maidson, TAPPI, Atlanta, p 191
Singhal PK, Gaur S, Talegaonkar L (1992) Relative contribution of different decay processes to the decomposition of the *E. crassipes* (Mart.) Solms. Aquat Bot 42:265–272
Stamp LD (1948) The land of Britain-its use and misuse. Longmans, Green, in conjunction with Geographical Publications, p 507
Struthers PH, Seiling DH (1950) Effects of organic ions in phosphorus precipitation by Fe and Al as influenced by pH. Soil Sci 69:205–208
Tarkow H, Feistm WC (1969) A mechanism for improving for improving the digestibility of lignocellulosic materials with dilute alkali and liquid ammonia. Adv Chem Ser 95:197
Umar KJ, Hassan LG, Dangoggo SM, Ladan MJ (2007) Nutritional composition of water spinach (*Ipomoea aquatica* Frosk.) leaves. J Appl Sci 7(6):803–809
Upendra NR, Sinha S, (2001) Distribution of metals in aquatic edible plants: *Trapa natans* (Roxb.) Makino and Ipomoea aquatica Frosk. Environ Monit Assess 70(3):241–252

Valmaseda M, Martinez MJ, Martinez AT (1991) Kinetics of wheat straw in solid state fermentation with *Tremetes vesicolour* and *Pleurotus ostreatus* lignin and polysaccharide alternation and production and production of related enzyme activities. Appl Microbiol Biotechnol 35:87

Wei Z, Chunsheng H, Li J, Christie P, Ju X (2012) Natural 15N abundance of tomato and soil amendments with urea and compost. J Food Agric Environ 10:287–293

Zafar SI, Sheeraj Q, Abdullah N (1889) Degradation of the lignocellulosic compound on wheat straw *Coriolus vesicolor* solid state fermentation under nitrogen starred condition. Biol Wastes 27:67

Zhu JY, Pan XJ (2010) Woody biomass pretreatment for cellulosic ethanol production: technology and energy consumption evaluation. Bioresour Technol 101(13):4992–5002

Chapter 5
Physical, Chemical and Biological Parameters for Compost Maturity Assessment: A Review

Rajinder Singh Antil, Dev Raj, Nuha Abdalla and Kazuyuki Inubushi

Abstract New livestock production system based on intensification of large farms produces huge amounts of organic waste materials in the form of farm, animal, domestic and agro-industrial waste all over the world without enough agricultural land for their direct application as fertilizers. Compost stability/maturity has become a critical issue for land application of composts because immature compost can be determinant for plant growth and the soil environment. Evaluation of maturity of compost has been widely recognized as one of the most important problems concerning composting process and application of this product to land. In order to provide and review the information found in the literature about composting of different organic waste materials, the first part of this chapter explains basic concepts of composting process. Then a summary of physical, chemical and biological parameters affecting stability and maturity of high quality compost prepared from organic waste materials.

Keywords Physical · Chemical and biological parameters · Organic wastes · Compost quality

5.1 Introduction

Composting of organic wastes is a bio-oxidative process involving the mineralization and partial humification of the organic matter, leading to a stabilized final product, free from phytotoxicity and pathogens (Zucconi and de Bertoldi 1987; Bernal et al. 2009). Composting is a controlled aerobic process carried out by successive

R. S. Antil (✉) · D. Raj · N. Abdalla
Department of Soil Science, CCS Haryana Agricultural University, Hisar 125004, India
e-mail: antilhau_56@rediffmail.com

D. Raj
e-mail: devraj_chauhan@rediffmail.com

N. Abdalla
e-mail: nuha75n@gmail.com

K. Inubushi
Graduate School of Horticulture, Chiba University, Matsudo, Chiba 271-8510, Japan

D. K. Maheshwari (ed.), *Composting for Sustainable Agriculture,* Sustainable Development and Biodiversity 3, DOI 10.1007/978-3-319-08004-8_5

microbial population combining both mesophilic and thermophilic activities, leading to the production of carbon dioxide, water, minerals and stabilized organic matter (Pereira-Neta 1987). Composting is one of the most effective means of recycling of organic wastes that can be used as a source of soil amendment and organic matter in agricultural land. The variability of the organic matter undergoing composting makes compost research challenging. In recent years, interest in composting has increased because of the social demand for an environmentally friendly waste treatment technology and for organic agricultural products. Composting is seen as not only an environmentally acceptable method of waste treatment, but also one of the more efficient methods of waste disposal which enables recycling of organic matter (Ko et al. 2008). The application of composted manure has increased over the years. This practice improves the quality of the crops and preserves the environment (Hoitink 2000; Ko et al. 2008). However, non-composted manure may have adverse effects on plant growth and/or seed germination (Hoekstra et al. 2002) because of their wide C:N ratio and production of phytotoxic substances such as phenolic and volatile fatty acids during organic matter decomposition (Kirchmann and Widen 1994). Mesophilic and thermophilic micro-organisms are involved in the composting and their succession is important in the effective management of composting process (Ishii et al. 2000). Many studies have shown that the composting process eliminates or reduces the risk of spreading of pathogens, parasites and weed seeds associated with direct land application of immature compost and leads to final stabilized product (Larney and Hao 2007; Bernal et al. 2009). The need to treat and dispose of organic wastes to reduce potential damage has made compost production and its agricultural application an attractive solution. The composting industry is poised for a new era of growth where the product must be evaluated both for safety and quality purposes. Composts prepared from different organic wastes differ in their quality, which further depends upon the composition of raw material and composting technology used for compost production (Ranalli et al. 2001). Quality of high grade compost generally relates to its fertilizer value. If unstable or immature compost is applied, it can induce anaerobic conditions as the micro-organisms utilize oxygen in the soil pores to break down the material (Mathur et al. 1993). Another problem associated with immature compost is the phytotoxicity due to the presence of organic acids during early stages of the composting process (Fuchs 2002; Cambardella et al. 2003). Compost quality is closely related to its stability and maturity. Maturity is associated with plant-growth potential or phytotoxicity (Iannotti et al. 1994), whereas stability is often related with the compost's microbial activity. However, both stability and maturity usually go hand in hand, since phytotoxic compounds are produces by the micro-organisms in unstable composts (Zucconi et al. 1985). Several parameters have been proposed for evaluating compost maturity and stability (Iglesias Jiménez and Pérez García 1992; Wu et al. 2000; Bernal et al. 2009; Raj and Antil 2011; Antil et al. 2012). However, there is no single method that can be universally applied to all types of compost due to variation of materials and composting technology (He et al. 1995; Benito et al. 2003; Chang and Chen 2010). Morel et al. (1985) proposed that the maturity of the compost may be assesses by the biological activity of the product, including total microorganisms count, monitoring biochemical parameters of microbial activity and analysis of biodegradable constituents. The changes in microbial properties during the composting process and their association with maturity must be carefully studied. The present chapter reviews the physical [color, odor, temperature, organic matter (OM) loss], chemical [pH, electrical conductivity (EC), water soluble C (WSC), C:N ratio, water soluble organic carbon (C_w):organic N (N_{org}) ratio, NH_4^+-N and NO_3^--N, humic acid (HA): fulvic acid (FA) ratio, humification index (HI) and cation exchange capacity (CEC):total organic carbon (TOC) ratio)] along with biological [seed germination index (GI) and microbiological parameters (bacterial, fungal and actinomycetes counts, O_2 and CO_2 respiratory, enzyme activities and microbial diversities (coliform, fecal coliforms and fecal enterococci population)] parameters affecting stability and maturity of high quality compost prepared from organic waste materials.

5.2 Concepts of Composting Process

While composting occurs, efficient composting requires the control of several factors to avoid nuisance problems of such as odour and dust and also for obtaining a quality agricultural product. The controlled conditions are basic for a composting procedure, distinguishing it from aerobic fermentation. Over the last decades, research has been focused on the study of the complex interaction amongst physical, chemical and biological factors that occurs during composting.

The precise details of the biochemical changes taking place during the complex processes of composting are still lacking. The phases which can be distinguished in the composting processes according to temperature are (Fig. 5.1):

a. Latent phase, which corresponds to the time necessary for the microorganisms to acclimatize and colonize in the new environment in the compost heap.
b. Growth phase, which is characterized by the rise of biologically produced temperature to mesophilic level.
c. Thermophilic phase, in which the temperature rises to the highest level. This is the phase where waste stabilization and pathogen destruction are most effective.
d. Maturation phase, where the temperature decreases to mesophilic and, subsequently, ambient level. A secondary fermentation takes place which is slow and favours humification; that is, the transformation of some complex organics to humic colloids closely associated with minerals and finally to humus. Nitrification reactions, in which ammonia is biologically oxidized to become nitrite (NO_2^-) and finally nitrate (NO_3^-), also take place (Metcalf and Eddy 1991).

The organic wastes are initially decomposed by the first level consumers such as bacteria, fungi (molds), and actinomycetes. Mesophilic bacteria are the first to appear. Thereafter, as the temperature rises, thermophilic bacteria, which inhabit all parts of the compost heap, appear. Thermophilic fungi usually grow after 5–10 days of composting. If the temperature becomes too high, i.e., greater than 65–70 °C, fungi, actinomycetes, and most bacteria become inactive and only spore forming bacteria can develop. In the final stage, as the temperature declines, members of the actinomycetes become the dominant group which may give the heap surface a white or gray appearance. After these stages the first level consumers become the food of

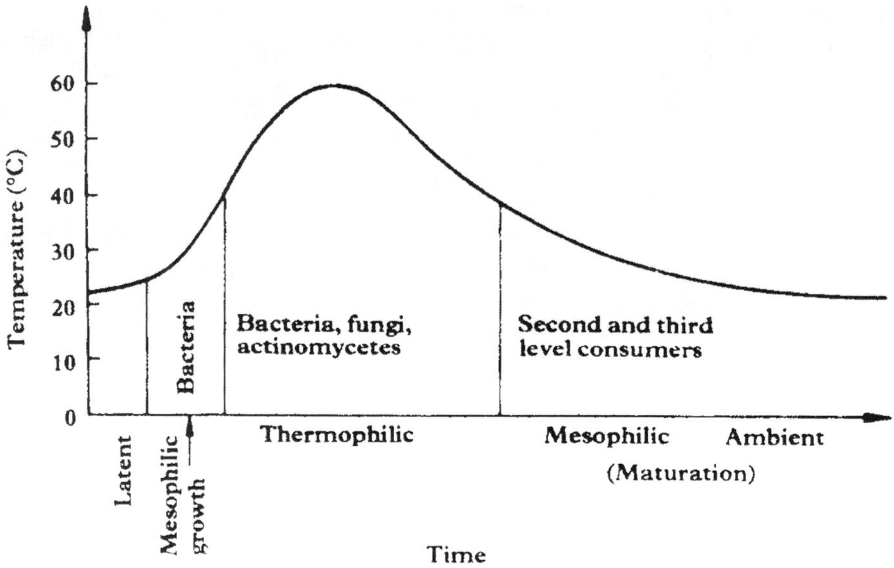

Fig. 5.1 Patterns of temperature and microbial growth in compost piles

second level consumers, such as mites, beetles, nematodes, protozoa and rotifers. Third level consumers, such as centipedes, rove beetles and ants, prey on the second level consumers.

5.2.1 Factors Affecting the Composting Process

Composting process is affected by several abiotic and biotic factors, which maintain its integrity on optimum and these factors are quite important during process to be maintained. These are discussed followed

5.2.1.1 Temperature

Temperature is the main factor that controls microbial activity during composting. Heating is essential to enable the development of a thermophilic population of micro-organisms which is capable of degrading the more recalcitrant compounds (natural and anthropogenic), and to kill pathogens and weed seeds (Boulter et al. 2000).

5.2.1.2 pH

A pH of 6.7–9.0 supports good microbial activity during composting. Optimum values are between 5.5 and 8.0 (de Bertoldi et al. 1983; Miller 1992). Usually pH is not a key factor for composting since most materials are within this pH range. However, this factor is very relevant for controlling N-losses by ammonia volatilization, which can be particularly high at pH > 7.5.

5.2.1.3 Aeration

Aeration has an indirect effect on temperature by speeding the rate of decomposition and therefore, the rate of heat production. The air requirement depends upon the type of waste (type of material, particle size), the temperature of the compost and the stage of the process. Air supply can be controlled to some extent by the use of a system of aeration. Under natural conditions warm air diffuses from the top of the windrow drawing fresh air into the base and sides (Hellmann et al. 1997). Aeration is further encouraged by periodic turning of the windrow. Alternatively, air may be actively forced into the pile, usually within a closed or in-vessel system with the aim of maximizing the rate of microbial decomposition. This is relatively costly, but is useful for materials that pose health risks. Forced aeration has also been used successfully on static piles giving a high degree of process control (Sasaay et al. 1997).

5.2.1.4 Microbial Activity

Microbial activity is influenced strongly by moisture content: activity decreases under dry conditions, and aerobic activity decreases under water-logged conditions due to the resulting decrease in air supply. The recommended optimum water content is 40–60 % on a mass basis (Epstein 1997). The microorganisms involved in composting develop according to the temperature of the mass, which defines the different steps of the process (Keener et al. 2000). Bacteria predominate early in composting, fungi are present during all the process but predominate at water levels below 35 % and are not active at temperatures > 60 °C. Actinomycetes predominate during stabilization and curing, and together with fungi are able to degrade resistant polymers.

5.2.1.5 Moisture Content

Changes in moisture content are related to aeration and temperature; in an aerated static pile system approximately 90 % of the heat loss is due to evaporation of water (Sasaay et al. 1997). Systems which actively encourage aeration can lead to desiccation and result in a decrease in the rate of decomposition in windrow composting (Itavaara et al. 1997). The compost must be kept aerobic to avoid the production of odours. Moisture is essential but if the compost is too wet then anaerobic conditions develop. Anaerobic conditions are also undesirable because of the loss of N by denitrification. There may also be build-up of organic acids, such as acetic acid, which can be toxic to plants.

5.2.1.6 C/N Ratio

The C/N is important in determining in general terms the rate of decomposition of organic materials. High C/N ratios make the process very slow as there is an excess of degradable substrate for the microorganisms. But with a low C/N ratio there is an excess of N per degradable C and inorganic N is produced in excess and can be lost by ammonia volatilization or by leaching from the composting mass. Then, low C/N ratios can be corrected by adding a bulking agent to provide degradable organic-C.

5.2.2 Advantages and Disadvantages of Composting

The advantages of composting organic waste materials with direct application are summarized below:

- Elimination of pathogens and weeds.
- Microbial stabilization.
- Reduction of volume and moisture.
- Removal and control of odours.
- Ease of storage, transport and use.
- Production of good quality fertilizers.

However, the disadvantages are:

- Cost of installation and management.
- Requirement for a bulking agent.
- Requirement for large areas for storage and operation.

5.2.3 Benefits of Composting

The benefits of composting can be divided in to two groups: to improve crop productivity and profitability; and to improve soil quality.

Improve crop performance and lower production costs through:

- Improved yields, product quality and storage life
- More efficient and reduced use of fertilizers and pesticides, including soil fumigants
- Better utilization of irrigation
- Increased crop resistance to pests and diseases.

Improve soil quality through:

- Better organic matter levels and organic cycles
- Increased available water to plants
- Increased nutrient availability and nutrient-holding capacity
- Improved structure
- Reduced soil-borne plant pathogens and pests

Table 5.1 Different maturity and stability parameters

Maturity parameters	Stability parameters
C:N ratio	Temperature
C_w/N_{org}	pH
HA/FA ratio	EC
HI	WSC
CEC	Total organic carbon/OM loss
CEC/TOC ratio	Microbial diversity population/activity
NH_4 and NO_3	Enzyme activity
O_2 and CO_2 respiratory	Germination index

5.3 Maturity and Stability Assessment for Compost Quality

Composts prepared from different organic waste materials differed in their quality and stability; and quality is closely related to stability and maturity. Hence, it is important to define maturity and stability. Compost maturity is the degree or level of completeness of composting. It is described by several physical, chemical and biological properties and therefore maturity is best assessed by measuring two or more parameters of compost. Compost stability refers to a specific stage or decomposition or state of organic matter during composting, which is related to the type of organic compounds remaining and the resultant biological activity in the material.

Several parameters have been proposed for evaluating compost maturity and stability (Iglesias Jiménez and Pérez García 1992; Bernal et al. 1998, 2009; Wu et al. 2000; Raj and Antil 2011; Raj and Antil 2012a, b; Antil and Raj 2012; Antil et al. 2012). However, there is no single method that can be universally applied to all types of composts due to variation of materials and composting technology. (He et al. 1995; Itavaara et al. 2002; Benito et al. 2003; Chang and Chen 2010). The parameters are grouped into maturity and stability shown in Table 5.1.

In addition to above parameters, the physical characteristics such as colour, odour, particle size, appearance of larvae of secondary consumers also show the decomposition stage, but give the little information with regard to degree of maturation. The maturity of compost, which may defined as the degree of compost stability in physical, chemical and biological properties is an important factor affecting successful application in agriculture and its impact on environment. The parameters proposed by several workers are presented in Table 5.2. Parameters or criteria proposed or indicated in the literature are grouped into physical parameters, chemical parameters and biological parameters.

5.3.1 Physical Parameters

Physical parameters are frequently used but they give only general information regarding maturity of compost.

Table 5.2 Maturity parameters established for composts of different sources

Parameter	Value	Reference
Colour (degree of darkness)	Dark brown or black colour	Sugahara et al. (1979)
Temperature	More or less constant after turning of material	Stickelberger (1975)
pH	Close to neutral value	Gray et al. (1971); Finstein and Miller (1985)
EC	Stable at the end	Avnimelech et al. (1996), Wu et al. (2000)
C/N	<20	Poincelot (1974); Cardenas and Wang (1980); Golueke (1981)
	<15	Juste (1980); Roig et al. (1988)
	<12	Iglesias Jiménez and Pérez García (1992); Bernal et al. (1998)
Water soluble C/N ratio	5–6	Chanyasak and Kubota (1981)
Cw/TOC	<0.70	Hue and Liu (1995)
NH_4 content	<0.04 %	Zucconi and de Bertoldi (1987)
NH_4/NO_3	<0.16	Bernal et al. (1998)
CEC	>60 meq/100 g	Harada and Inoko (1980)
CEC/TOC	>1.70	Roig et al. 1988
HA/FA	>1.9	Saviozzi et al. (1988); Iglesias Jiménez and Pérez García (1992)
HI	>30	Raj and Antil (2011), Bernal et al. (1998)
GI	>50 %	Zucconi et al. (1981)
	>80 %	Tiquia et al. 1996; Sellami et al. 2008
	>110 %	Ko et al. (2008)
CO_2 production rate	<120 mg CO_2/kg/h	Hue and Liu (1995)
Microbial diversity bacterial, fungal and actinomycetes counts)	Decrease in bacterial and fungal counts, and increase in actinomycetes counts and stable at the end of composting	Petkova and Kostov (1996); Antil and Raj (2012)
Micro-organism counts (total coliform, faecal coliforms and faecal enterococci)	<500 MPN/g	Vuorinen and Saharinen (1997); Dahshan et al. (2013)
Enzyme activities cellulase, xylanase and protease)	Maximum between 30 and 60 days of composting	Goyal et al. (2005)

5.3.1.1 Colour and Odour

During composting of organic wastes, a gradual darkening or melanisation of the material takes place. The final product, after a sufficiently long period of maturation, is dark brown or almost black colour (Iglesias Jiménez and Pérez García 1992). Sugahara et al. (1979) proposed a simple technique to determine the maturity of compost by measuring the degree of darkness of composting material. It is also possible to monitor it visually the gradual process of compost darkening. In general unpleasant odour emission takes place during first thermophilic phase and then starts decreasing, with the maturity of compost. At the end of composting process, when optimal maturation is achieved, the unpleasant odour should be absent in a compost heap even on its turning of the material (Haug 1980; Vander Hoeck and Oosthoeck 1985). Although colour and odour are the simplest criteria to evaluate the maturity and stability of the compost but for confirmation some physical, chemical and biological parameters may also studied.

5.3.1.2 Temperature

Temperature evolution is an indication of microbial activity during the composting process. The temperature in the compost heap increased to thermophilic range (60–70 °C) during the first few days and then decreased gradually to a constant temperature and finally reached to ambient level (Golueke 1981; Satisha and Devarajan 2007; Raj and Antil 2011, 2012b). Stickelberger (1975) stated that compost is matured enough when its temperature remains more or less constant and does not vary with the turning over the material. Tiquia et al. (1996) also considered the temperature an important and simple maturity parameters and found positive correlation with other parameters during composting of spent pig manure saw dust litter. However, Iglesias Jiménez and Pérez García (1992) reported that exclusive determination of temperature in pilling composting plant cannot be considered in all cases the conclusive criterion to estimate the stability of compost. The rise in temperature to the thermophilic range is very necessary for weed seed and pathogen destruction in the final compost and regular monitoring of temperature in the composting pile is required to ensure the proper decomposition of organic material.

5.3.1.3 Weight Loss/Organic Matter Loss

The weight loss determination is the simplest procedure to measure the mineralization rate of organic matter (OM) during composting. The cumulative loss of organic matter increased with composting time in all types of composts and also measured in the form of weight loss. The weight loss/OM loss was found highly correlated with C:N ratio and other maturity parameters (Chefetz et al. 1996; Raj and Antil 2012b). However, thorough examination of maturity parameters involving a long term study of a single compost pile may find some correlation between them and curing time of compost (Adani et al. 1995; Wu et al. 2000).

5.3.2 Chemical Parameters

Chemical methods are widely used to assess the compost maturity. These parameters are more reliable than the physical parameters.

5.3.2.1 pH

The pH is a good indication showing the development in different stages of composting. The pH dropped slightly at the beginning of the composting process due to production of organic acids. Soon after, on utilization of these acids as substrates by other aerobic microbes, the pH increased, during the cooling and maturation stages, following lowering in pH and reached a value close to neutral (Satisha and Devarajan 2007; Ko et al. 2008). This trend of pH could be used to monitor the stabilization and maturation of compost (Gray et al. 1971; Wu et al. 2000; Raj and Antil 2012b).

5.3.2.2 EC

The EC is a measure of dissolved salts in the compost. This measure is significant because it reflects the salinity of the compost, and overly saline compost is likely harmful to plants. The sum of soluble salts in the water extracts is increased with the maturation of compost because of release of organic acids and soluble salts during organic matter decomposition indicating the stability of compost (Avnimelech et al. 1996; Wu et al. 2000).

5.3.2.3 C:N Ratio (Solid Phase)

This is the criteria traditionally used to determine maturity of compost. The relevance of C:N ratio relies on the fact that a decrease in ratio implies in the degree of humification of organic matter. As the decomposition progressed due to losses of carbon mainly as carbon dioxide, the carbon content of the compostable material decreased with time and N content per unit material increased which resulted in the decrease of C:N ratio. The researchers reported that a C:N ratio below 20 was assumed to be indicative of maturity compost (Golueke 1981), a ratio of 15 or less is preferable (Morel et al. 1985; Bernal et al. 2009). On the other hand, some researchers (Chen and Inber 1993; Sellami et al. 2008) reported that C:N ratio alone is not a sufficient criteria to determine the compost maturity and it is very necessary to associate it with some other chemical and microbial parameters to establish compost maturity (Goyal et al. 2005).

5.3.2.4 C:N Ratio (Water Extract)

Composting is a biochemical transformation of organic matter by microorganisms whose metabolism occurs in water soluble phase. Therefore, a study of the changes occurring in the soluble organic matter can be useful for the assessing maturity of compost. A C_w:N_{org} ratio of 5–6 was established by Chanyasak and Kubota (1981) as an essential indicator of compost maturity. However, this ratio is sometimes difficult to evaluate since the concentration of organic-N in water extract of mature samples is usually very low. For this reason, Hue and Liu (1995) suggested using the Cw:N_{org} ratio as a suitable parameter for assessing compost maturity, proposing a value of < 0.70 as a new index of compost stability. However, Bernal et al. (1998) found that some immature compost reached this value at thermophilic stage, thus they proposed a limit <0.55 of Cw:N_{org} to describe well matured and stabilized composts. Raj and Antil (2011) also confirmed this value as maturity index of composts prepared from farm and agro-industrial waste materials.

5.3.2.5 Water Soluble Carbon (WSC)

The WSC represents the most easily biodegradable C fraction during the composting process because it consists of sugars, organic acids, amino acids and phenols, apart from the soluble fraction of fulvic acids (Garcia et al. 1991). The concentration of WSC declined with composting time in all types of composts. The biodegradable C fractions were consumed first by microbes, which leads the decomposition of complex organic compounds resulting the release of CO_2 at the end and some of them polymerizes with nitrogenous compound producing humic substances. Eggen and Vethe (2001) and Garcia et al. (1991) established the 0.5% value of WSC as a maximum content above which compost could be considered as mature. Other values of this parameter were those suggested by Hue and Liu (1995) (WSC < 1%) and by Bernal et al. (1998) (WSC < 1.7%).

5.3.2.6 NH_4^+ and NO_3-N Concentration

Compost maturity can also be defined in terms of nitrification. When the NH_4^+-N concentration decreases and NO_3-N concentration increases in the composting material is considered ready to be used as compost (Finstein and Miller 1985). A high level of NH_4^+ concentration indicates an un-stabilized material. The absence or decrease in NH_4^+-N is an indication of both good quality and completion of maturation process (Tiquia et al. 1997). Zucconi and de Bertoldi (1987) suggested the maximum limit of NH_4^+-N content to be below 0.04% for mature compost. However, Raj and Antil (2011, 2012b) did not found this limit as an indication of maturity of composts prepared from different organic wastes. They suggested that decrease in NH_4^+-N concentration and increase in NO_3-N concentration indicates the maturity of compost. Bernal et al. (1998) suggested that the ratio of NH_4^+: NO_3-N < 0.16 is an indication of maturity of composts prepared from wide range of organic wastes.

5.3.2.7 CEC (Ash Free Material Basis)

The CEC of the composts increased with the advancement of composting time due to decomposition of organic matter. The humification process produces functional groups influenced by the increased oxidation of organic matter, which leads to rise in CEC. Consequently, measurement of CEC should be considered useful for estimating the degree of maturity (Harada and Inoko 1980). They found a highly significant negative correlation between the CEC and C:N ratio and proposed a value >60 cmol (P^+)/kg (on ash free material basis) to determine the degree of maturity of composts. Later on several workers validated this criterion for assessment of maturity of different composts (Estrada et al. 1987; Iglesias Jiménez and Pérez García 1992; Namkoong et al. 1999). However, Bernal et al. (1998) could not found CEC>60 cmol (P^+)/kg (ash free basis) a valid criteria for maturity of composts prepared from the combination of agricultural and sewage sludge waste. This was also supported by Raj and Antil (2011, 2012b) for composts prepared from farm and agro-industrial wastes. Earlier, Roig et al. (1988) reported that correlation for CEC/TOC versus Ct/Nt (t=total), seems to be very useful for estimating the degree of maturity of cattle, sheep, chicken and rabbit manure. They suggested that a value of CEC/TOC over 1.7 reflect a good humification level of these waste manures. Iglesias Jiménez and Pérez García (1992) reported that numerical value of CEC is influenced by nature of the original material and suggested that CEC/TOC (total organic carbon) ratio can be used as an index value for assessment of maturity of composts as also evidenced by Raj and Antil (2011) who found this criteria highly correlated with other parameters and suggested CEC/TOC>1.7 as maturity index.

5.3.2.8 Humification Parameters

The humified fraction of the soil organic matter is the most important one responsible for organic fertility functions in the soil (Bernal et al. 2009). So the evaluation of the humification degree of the OM during composting is a criterion for compost maturity. During composting, humic substances (alkali-extractable organic-C, C_{EX}) are produced and humic acid like organic-C (C_{HA}) increases, while fulvic acid like organic-C (C_{FA}) and water extractable organic-C decreases due to microbial degradation (Singh et al. 1987; Singh and Amberger 1990; Raj and Antil 2012b). Some indices used for evaluation of the humification level in the material during composting include (Roletto et al. 1985; Senesi 1989)

- Humification ratio (HR): $C_{EX}/Corg \times 100$
- Humification index (HI): $C_{HA}/Corg \times 100$
- Percent humic acids (PHA): $C_{HA}/C_{EX} \times 100$
- Humic acid to fulvic acid ratio: C_{HA}/C_{FA}

Since maturation also implies the formation of some humic like substances, the degree of OM humification is generally accepted as a criterion of maturity and above parameters can be used as maturity index (Bernal et al. 2009). Saviozzi et al. (1988) proposed HA:FA ratio to judge the matutity of compost. Iglesias Jiménez and Pérez García (1992) adopted a ratio of 1.9 to determine the maturity of composts prepared from city waste and sewage sludge compost. Raj and Antil (2012b) proposed HI>30 as a maturity index for the composts prepared from agro-industrial and farm wastes composts. Numerous chemical, physico-chemical and spectroscopic methods such as: elemental and functional groups composition, ratio of absorbance measured at 465 and 665 nm (E4/E6), molecular weight distribution, electrophoresis and electrofocusing, pyrolysis-gas chromatography–mass spectrometry (GC-MS), infrared and Fourier transformed–infrared (FT-IR) spectroscopy, electron spin resonance (ESR) spectroscopy and fluorescence spectroscopy (Moral et al. 2009). Amongst these methods, advanced techniques such as NMR, FT-IR and pyrolysis have been employed to achieve a better understanding of the structural changes of the OM during composting and hence to evaluate composting efficiency and compost maturity (Chen 2003). However, these humification parameters depend on the type of waste used for composting and their same value cannot be used for all type of composts.

5.3.3 Biological Parameters

Biological methods are widely used to assess the compost maturity. These parameters are comparatively more reliable than the other parameters.

5.3.3.1 Germination Test

The application of immature compost to arable soil may inhibit seed germination and or reduce the root length of seedlings due to the creation of reducing conditions in the soil and the presence of phytotoxic compounds in the compost. Based on this fact the germination test has been developed to determine the degree of maturity, GI was calculated using the following expression (Zucconi et al. 1981):

$$\text{GI} = \text{Percent seed germination} \times \text{mean root length}$$

They used cress seeds (*Lepidium sativum*) because of their quick response for determining the presence of phytotoxic substances in the compost. The proposed value of GI above 50 % is an index to judge maturity of compost. Some workers also studied the plant growth test (assessment of top growth and sometimes root biomass) and found that growth of plants inhibited by immature compost (Inbar et al. 1993; Iannotti et al. 1994; Chefetz et al. 1996). Zucconi and de Bertoldi (1987) discussed the differences between the seed germination and growth tests. Germination tests provide an instant picture of phytotoxicity, whereas growing test will be affected by continuing changes in the stability or maturity of the compost tested: there may be damaging effects on growth in the earlier stages, but beneficial effects later on, with different conclusions depending up on time of assessment. According to Zucconi et al. (1981), a GI value of 80 % has been used as an indicator of disappearance of phytotoxicity in composts. Tiquia et al. (1996) used this value not only as an indication of disappearance of phytotoxicity, but also an indication of maturity of compost. However, Antil and Raj (2012) reported a lower value of GI (>70 %) as a maturity index of composts prepared from wide range of organic wastes. Ko et al. (2008) reported that the compost having GI value >110 % was considered mature compost because the GI values reported by previous researchers were not a suitable threshold value for determining the maturity of composted animal manure. However, the results obtained using GI should be interpreted with caution, because the GI was affected by the type of seed used and applied extraction rates (Bernal et al. 1998; Wu et al. 2000; Tang et al. 2006).

5.3.3.2 Oxygen and CO_2 Respirometry

The aerobic respiration rate was selected as the most suitable parameters to assess the aerobic biological activity and hence stability. The respiration can be measured in several ways: carbon dioxide evolution, oxygen consumption and self-heating, which are indicative of the amount of degradable OM still present and which are related inversely to stabilization (Zucconi and de Bertoldi 1987). Immature compost has a strong demand for oxygen and high CO_2 production rates, due to high microbial activity as a consequence of the abundance of easily biodegradable compounds in the raw material. For this reason, O_2 consumption and CO_2 production are indicative of compost stability and maturity (Iannotti et al. 1994; Hue and Liu 1995; Barrena-Gómez et al. 2006). Hue and Liu (1995) set the limit of the CO_2 production rate for compost maturity at <120 mg CO_2 $kg^{-1}h^{-1}$. Wang et al. (2004) used a respiration rate of <1 mg CO_2-C g^{-1} dwd^{-1} to define a highly stabilized compost. Cooperband et al. (2003) suggested a NO_3-N/CO_2-C g^{-1} ratio >8 per day as an index for compost maturity. The relationship between CO_2 respiration and phytotoxicity of immature compost was studied by García-Gómez et al. (2003) and found that the CO_2-C evolved correlated with plant growth and immature compost caused N-immobilisation in the soil, leading to plant N-deficiency.

5.3.3.3 Microbial Population/Count

The decomposition of organic matter is a microbial process and directly related with total microbial count and their activity during composting of organic wastes (Morel et al. 1985). This method is based on the initial hypothesis that the microbial population stabilized on the maturity of compost. Some important studies in the evolution and quantification of different group of microorganisms involved in the process of composting have been carried (Albonetti and Massari 1979; De Bertoldi and Zucconi 1980). The microbial population increased at thermophilic stage and then decreased at maturation phase and become constant at the end of composting indicating the stable nature of composting material (Petkova and Kostov 1996; Tiquia et al. 1996). A decrease in bacterial and fungal counts, and increase in actinomycetes counts and stable at the end of composting appear as useful indicator for establishing biological stabilization and optimum degree of compost maturity (Antil and Raj 2012). A gummy whitish appearance in composting materials indicates the presence of actinomycetes. For compost hygenization, the average number of total coliform, faecal coliforms and faecal enterococci densities should be less than <500 MPN/g (Vuorinen and Saharinen 1997; Dahshan et al. 2013).

5.3.3.4 Enzyme Activities

Various hydrolytic enzymes (cellulase, xylanase and protease) are believed to control the rate at which different substrates are degraded. Enzymes are the main mediators of various degradative processes (Tiquia et al. 1996). Maximum activities of cellulase, xylanase and protease observed between 30 and 60 days of composting can be taken as index of compost maturity (Goyal et al. 2005).

5.4 Compost Quality Standards

Different official and private organisations (European commission 2001; TMECC 2002; BOE 2005; BSI 2005; Ge et al. 2006) proposed many compost quality standards by taking into consideration compost properties viz. foreign matter (inert contamination), potentially toxic elements (organic contamination and heavy metals), sanitisation (pathogens and phytopathogens), maturity and stability, weed seeds, water, OM and nutrient content (Bernal et al. 2009). At present, there is a need for standardization of such criteria at the international level.

5.5 Conclusion

The composting of organic waste materials has been demonstrated to be an effective method of producing end-products which are stabilized, ensuring their maximum benefit for agriculture. Composts are valuable products which can be used as sources of soil amendment and organic matter in agricultural land. Composts prepared from different organic wastes differ in their quality and stability. Several indices based on physical, chemical and biological parameters have been used for compost by different authors. Despite all the proposed methods to establish the degree of maturity and stability of composts, no single method can be universally applied to all composts due to wide variations in composition of raw material used for composting and composting technology. Hence, compost maturity should be assessed by measuring two or more parameters. However, it is necessary to standard the criteria used by official institutes from different countries.

References

Adani F, Genevini PL, Tambone F (1995) An new index of organic matter stability. Compost Sci Util 3(2):25–37

Albonetti SG, Massar G (1979) Microbiological aspects of a municipal waste composting system. Eur J Appl Microbiol Biotechnol 7:91–98

Antil RS, Raj D (2012) Chemical and microbiological parameters for the characterization of maturity of composts prepared from farm and agro-industrial wastes. Arch Agron Soil Sci 58:833–845

Antil RS, Raj D, Narwal RP, Singh JP (2012) Evaluation of maturity and stability parameters of composts prepared from a wide range of organic wastes and their response to wheat. Waste Biomass Valoriz 4:95–104

Avnimelech Y, Bruner M, Ezrony I, Sela R, Kochba M (1996) Stability index for municipal solid wastes compost. Compost Sci Util 4(2):13–20

Barrena-Gómez R, Vázquez Lima F, Sánchez Ferrer A (2006) The use of respiration indices in the composting process: a review. Waste Manage Res 24:37–47

Benito M, Masaguer A, Moliner A, Arrigo N, Palma RM (2003) Chemical and microbiological parameters for the characterization of the stability and maturity of pruning waste compost. Biol Fertil Soils 37:184–189

Bernal MP, Paredes C, Sanchew-Monedero MA, Cegarra J (1998) Maturity and stability parameters of composts prepared with a wide range of organic wastes. Bioresour Technol 63:91–99

Bernal MP, Alburquerque JA, Moral R (2009) Composting of animal manures and chemical criteria of compost maturity assessment: a review. Bioresour Technol 100:5444–5453

BOE (2005) Real Decreto 824/2005, de 8 de julio, sobre productos fertilizantes. Boletín Oficial del Estado 171:25592–25669

Boulter JI, Boland GJ, Trevors, JT (2000) Compost: a study of the development process and end-product potential for suppression of turfgrass disease. World J Microbio Biotech 16:115–134

BSI (2005) PAS 100:2005, Specification for composted materials. British Standard Institution, London

Cambardella CA, Richard TL, Russell A (2003) Compost mineralization in soil as a function of composting process conditions. Eur J Soil Biol 39:117–127

Cardenas RR, Wang LK (1980) Composting process. Handbook of environmental engineering, vol II. The Human Press, New York, pp 269–327

Chang JI, Chen YJ (2010) Effects of bulking agents on food waste composting. Bioresour Technol 101:5917–5924

Chanyasak V, Kubota H (1981) Carbon/organic nitrogen ratio in water extract as measure of compost degradation. J Ferment Technol 59:215–219

Chefetz B, Hatcher PG, Hadar Y, Chen Y (1996) Chemical and biological characterization of organic matter during composting of municipal solid wastes. J Environ Qual 25:776–785

Chen Y (2003) Nuclear magnetic resonance, infra-red and pyrolysis: application of spectroscopic methodologies to maturity determination of composts. Compost Sci Utiliz 11:152–168

Chen Y, Inbar Y (1993) Chemical and spectroscopic analyses of organic matter transformation during composting in relation to compost maturity. In: Hoitink HAJ, Keener HM (eds) Science and engineering of composting: design, environmental, microbiology and utilization aspects. Renaissance Publications, Ohio, pp 551–600

Cooperband LR, Stone AG, Fryda MR, Ravet JL (2003) Relating compost measures of stability and maturity to plant growth. Compost Sci Util 11:113–124

Dahshan H, Abd-Elall AMM, Mahdy A, Abd-El-Kader, Megahed AM (2013) Field-scale management and evaluation of recycled cattle manure solids from livestock in Nile Delta ecosystem. African J Agric Res 8:2829–2836

de Bertoldi M, Zucconi F (1980) Microbiologia della transformazione deirifiuti solidi unbani in compost e loro utilizzazione in agricoltura. Ingegneria ambientale 9:209–216

de Bertoldi M, Vallini G, Pera A (1983) The biology of composting: a review. Waste Manage Res 1:157–176

Eggen T, Vethe O (2001) Stability indices for different composts. Compost Sci Util 9:19–26

Epstein E (1997) The science of composting. Technomic Publishing Company Inc., Lancaster

Estrada J, Sana J, Cequiel RM, Cruanas R (1987) Application of a new method for CEC determination as a compost maturity index. In: De Berfoldi M, Feerranti MP, Hermite PL, Zucconi F (eds) Compost: production, quality and use. Elsevier, London pp 334–340

European Commission (2001) Working document: biological treatment of biowaste, 2nd draft, pp 22

Finstein MS, Miller FC (1985) Principles of composting leading to maximization of decomposition rate, odor control and cost effectiveness. In: Gasser JKR (ed) Composting of agricultural and other wastes. Elsevier Applied science Publications, Barking, pp 13–26

Fuchs JG (2002) Practical use of quality compost for plant health and vitality improvement. In: Insam H, Riddech N, Klammer S (eds) Microbiology of composting. Springer, Heiderlberg, pp 435–444

Garcia C, Hernandez T, Costa F (1991) Changes in carbon fractions during composting and maturation of organic wastes. Environ Manage 15:433–439

García-Gómez A, Bernal MP, Roig A (2003) Carbon mineralisation and plant growth in soil amended with compost samples at different degrees of maturity. Waste Manage Res 21(2):161–171

Ge B, McCartney D, Zeb J (2006) Compost environmental protection standards in Canada. J Environ Eng Sci 5:221–234

Golueke CG (1981) Principles of biological resources recovery. BioCycle 22:36–40

Goyal S, Dhull SK, Kapoor KK (2005) Chemical and biological changes during composting of different wastes and assessment of compost maturity. Bioresour Technol 96:1584–1591

Gray KR, Sherman K, Biddlestone AJ (1971) A review of composting. Part I. Microbiology and biochemistry, process. Biochemistry 6:32–36

Harada Y, Inoko A (1980) Relationship between cation exchange capacity and degree of maturity of city refuse composts. Soil Sci Plant Nutr 26:353–362

Haug RT (1980) Compost engineering: principals and practice. Ann Arbor Science, Michigan

He XT, Logan TJ, Traine SJ (1995) Physical and chemical characteristics of selected US municipal solid waste composts. J Environ Qual 24:543–552

Hellmann B, Zelles L, Palojarvi A, Bai Q (1997) Emissions of climate-relevant trace gases and succession of microbial communities during open-windrow composting. Applied Environ Microbiol 63:1011–1018

Hoekstra NJ, Bosker T, Lantinga EA (2002) Effects of cattle dung from farms with different feeding strategies on germination and initial root growth of cress (*Lepidium sativum* L). Agric Ecosyst Environ 93:189–196

Hoitink HAJ (2000) Trends in treatment and utilization of solid wastes through composting in the United States. In: Warman PR, Taylor BR (eds) Proc International Composting Symp, vol 1. CBA Press Inc., Nova Scotia, pp 1–13

Hue NV, Liu J (1995) Predicting compost stability. Compost Sci Util 3:8–15

Iannotti DA, Grebus ME, Toth BL, Madden LV, Hoitink HAJ (1994) Oxygen respirometry to assess stability and maturity of composted municipal solid waste. J Environ Qual 23:1177–1183

Iglesias Jiménez E, Pérez García V (1992) Determination of maturity indices for city refuse composts. Agric Ecosyst Environ 38(4):331–343

Inbar Y, Hadar Y, Chen Y (1993) Recycling of cattle manure: the composting process and characterization of maturity. J Environ Qual 22:857–863

Ishii K, Fukui M, Takii S (2000) Microbial succession during a composting process as evaluated by denaturing gradient gel electrophoresis analysis. J Appl Microbiol 89:768–777

Itavaara M, Vikman M, Venelampi O (1997) Windrow composting of biodegradable packaging materials. Compost Sci Util 5:84–92

Itavaara M, Venelampi O, Vikman M, Kapanen A (2002) Compost maturity—problems associated with testing. In: Insam H, Riddech N, Klammer S (eds) Microbiology of composting. Springer, Heidelberg, pp 373–382

Juste C (1980) Advantages et inconvenient de l' utilization des composts d' ordures menageres comme amendment organique des soils ou supports de culture International conference on compost, 22–26 Jan, Madrid, Spain. Min Obras Publicas

Keener HM, Dick WA, Hoitink HAJ (2000) Composting and beneficial utilization of composted by-product materials. In: Dick WA (ed) Land application of agricultural, industrial, and municipal by-products. Soil Science Society of America Inc., Madison, pp 315–341

Kirchmann H, Widen P (1994) Separately collected organic household wastes. Swedish J Agric Res 24:3–12

Ko HJ, Kim KY, Kim HT, Kim CN, Umeda M (2008) Evaluation of maturity parameters and heavy metal contents in composts made from animal manure. Waste Manage 28:813–820
Larney FJ, Hao X (2007) A review of composting as a management alternative for beef cattle feedlot manure in southern Alberta, Canada. Bioresour Technol 98:3221–3227
Mathur SP, Owen G, Dinel H, Schnitzer M (1993) Determination of compost biomaturity. 1. Literature review. Biol Agric Hortic 10:87–108
Metcalf, Eddy Inc. (1991) Wastewater engineering: treatment, disposal and reuse, 3rd edn. McGraw-Hill, New York
Miller FC (1992) Composting as a process based on the control of ecologically selective factors. In: Metting FB Jr (ed) Soil microbial ecology, applications in agricultural and environmental management. Marcel Dekker Inc., New York, pp 515–544
Moral R, Paredes C, Bustamente MA, Marhuenda-Egea F, Bernal MP (2009) Utilisation of manure composts by high-value crops: safety and environmental challenges. Bioresour Technol 100:5454–5460
Morel JL, Colin F, Germon JC, Godin P, Juste C (1985) Methods for the evaluation of the maturity of municipal refuse compost. In: Gasser JKR (ed) Composting of agricultural and other wastes. Elsevier Applied Science Publications, Barking, pp 56–72
Namkoong W, Hwang EY, Cheong JG, Choi JY (1999) A comparative evaluation of maturity parameters for food waste composting. Compost Sci Util 7:55–62
Pereira-Neta JT (1987) On the treatment of municipal refuse and sewage sludge using aerated static pile composting—a low-cost technology approach. Ph.D. dissertation, Leeds University, UK
Petkova G, Kostov O (1996) Microbiological processes under vine-twig composting. Pochvo Agrok Ekol 31(5):25–28
Poincelot RP (1974) A scientific examination of the principals and practice if composting. Compost Sci 15:24–31
Raj D, Antil RS (2011) Evaluation of maturity and stability parameters of composts prepared from agro-industrial wastes. Bioresour Technol 102:2868–2873
Raj D, Antil RS (2012a) Phytotoxicity evaluation and response of wheat to agro-industrial waste composts. Arch Agron Soil Sci 58:73–84
Raj D, Antil RS (2012b) Evaluation of maturity and stability parameters of composts prepared from farm wastes. Arch Agron Soil Sci 58:817–832
Ranalli G, Botturea G, Taddei P, Garavni M, Marchetti R, Sorlini G (2001) Composting of solid and sludge residues from agricultural and food industries. Bioindicators of monitoring and compost maturity. J Environ Sci Health 36:415–436
Roig A, Lax A, Gegarra J, Costa T, Harnandez MT (1988) Cation exchange capacity as a parameter for measuring the humification degree of manures. Soil Sci 146:311–316
Roletto E, Barberis R, Consiglio M, Jodice R (1985) Chemical parameters for evaluating compost maturity. BioCycle 9:46–48
Sasaay AA, Lasaridi K, Stentiford E, Budd T (1997) Controlled composting of paper sludge using the aerated static pile method. Compost Sci Util 5:82–96
Satisha GC, Devarajan L (2007) Effect of amendments on windrow composting of sugar industry pressmud. Waste Manage 27:1083–1091
Saviozzi A, Levi-Minzi R, Riffaldi R (1988) Maturity evaluation of organic wastes. BioCycle 29:54–56
Sellami F, Hachicha S, Chtourou M, Medhioub K, Ammar E (2008) Maturity assessment of composted olive mill wastes using UV spectra and humification parameters. Bioresour Technol 99:6900–6907
Senesi N (1989) Composted materials as organic fertilisers. Sci Total Environ (81/82):521–542
Singh CP, Amberger A (1990) Humic substances in straw compost with rock phosphate. Biol Waste 31:165–174
Singh CP, Ruhal DS, Singh MS (1987) The solubilization of low grade rock phosphate with anaerobic digested slurry. Int Biodeterior 23:249–256
Stickelberger D (1975) Survey if city refuse composting. In: Organic matters as fertilizers. Swedish International Development Authority, FAO, Soils Bulletin. 27, Rome, pp 185–209

Sugahara K, Harada Y, Inoko A (1979) Color change of city refuse during composting process. Soil Science Plant Nutr 25:197–208

Tang JC, Maie N, Tada Y, Katayama A (2006) Characteristics of the maturing process of cattle manure compost. Process Biochem 41:380–389

Tiquia SM, Tam NFY, Hodgkiss IJ (1996) Effects of composting on phytotoxicity of spent pig-manure sawdust litter. Envrion Pollut 93:249–256

Tiquia SM, Tam NFY, Hodgkiss IJ (1997) Effect of turning frequency on composting of spent pig-manure sawdust litter. Bioresour Technol 62:37–42

TMECC (2002) Test methods for the examination of composting and compost. US Composting Council, Bethesda

Vander Hoeck KW, Oosthoeck J (1985) Composting odour emission and odour control by bio-filtration. In: Gasser JKR (ed) Composting of agricultural and other wastes. Elsevier Applied Science Publishers, London, pp 271–279

Vuorinen AH, Saharinen MH (1997) Evolution of microbiological and chemical parameters during manure and straw co-composting in a drum composting system. J Agric Ecosyst Environ 66:19–29

Wang P, Changa CM, Watson ME, Dick WA, Chen Y, Hoitink HAJ (2004) Maturity indices for composted dairy and pig manures. Soil Biol Biochem 36:767–776

Wu L, Ma LQ, Martinez GA (2000) Comparison of methods for evaluating stability and maturity of biosolids composts. J Environ Qual 29:424–429

Zucconi F, de Bertoldi M (1987) Compost specification for the production and characterization of compost from municipal solid waste. In: De Berfoldi M, Feerranti MP, Hermite PL, Zucconi F (eds) Compost: production, quality and use. Elsevier, Barking, pp 30–50

Zucconi F, Pera A, Forte M, de Bertoldi M (1981) Evaluating toxicity of immature compost. Bio-Cycle 22:54–57

Zucconi F, Monaco CA, Forte M, de Bertoldi M (1985) Phytotoxins during the stabilization of organic matter. In: Gasser JKR (ed) Composting of agricultural and other wastes. Elsevier Applied Science Publishers, Barking, pp 73–85

Chapter 6
Thermophilic *Bacilli* and their Enzymes in Composting

Abhishek Bhattacharya and Brett I. Pletschke

Abstract Thermophilic *Bacilli* play a key role in the biocomposting process. Methods for analysing microbial diversity and community structure during composting have been especially helpful in characterising this process more completely. Thermophilic *Bacilli* account for the most prevalent group of thermophiles present in compost and the role of these bacteria in composting is highlighted. The spectrum of thermophilic *Bacilli* and the progression of their enzymes during the composting process are discussed, with a special emphasis on cellulolytic thermophilic *Bacilli* and cellulases. Finally, the ecological and economic impacts, as well as future prospects of composting are also discussed.

Keywords Bacillus · Cellulases · Composting · Enzymes · Thermophilic

6.1 Introduction

Decomposition of plant and animal matter in a naturally occurring or man-made environment, resulting into a nutritionally rich product is a basic definition of composting. The natural process of composting involves the decomposition of large amounts of organic material on a habitual basis by naturally occurring micro-flora, resulting into the formation of humus and is an excellent route of nutrient turnover in the natural ecosystem. The process of humus formation is a slow process which can be accelerated by creating conditions ideal for growth and activity of microbial decomposers. Decomposition of organic matter at an accelerated rate by a mixed population of microorganisms in a warm, moist, aerobic environment defines the process of composting (Rawat et al. 2005). The process can be defined as either 'passive' or 'active', involving complex physico-chemical interactions between the

B. I. Pletschke (✉) · A. Bhattacharya
Department of Biochemistry and Microbiology,
Rhodes University, PO Box 94, 6140, Grahamstown, South Africa
e-mail: b.pletschke@ru.ac.za

D. K. Maheshwari (ed.), *Composting for Sustainable Agriculture*, Sustainable Development and Biodiversity 3, DOI 10.1007/978-3-319-08004-8_6

organic matter and decomposers, and results in the formation of a range of end products or compost. The compost has many interesting and useful properties and can be used as a fertiliser, a substitute for peat in horticulture, a natural pesticide, a soil conditioner (as it improves soil structure, texture, aeration and water retention), and a microbial additive to increase enzyme activity as humus (Odlare 2005). There are several advantages to the process of composting over incineration of waste material. Composting is economical, safe, and environmentally favourable. Decomposition of organic waste by composting instead of incineration avoids the need of oil and also mitigates the release of CO_2 and other greenhouse gasses into the atmosphere.

Composting is environmentally safe since it does not produce harmful substances such as nitrogen oxides (NOx), sulfur oxides (SOx), and dioxins. Composting kills pathogenic microbes and viruses, and also seeds of weeds. Since incineration is now generally prohibited in many countries, composting has become more important for decomposing organic wastes than before.

Changes in physico-chemical conditions of compost drive the changes in microbial community structures and composition. Compost has a defined thermophilic microflora which constitutes the predominant component and provides selectivity to compost. Thus, the optimisation of compost quality is directly linked to composition and succession of microbial communities in the composting process. Heat emitted from fermentation processes of microorganisms causes the inside of compost piles to become hot, and thus acts as one of the important environments for isolating thermophiles (Finstein and Morris 1975). According to the literature, the inside temperature reaches up to 75–80 °C (Saiki et al. 1978). The isolation of many moderate thermophiles belonging to the genera *Geobacillus, Bacillus*, and *Clostridium* and related species has been reported. Methanogens, including thermophilic methanogens (which belong to Archaea) have also been isolated from traditional compost.

6.2 Thermophiles

Of the enormous range of temperatures known (from 0 to approx. 3×10^9 K), only a small fraction is compatible with life. However, significant attention towards exploring the biodiversity of life at elevated temperatures has established that microbial life can prevail or thrive in the upper as well as lower temperature limits. Life under extremes of temperature creates a series of challenges, at low temperatures the structural conformation of the cell is disintegrated due to formation of ice crystals; while at the other extreme the denaturation of biomolecules and cell components bring about the destruction of cells at high temperatures. A wide variety of microorganisms that thrive under these denaturing conditions at elevated temperatures have, however, been discovered and can overcome these challenges. One such group of microorganisms includes the thermophiles. The word "thermophile" has been derived from two Greek words "thermotita" (meaning heat) and "philia" (meaning love). Thermophiles are heat-loving organisms, which not only tolerate high temperatures but also usually require these for their growth and survival. A thermophile as defined by Brock (1978) is "an organism capable of living at temperatures at or near the maximum for the taxonomic group of which it is part." Cellular structure and activities are affected by various factors including temperature. Thermophilic microbes have developed special attributes in their structural make-up in terms of its major components, including proteins, nucleic acids and lipids, which allow them to prevail and thrive at high temperatures. A significant increase in the proportion of nonpolar, hydrophobic amino acids (isoleucine, leucine, methionine, phenylalanine, and valine) indicates the role of hydrophobicity for stabilization of thermophilic proteins (Lieph et al. 2006). A reduction in the glycine content and extensive ionic interactions that form a network over the surface of the molecule assists the compacted protein to resist unfolding at high temperature (Ladenstein and Antranikian 1998). A higher content of arginine and a reduction in the total number of thermally unstable residues, such as Cys, Lys, Met, Asn and Gln (Cicicopol et al. 1994), have been associated with thermostable proteins. An increase in ion-pair content, or a change from monomeric to oligomeric structure (Cicicopol et al. 1994) a decrease in the length of surface loops that connect elements of secondary structure (Thompson and Eisenberg 1999), additional networks of hydrogen bonds (Jaenicke and Bohm 1998), an increase in disulfide bond formation (Beeby et al. 2005), and an exchange of amino acids to increase helix propensity of residues in alpha–helices and the presence of noncovalent, ionic bonds called salt bridges on a protein's surface, are some of the known factors that have been shown to play a major role in maintaining the biologically active structure of thermophilic proteins (Das and Gerstein 2000).

6.2.1 A Brief Classification of Thermophilic Bacilli

The thermophilic *Bacilli* are a phenotypically diverse group, that are Gram-positive and/or Gram-variable, aerobic or facultative anaerobic, endospore-forming, rod-shaped bacteria. This group has undergone considerable reclassification, as the advancement in molecular biology has resulted in a shift in the identification pattern, from the basis of phenotypic characterization towards nucleotide sequence analysis of conserved genes, particularly, by 16s rRNA gene analysis. According to Ludwig et al. (2007) and Logan et al. (2009), the thermophilic *Bacilli* are categorised under the phylum *Firmicutes* and class *Bacilli* which is further subdivided into order *Bacillales* and *Lactobacillales*. The order *Bacillales* consists of nine families, namely: (I) Bacillaceae, (II) Alicyclobacillaceae, (III) *Listeriaceae*, (IV) *Paenibacillaceae*, (V) Pasteuriaceae, (VI) *Planococcaceae*, (VII) *Sporolactobacillaceae*, (VIII) *Staphylococcaceae*, (IX) *Thermoactinomycetaceae*. The order *Lactobacillales* is composed of six families, including: (I) *Lactobacillaceae*, (II) *Aerococcaceae*, (III) *Carnobacteriaceae*, (IV) *Enterococcaceae*, (V) *Leuconostocaceae* and (VI) *Streptococcaceae*. The genus *Bacillus, Paenibacillus* and *Streptococcus* are extensively studied due to their vast occurrence and their ability to utilize diverse carbon compounds as energy source (Cihan et al. 2012). *Bacillus* and related genera have been shown to produce a variety of enzymes with industrial importance owing to their versatility and performance over a broad pH and temperature range (Cihan et al. 2012).

6.3 Methods for Analysing Microbial Diversity and Community Structure during Composting

Compost bears a rich microbial diversity and community structure during process. Several methods including advantageous molecular method though employed in analysis of microbial community dynamics in composting processes, among few are discussed-

6.3.1 Enrichment Methods

Enrichment is a culture dependent approach and has been traditionally used for cultivation of selective microbial communities. The compost provides a complex nutritional source that can support different types of microbes with unique growth requirements. Thus, the enrichment process uses different incubation and cultivation conditions that can support the group of multiple communities with specific growth requirements. Enrichment methods have been successfully used for the enumeration and observation of microbial community dynamics during composting of animal manures (Tiquia et al. 2002), municipal solid (Hassen et al. 2001), food waste (Ryckeboer et al. 2003) etc. These types of methods can only account for 0.1–10.0 % fraction of the total microbial diversity and thus have been replaced with a number of cultivation independent approaches.

6.3.2 Biochemical Methods

The analysis of community structure using biochemical identification systems such as API and Biolog were introduced by Garland and Mills (1991). It is termed as the Sole-Carbon-Source Utilization (SCSU) and also known as Community Level Physiological Profiling (CLPP). The method is based on the metabolic machinery of the microbes and their ability to utilize different carbon sources. However, this method is limited by the availability of kits that can be used to decipher the physiological and metabolic diversity (Derry et al. 1998). Optimization of substrate combinations for the study of environmental isolates is limited that often results in problems leading to invalid identification of microbial species.

6.3.3 Phospholipid Fatty Acid (PLFA) Analysis

Microbial characterization based on fatty acid composition has been extensively used. The method is quantitative, robust, and reproducible. The method analyses the differences in the fatty acid composition of the microbes, especially in the range of C2 to C24, and is independent of plasmids, mutations, or damaged cells. It is also known as fatty acid methyl ester (FAME) analysis. This method has been used to monitor the microbial community succession in a number of thermophilic composting processes. The process is limited by the fact that it does not identify the species; rather it provides a description of microbial communities based on functional groups (Eiland et al. 2001).

6.3.4 *Molecular Methods*

The last decade has seen an upsurge in the use of molecular phylogenetic techniques for analysis of microbial diversity, community structure analysis and community succession monitoring.

6.3.4.1 Mole Percentage Guanine + Cytosine (mol% G + C)

Mol% G + C is determined by the thermal denaturation of DNA, within bacteria this value ranges from 25 % up to 75 %, although it remains constant for a certain species. The method is not influenced by the anomalies associated with PCR and has been used to uncover rare members in microbial populations (Tiedje et al. 1999).

6.3.4.2 Nucleic Acid Hybridization

Nucleic acid hybridization provides a powerful tool for the qualitative and quantitative assessment of bacterial ecology and community structure. The method uses oligonucleotide or polynucleotide's probes obtained from sequences available in the library specific to a community or species and can be tagged with markers at the 5′ end (Goris et al. 2007). Dot-blot hybridization and cellular level hybridization have been successfully used to provide valuable information regarding the distribution of microbial communities in a natural environment. Fluorescent In-situ hybridization (FISH) is a very popular method to study bacterial diversity and community structure (Fakruddin and Mannann 2013).

6.3.4.3 DNA Re-association

The diversity of the microbial communities of an environment can be studied based upon the kinetics of DNA re-association, which reflects the variety of sequences present in the environment and thus provides a measurement that reflects the genetic complexity of the microbial community (Torsvik et al. 1996). The rate of annealing or re-association of the total DNA will depend upon the diversity of the microbial community and a decrease in the rate of association will indicate the complexity of the microbial community in a given environment (Theron and Cloete 2000).

6.3.4.4 Restriction Fragment Length Polymorphism (RFLP)

This method relies on the DNA polymorphism and has been applied to estimate the diversity and community structure in different environments (Moyer et al. 1996). This method, in combination with DNA hybridization and electrophoresis, provides a powerful tool for strain typing and also for the determination of intra-species variation. The method has been successfully used for the detection of specific phylogenetic groups within a community; however the complexity of the banding patterns limits the application of this method for analysing bacterial diversity (Tiedje et al. 1999).

6.3.4.5 Terminal Restriction Fragment Length Polymorphism (T-RFLP)

T-RFLP is an extension of RFLP/ARDRA and addresses their shortcomings; it provides an easy and powerful tool for microbial community analysis in various environments. In this method, one set of the PCR primer is labelled with a fluorescent dye, such as TET (tetracholoro-6-carboxyfluoroscein) or 6-FAM (phosphoramidite fluorochrome 5-carboxyfluorescein). The generated Amplicons are digested with restriction enzymes and electrophoresed using an agarose gel. The labelled fragments termed as T-RF's are separated and analysed using an automated sequencer. Each unique fragment length is described as an Operational Taxonomic Unit (OTU) and the frequency of each OTU can be counted and used as a measure of diversity and richness of the microbial community (Liu et al. 1997). This method has been applied to determine bacterial diversity and community succession during composting in different environments. T-RFLP has been considered a highly effective tool for comparing the relationships between different environmental samples and also in the measurement of spatial and temporal changes in bacterial communities with five times greater success rate than DGGE (Tiedje et al. 1999).

6.3.4.6 Denaturant Gradient Gel Electrophoresis (DGGE)/Temperature Gradient Gel Electrophoresis (TGGE)

DGGE or TGGE allows the separation of DNA fragments with the same length but with different base pair sequences. The method involves the amplification of the 16S or 18S rRNA gene and separation of the partially melted DNA molecules in acrylamide gels containing a linear gradient of DNA denaturants (urea and formamide). The variation in the nucleotide sequences within the DNA generates different fragments based on the differences in melting points. DNA sequences with a difference of only one base pair can be separated using DGGE. TGGE is based on the same principle as DGGE, but in this method a temperature gradient is employed in place of chemical denaturants. The method is cost-effective, reproducible, reliable and rapid (Muyzer 1999). DGGE has been used to monitor microbial community profiles and succession in a variety of composts (Ishii et al. 2000; Mühling et al. 2008).

6.3.4.7 Single Strand Conformation Polymorphism (SSCP)

The method was originally used for the detection of novel polymorphisms or point mutations in DNA (Peters et al. 2000) and has been adapted to analyse the succession of microbial communities. Similar to DGGE, this method allows the separation of DNA molecules with the same length and different DNA fragments; however the separation on polyacrylamide gel is based on the differences in motilities that originate from the presence of secondary structures (heteroduplex) in the DNA molecules. The conformational changes in the secondary structure are a unique characteristic of each DNA molecule, and this method has been used to track changes in bacterial populations throughout the composting process, thereby allowing the assessment of bacterial diversity and also in the identification of important microbes (Alfreider et al. 2002).

6.3.4.8 Pyrosequencing

Pyrosequencing determines the order of nucleotides in DNA based on the sequencing by synthesis principle and relies on the detection of pyrophosphate release on nucleotide incorporation rather than chain termination with dideoxynucleotides. This method has several advantages over other molecular methods, as it allows the determination of the exact sequence by providing the same accuracy as conventional sequencing methods. It dispenses with the need for labelled nucleotides, labelled primers, and electrophoresis and allows the sequencing of large number of samples in a short period of time. In the context of analysing bacterial diversity, it allows for the study of a large number of samples from a given population and thus counters the deficiency associated with limited analysis of a small number of clones and a few different samples for the determination of a bacterial community (Lauberet et al. 2009; Fakruddin et al. 2012).

6.3.4.9 Illumina-based High Throughput Microbial Community Analysis

In this method, DNA molecules and primers are first attached on a slide and amplified with polymerase, so that local clonal DNA colonies, or otherwise termed "DNA clusters" are formed. To determine the sequence, four types of reversible terminator bases (RT-bases) are added and non-incorporated nucleotides are washed away. This technology supports both single read and paired-end libraries. This method offers a short-insert paired-end capability for high-resolution genome sequencing as well as long-insert paired-end reads using the same robust chemistry for efficient sequence assembly, de novo sequencing, large-scale structural variation detection etc. The combination of short inserts and longer reads increases the ability to fully characterize any genome (Caporaso et al. 2012).

6.4 Role of Thermophilic *Bacilli* in Composting

On the basis of temperature, composting can be divided into two types (a) hot composting—this is the most preferred type, as it is rapid with three different phases and is carried out under a more or less controlled environment, and (b) cold composting—this method requires longer time periods and may last for many years.

Thermophilic composting is the most widely used procedure for the conversion of waste/undesired material into important by-products and involves three different phases, (a) mesophilic, or the moderate temperature phase (~40 °C), that lasts up to 2–3 days, this phase is characterized by an exponential increase in microbial biomass with the rapid breakdown of soluble and easily biodegradable substrates. The microbial activity results in heat generation causing an increase in temperature. A succession of microbial community from mesophiles to thermophiles is observed during this stage, (b) the thermophilic phase, usually lasts for several days, which is determined by the physico-chemical and the nutritional parameters of the compost under study. The temperature in large piles and/or man-made compost systems can reach up to 55–80 °C during this stage, however in small bioreactors the temperature only reaches up to 45–50 °C. The temperature during the thermophilic phase has a major influence on the microbial community structure, although the nature of the available nutrients and the corresponding physico-chemical factors also influence the community structure. The microbial population of thermophiles are dominated by the bacterial population; however the role of thermophilic fungi cannot be overlooked. Irrespective of the nature and the type of compost, thermophilic *Bacilli* account for the most prevalent group of thermophiles and are phenotypically represented by the diverse members of the genera *Bacillus*, and play a vital role in the breakdown of complex substrates including, fats, proteins, cellulose, hemicellulose, pectins and lignins. This also assists in the eradication of many pathogenic microbes and weeds. The exhaustion of nutrients results in a decrease of the microbial load and the thermophilic phase moves onto (c) 'curing' or maturation phase, characterized by the presence of mesophiles and slow breakdown of the organic matter over long periods with humus as the final product. An important aspect of thermophilic *Bacilli* during the entire process of composting is their remarkable ability to utilize diverse compounds as carbon source and also the spectrum of their thermostability, as many representatives of the class *Bacilli* are able to grow both at 30 and 50 °C, which allows them to survive and remain metabolically active over all three phases.

6.5 The Spectrum of Thermophilic *Bacilli*

Thermophilic bacterial strains have been isolated from almost every type and nature of compost, including heterotrophs, autotrophs and mixotrophs. Beffa et al. (1996) reported the presence of a large number of heteroteophic spore forming bacteria notably *Bacillus* spp. during the thermophilic phase of garden manure using the enrichment culture technique. Blanc et al. (1997) used amplified ribosomal DNA analysis (ARDRA) to deduce the predominance of *Bacillus pallidus*, *B. stearothermophillus* and *B. thermodenitrificans* in hot composts. Dees and Ghiorse (2001) reported the abundance of thermophilic bacterial strains (76.1 %) in garden and domestic composts. The RAPD (randomly amplified polymorphic DNA) profiles indicated the abundance of two strains notably, *B. licheniformis* and *B. thermodenitrificansi*. Other strains, including *B. sporothermodurans* and *B. thermosphaericus,* were present at relatively lower numbers (~ 10.0 %). The presence of *Anurenibacillus* and *Brevibacillus* was also detected, although these strains are not characteristically associated with garden type composts. Ishii et al. (2000) used DGGE profiling to identify the members of genus *Bacillus* as the predominant representatives during the thermophilic phase of garbage compost, however sequences related to genus *Virgibacillus* and *Gracilibacillus* were also identified. The thermophilic phase between 50–60 °C was characterized by the abundance of heterogenic members belonging to the genus *Bacillus*. Hassen et al. (2001) used the enrichment culture technique and API biomurex strips for isolation and identification of microbes from a municipal waste. The members of the genus *Staphylococcus* were found to dominate the mesophilic phase and early thermophilic phase notably: *S. capitis*, *S. cohnii*, *S. sciuri*, and *S. similans*. The presence of the last species was observed only during the thermophilic phase. Schloss et al. (2003) used ARISA (automated 16S-23S rRNA intergenic spacer amplification) to observe the community succession in a compost heap. They detected the predominance of sequences related to genus *Lactobacilli* during the first 60 h period, while between 60–96 h a shift in the bacterial population was observed, with members of the genus *Bacillus* appearing as the dominant flora. The SSCP profiling of samples collected from thermophilic compost containing corn, wood chips and manure indicated the predominance of five different species of *Bacillus* (Peters et al. 2000). DGGE analysis of the microbial community from a composting bin system indicated the presence of a large number of bands corresponding to the genus *Bacilli*. *In situ* FISH hybridization studies indicated that 30 % of the sequences were related to *B. licheniformis* (Rawat and Johri 2013). The 16S rRNA gene sequence analysis of samples obtained from spent mushroom compost indicated the presence of Gram-positive bacteria mostly associated with the genera *Bacillus*, *Paenibacillus,* and *Staphylococcus*. Novel bacterial isolates belonging to the genera *Bacillus* were also identified (Ntougias et al. 2004). Tiago et al. (2004) analysed the microbial community in municipal sludge compost using the enrichment culture technique and API biomurex system, RAPD was used for determining the genetic diversity of the genus *Bacillus*. *B. cereus*, *B. licheniformis,* and *B. gelatini* were distinguished based on RAPD profiles. All the *Bacillus* sp. except for *B. cereus* group, were able to grow under both mesophilic and thermophilic conditions. Partanen et al. (2010) used a culture independent approach by analysing the 16S rRNA gene sequences of sample(s) from both pilot scale and large scale municipal waste compost and reported the abundance of members associated with the genera *Lactobacillus* during the mesophilic and early thermophilic stages. The thermophilic phase in the pilot scale and the full scale compost was characterized by the abundance of diverse sequences related to *Bacillus* sp. In the pilot scale, the *Bacillus* sp. appeared quite early in thermophilic phase and was identified throughout the phase. However, in full scale composting, sequences related to *Bacillus* sp. were identified only during the later stages of the composting process. The 16S rRNA gene sequence analysis of thermophilic bacterial isolates obtained from hot manure compost indicated the presence of four separate clusters belonging to four genera (i) *Geobacillus*, (ii) *Bacillus*, (iii) *Ureibacillus*, and (iv) *Anureinibacillus*. PCR-RFLP and 16S-ITS-23S region analysis were used for genetic characterization. Four strains; *Geobacillus thermodenitrificans*, *Bacillus smithii*, *Ureibacillus suwonensis* and *Anureibacillus thermoaerophilus* were found to be prevalent and were biochemically characterized. Other *Bacilli* including *B. coagulans*, *B. subtilis*, *G. stearothermophillus*, *G. kaustophilus*, *U. koreensis*, *A. anureinilyticus* and *Brevibacillus brevis* were also identified (Charbonneau et al. 2012). The dynamics of the microbial community and its succession on hot chicken manure compost was characterized using DGGE. The initial abundance of bacterial community shifted from *Acrobacter* sp. to *Bacillus* sp. TP-84 from 5th to 30th day of the composting. Emergences of sequence related to another *Bacillus* sp. MSP06G was observed after 15 days of composting. The higher number of *Bacillus* sp. SCSSS08 was exclusively observed during the later phases of the composting. The study reflected that *Bacillus* spp. are the predominant contributors towards the hydrolytic degradation of organic matter during the thermophilic phase (He et al. 2013). Amore et al. (2013) reported the predominance of *Bacillus* sp. from industrial waste based compost using 16S rRNA gene identification.

6.6 Enzymes and Enzymatic Processes Associated with Composting

Composting is a highly complex process that mainly depends on the metabolic machinery of the microbes and involves different enzymes and enzymatic processes. The enzymes involved in any composting process vary depending on the composition of the organic matter in the compost, the physico-chemical conditions that are prevalent during the composting and the changing dynamics of microbial community succession. Although a lot of focus has been targeted towards the study of microbial community structure and succession, only a few studies have focused on the type, functionality, and progression of the enzymes during composting process.

6.6.1 Proteases

The role and importance of proteases during composting has been acknowledged by different studies (He et al. 2013). Proteases have been associated with the nitrogen cycle and hydrolysis of proteins and have usually been identified during the initial phase of the composting process, most likely because of the availability of oligo-and polypeptides in the initial mixture (Castaldi et al. 2008). An analysis of enzymatic activities associated with municipal solid waste compost indicated an increase in activity from day 1 to a peak on day 7, followed by a gradual decrease thereafter (Castaldi et al. 2008). Raut et al. (2008) reported an increase in protease activity up to the 9th–12th day of the experiment, followed by a gradual decrease—this observation was made with a properly aerated compost pile. Interestingly, with an un-aerated compost pile, an increase in protease activity up to the last day (21st) of the experiment was observed. The protease activity in dairy manure with rice chaff was determined with three different piles. In piles 1 and 2, the thermophilic stage was reached after 7 days, while the peak temperature reached after 17 days in pile 3. The protease activity was highest at the initial stages in pile 1 and 2, followed by a gradual decrease in activity; while the highest protease activity associated with pile 3 was reached after 10 days (Liu et al. 2011). The protease activity was associated with a number of *Bacillus* sp. during the early phase of animal manure composting. Proteases from *B. stearothermophillus* were found to be activated in the presence of calcium and ferrous ions during the aerobic degradation of sludge compost (Kim et al. 2002).

6.6.2 Cellulases and Hemicellulases

Cellulases and hemicellulases include a wide array of enzymatic machinery which plays significant roles in the degradation of organic matter, particularly plant matter and lignocellulosic residues. These enzymes have decisive influence on the carbon cycle and also affect the overall nutrient cycle during composting. More than any other substrate, the breakdown of cellulose limits the formation of compost (Goyal et al. 2005). He et al. (2013) reported the highest cellulase and β-glucosidase activity associated with the thermophilic *Bacilli* (predominantly *Bacillus* sp.) on days 12–14, over the 45 days composting period. Goyal et al. (2005) studied the changes in cellulase and xylanase activity associated with the breakdown of different compost types including sugarcane trash + cattle dung (4:1), sugarcane trash + cattle dung (1:1), press mud, poultry waste and water hyacinth. An increase in cellulase activity was observed from the initial period up to 30 days followed by a sequential decrease until the end of the study period (90 days). The highest cellulase activity was observed with water hyacinth, followed by sugarcane trash + cattle dung (1:1), press mud, poultry waste and sugarcane trash + cattle dung (4:1). The xylanase activity showed a steady increase from the initial period and the highest activity was recorded on day 60 of the composting period, followed by a decrease in activity towards the end of the study period. The highest xylanase activity was observed with water hyacinth followed by press mud, sugarcane trash + cattle dung (1:1), poultry waste and sugarcane trash + cattle dung (4:1). The thermophilic stage was reached after 14 days and thermophilic bacteria count was highest during this stage. The thermophilic bacteria count was the highest in sugarcane trash + cattle dung (1:1), followed by poultry waste and press mud piles. The differences in the phasing of the cellulases and xylanase activity are also an indication of change in the structure of the microbial community and indicate the emergence of thermophilic cellulolytic bacteria followed by their xylanolytic counterparts. The study also indicates the relative abundance of mesophilic bacterial strains from days 40–60 of composting, suggesting that cellulase activity is mostly associated with thermophilic *Bacilli*; however, many members of the order Bacillales are known to be thermoduric and grow under thermophilic and mesophilic conditions. The composting of municipal solid waste is reported by Castaldi et al. (2008). In their study, a decrease in cellulase activity was observed during the first week followed by an increase in the second week and again a decline after third week, finally becoming stable after 70 days. The β-glucosidase enzyme is a key component in the carbon cycle, the enzyme catalyses the termination reactions of the glucose chains to release glucose. An increase in β-glucosidase activity was observed during the first two weeks followed by a sharp decrease until day 70, which stabilized over the course of the last month of the process. The decrease in the activity of both cellulase and β-glucosidase during the second part of composting could be attributed to an increase in the lignin content and a gradual decrease in the availability of accessible cellulose content. The authors indicated that change in enzyme activity is associated with a corresponding change in organic matter content and microbial community succession during the composting process. Raut et al. (2008) reported an increase in cellulase activity that reached a maximum on day 12 of municipal sludge waste composting (aerated with added glucose) followed by a decline in activity until the end of the composting period (21 days). However, in the normal compost (without aeration) an increase in cellulase activity until the end of the composting period (21 days) was observed, indicating the availability of cellulose. The increase in cellulase activity in normal compost corresponds to an increase in the biomass of thermophilic *Bacilli* towards the end of the composting period. Wei et al. (2012) used microbial rDNA abundance as a tool to study the dynamics of microbial community change along with their secreted enzymes involved in the composting of a mix of yellow poplar wood-chips and mown lawn grass dippings (85:15 in dry-weight). The changes in the compositional data of the composed materials indicated a 50 and 42 % decrease in cellulose and total hemicellulose concentration, respectively, over a period of 27 weeks. During composting predominance in cellulase activity was observed in the later stages (24 weeks). In contrast, the measured hemicellulase activities, mainly α-arabinosidase and β-galactosidase, were higher in the earlier stages of composting (3 weeks). Parallel studies with light and fluorescence microscopy indicated the exposure of cellulose on the surface of the woody substrates during the later stages of the composting.

6.6.3 Other Enzymes Involved During Composting

Urease is involved in the hydrolysis of urea to ammonium and carbon dioxide and is closely related to the nitrogen cycle. Castaldi et al. (2008) observed a strong correlation between urease, protease and β-glucosidase activity. Urease activity increased during the first 3 weeks of composting and was the highest on day 21, followed by a steep decline in activity in the fourth week and then stabilized after 70 days of composting. The initial increase in urease activity could be attributed to the availability of water soluble nitrogen due to the action of proteases. The sharp decline in urease activity may be due to the accumulation of nitrates in composting mixture. Cayuela et al. (2008) reported very low urease activity during the entire duration of semi-solid olive mill waste composting, which was attributed to the low concentration of nitrogen and also to the low levels of available substrates released during mineralization. Liu et al. (2011) reported the presence of significant urease activity during the second week of composting of daily manure with rice chaff. Correlation analysis indicated a strong correlation with water soluble nitrogen and ammonium ions. A similar result was also reported by Garcia et al. (1995), wherein an increase in urease activity was directly correlated to the availability of water soluble nitrogen. Dehydrogenase activity is used as measure of overall microbial activity and reflects the amount of microbial biomass involved in the process of respiration and metabolic processing (Castaldi et al. 2008). Phosphatase activity has been reported during the early phase of composting followed by decline in activity (Raut et al. 2008); however, Ros et al. (2006) reported an increase in alkaline phosphatase activity during the beginning of composting, reaching a maximum by the end of the process. Phosphatase activity could be related to the amount and availability of organic phosphate compounds present in the composting mixture. This enzyme has high agronomic value because it hydrolyses compounds of organic phosphorous and transforms them into different forms of inorganic phosphorus which is assimilated by plants. The phosphatase enzyme is also considered a general microbial indicator (Raut et al. 2008). Phosphatase is a key enzyme in the phosphorus cycle and is induced by the presence of carbohydrate derived structures that are degraded to glucose by enzymatic action. This enzyme is only synthesized by microbes, and is not released by plants and/or organic matter residues Alkaline phosphatase is thus a relevant enzyme for characterization of the composting process (Raut et al. 2008; Fig. 6.1).

6.7 Cellulolytic Thermophilic *Bacilli* and Cellulases from Compost

Amore et al. (2013) reported the isolation of novel cellulolytic bacterial strains from industrial waste based compost. The potent bacterial strains were identified as *B. licheniformis* strain 1, *B. subtilis* subsp. *subtilis* strain B7B, *B. subtilis* subsp. *spizizenii* strain 6 and *B. amyloliquefaciens* strain B13C. *B. amyloliquefaciens* strain B13C was found to be thermotolerant as growth was observed in the range of 28–47 °C. The enzyme was purified to homogeneity and SDS-PAGE indicated the presence of a 55.0 kDa protein. LC-MS analysis of the proteome corresponded to the presence of a peptide belonging to the GH5 family of endoglucanases. The endoglucanase exhibited broad temperature optima (50–70 °C) and retained 90 % activity following 144 h incubation at 40 °C. The kinetic parameters including K_m (9.95 mg/ml) and V_{max}(284 μmol/min) were determined for the purified enzyme. Acharya et al. (2012) reported the isolation of 10 thermophilic and 15 thermotoler-

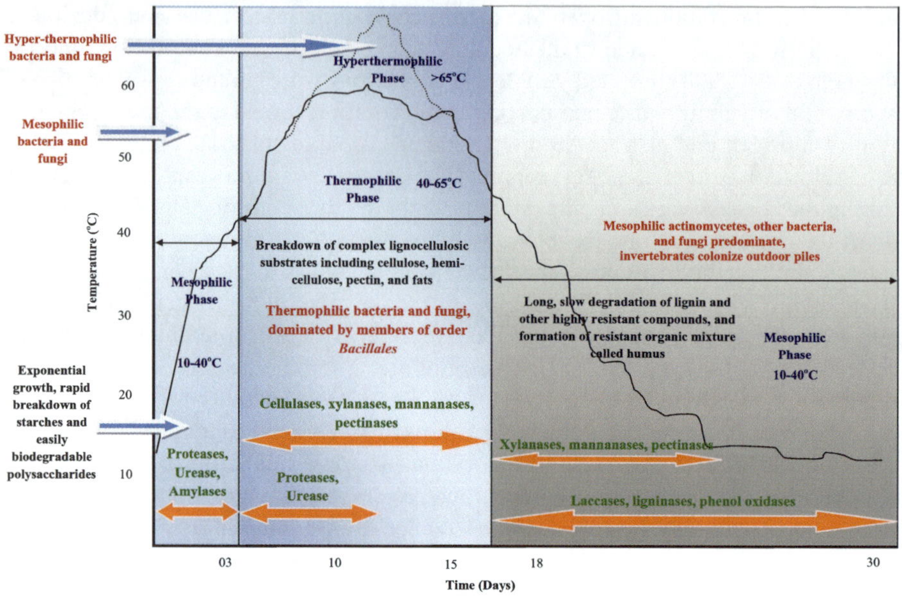

Fig. 6.1 A schematic illustration of the enzymes and microbial community prevailing during the three different phases of composting (modified from cwmi.css.cornell.edu/chapter1.pdf)

ant bacterial strains from hot compost in Nepal. The isolates were presumptively identified as *B. subtilis*, *B. licheniformis*, *Bacillus* sp., *Geobacillus* sp., and *Paenibacillus* sp., based on phenotypic and biochemical parameters. Highest activity on carboxymethylcellulose (CMC) was observed with *B. subtilis*. The partially purified enzyme was stable in the pH range 6.6–9.0 and showed optimum activity at pH 7.2 and 50 °C. Members of thermophilic *Bacilli* belonging to the genus *Bacillus* were isolated and identified from garden compost at Rhodes University, Grahamstown (Mayende et al. 2006). The authors reported the presence of cellulases from *B. subtilis* and *Bacillus* sp. that were found to perform optimally at 70 and 80 °C, respectively. *Paenibacillus cookii* strain B39 was isolated and identified from poultry manure compost in Taichung, Taiwan (Wang et al. 2008). The cellulases from strain B39 was purified using ion-exchange chromatography with a specific activity of 54.94 U/mg proteins. SDS-PAGE indicated the presence of a 148 kDa polypeptide; activity staining was used to confirm the presence of cellulases. The optimum CMCase activity was observed at 60 °C and pH 6.5. Substrate specificity studies revealed that these cellulases could also hydrolyse Avicel and filter paper in addition to CMC. The presence of Avicelase activity in bacterial cellulase is quite rare and makes this cellulase important for industrial application. In the presence of calcium ions the hydrolytic effect towards all the substrates was enhanced two-fold. The enzyme was also found to be activated in the presence of copper ions but was inhibited in presence of EDTA, $HgCl_2$ and SDS. Ng et al. (2009) reported the isolation of over 100 *Geobacillus* sp. strains from rice straw compost supplemented with pig manure. The genomic library of *Geobacillus* sp. 70PC53 was screened for presence of cellulase activity. The 1176-kb ORF encoding a putative endo-glucanase of 392 amino acids was associated with high cellulase activity and was designated as CelA. The recombinant CelA was over-expressed (85 mg/l) in *E. coli*. Activity staining was used to confirm the enzymatic activity of purified CelA with a molecular weight of 43 kDa. The cellulase was found to be active over a broad temperature range (45–75 °C), retained over 80 % activity for 4 h at 75 °C and was optimally active at 65 °C and pH 5.0. The enzyme retained 80 % activity after 16 h incubation over a broad pH range from 5.0 to 9.0. The endo-glucanase activity was found to be inhibited by Cu^{2+} and Zn^{2+} ions; however Mn^{2+}, Co^{2+}, Ca^{2+} stimulated the enzyme activity. The authors also reported a ten-fold greater specific activity for CelA endoglucanases when compared to the commercially available *T. reesei* endo-glucanases. The CelA endo-glucanase showed resemblance to the GH5 family of glycosyl hydrolases; it has been considered a novel enzyme, as the highest sequence identity of CelA was 45.1 % to that of the CelN from *Pectobacterium atrosepticum*. The unique properties of this cellulase, including a broad range of thermostability and pH stability, make it an ideal candidate for application in biofuel and food industry. Yang et al. (2010) reported the isolation and identification of a novel cellulase producing *Bacillus subtilis* strain 115 from long-term thermal compost containing rich cellulose materials in a factory of Zhengzhou in China. The cellulases exhibited optimum activity at pH 6.0 and 60 °C. The enzyme was thermostable as it retained more than 90 % activity at 65 °C after 2 h incubation. The cellulase gene was cloned and expressed in *E. coli* BL21 (DE3) and purified using a Sephadex G-100 column. Molecular weight of the purified protein was found to be 52 kDa using SDS-PAGE electrophoresis. Cloning and expression resulted in three-fold higher cellulase production with no change in properties compared to the wild-type protein. Zambare et al. (2011) reported the isolation of a thermophilic microbial consortium (TMC) producing cellulolytic and xylanolytic enzymes from yard waste following enrichment with CMC and birchwood xylan. High titres of cellulases and xylanases were observed following production of the TMC cellulases on lignocellulosic substrates (corn stover and prairie grass). Characterization studies with extracellular TMC cellulases showed the presence of pH optima peaks at 4.0, 7.0, and 10.0. The temperature optimum for cellulase was found to be at 60 °C, although the enzyme was active over a broad range of temperature (40–80 °C). The TMC cellulases retained 98 % activity after 1 h incubation at 50 °C, and 77 % after incubation at 60 °C for 3 h. SDS-PAGE of the crude TMC cellulases resulted in several bands with different molecular weights, however zymogram analysis revealed the presence of three bands corresponding to 60, 35 and 27 kDa proteins that exhibited cellulase activity. The substrate specificities of the TMC cellulases were the highest for Avicel followed by microcrystalline cellulose, filter paper, pinewood saw dust and CMC. The authors also reported the higher hydrolytic activities of the TMC cellulases compared to the cellulase produced by isolated strains from the microbial consortium. They also suggested the similarity of cellulases in TMC to that of *Bacillus* sp. reported elsewhere (based on the enzyme characteristics). The versatility and high activities of TMC cellulases on different lignocellulosic substrates suggests a robust and compatible system for the cost-effective breakdown of lignocellulosic biomass.

Eida et al. (2012) reported the isolation and characterization of cellulose decomposing bacteria from saw dust and coffee residue composts. A number of thermophilic bacteria belonging to the genus *Paenibacillus*, *Cohnella*, *Streptomyces* and *Microbiospora* were identified using 16S rRNA gene analysis. The two strains belonging to *P. woosongensis* were found to produce many hydrolytic enzymes, including cellulases, xylanases, β-glucanase, and mannanase. Kim et al. (2012) reported the isolation of hundreds of cellulolytic bacteria based on the screening of different compost samples from Jeju Island, South Korea. Based on qualitative screening three potential strains with high cellulolytic activity were identified as *Bacillus subtilis* based on 16S rRNA gene sequencing. CMCase and Avicelase activities were detected in the extracellular fraction, while β-glucosidase activity was found to be cell bound. Shu-bin et al. (2012) reported the isolation and identification of *B. subtilis* strain Pa5 from soil samples rich in rotting rice straw. The optimal pH and temperature for both CMCase and cellobiase were observed at pH 7.0 and 50 °C. The enzyme exhibited a broad range of thermostability (30 to 50 °C) and pH stability (5.0–8.0). Nizamudeen and Bajaj (2009) reported the isolation of numerous cellulolytic bacteria from different lignocellulosic sources. They identified a highly thermotolerant and alkali-tolerant endoglucanase producing *Bacillus* sp. NZ. Maximum enzyme production was observed after 72 h at 45 °C and pH 9.0. High endo-glucanase activity was observed with wheat straw, filter paper and saw dust as lignocellulosic substrates. The enzyme was active over a broad pH (5–10) and temperature (50–100 °C) range. The enzyme exhibited maximum activity at pH 9–10 and at two different temperatures 50 and 90 °C. The enzyme was highly stable for 30 min from 60 to 90 °C. Thermophilic bacterial strains and their hydrolytic enzymes produces during different stages of composting process is tabulated in Table 6.1.

6.8 Future Prospects and Recommendations: The Ecological and Economic Impact of Composting with Special Reference to Agricultural Residues

Agricultural crop residues including field residues and processing residues are renewable and abundant resources that contain large amounts of untapped energy. The crop residues (including lignocellulosic biomass) in the form of hot composts offer an intricate ecosystem that provides the natural selection for the isolation of thermophilic cellulolytic and lignocellulolytic bacteria. The isolation of *Bacillus* and *Geobacillus* sp. reported above provides ample support towards the application of composting for the generation of biofuels. The use of engineered strains as special additions during the thermophilic phase of the composting process can be used as an effective strategy for consolidated bioprocessing. Rice straw, wheat straw, corn stover and sugarcane bagasse are the major agricultural wastes in terms of quantity of biomass available and it is estimated by Kim and Dale (2004) that the potential production of 49.1 GL year^{-1} of bioethanol from 73.9×10^6 t of dry waste crops available in the world. The lignocellulosic biomass from the seven major crops ac-

Table 6.1 Summary of major thermophilic bacterial strains and their secreted enzymes in the course of different composting processes

Compost type	Major bacterial groups	Enzyme(s)	Reference
Municipal solid waste	*Bacillus* sp. and other thermophiles	Proteases	Raut et al. (2008)
Diary manure with rice	*Bacillus* sp. and other thermophiles	Proteases	Liu et al. (2011)
Sludge compost	*Bacillus* sp. *B. stearo-thermophillus*, and other thermophiles	Proteases	Kim et al. (2002)
Municipal solid waste	Majority mesophiles and some thermophiles	Urease	Castaldi et al. (2008)
Daily manure and rice chaff	Majority mesophiles and some thermophiles	Urease	Liu et al. (2011)
Municipal solid waste	Majority mesophiles and some thermophiles	Phosphatase	Raut et al. (2008)
Municipal solid waste	*Bacillus* sp. *B. stearo-thermophillus*, and other thermophiles	Cellulases	Raut et al. (2008)
Municipal solid waste	*Bacillus* sp. and other thermophiles	Cellulase and β-glucosidase	He et al. (2013)
Sugarcane trash + cattle mud, press mud poultry waste, water hyacinth	Majority thermophiles of the order *Bacillales* and some mesophiles	Cellulases	Goyal et al. (2005)
Sugarcane trash + cattle mud, press mud poultry waste, water hyacinth	Majority mesophiles belonging to *Bacillus* and *Paenibacillus* and some thermophiles	Xylanases	Goyal et al. (2005)
Municipal solid waste	Thermophilic *Bacilli* and other groups of thermophiles	Cellulases and β-glucosidase	Castaldi et al. (2008)
Mix of yellow poplar wood chips and mown lawn grass	Thermophilic *Bacilli* and other groups of thermophiles		
	Cellulases	Wei et al. (2012)	
Mix of yellow poplar wood chips and mown lawn grass	Thermophilic *Bacilli* and other groups of thermophiles	Hemicelluases	Wei et al. (2012)
Industrial waste compost	*B. licheniformis, B. subtilis* sub *subtilis, B. subtilis* sub *zizenii, B. amyloliquefaciens*	Cellulases	Amore et al. (2013)
Hot compost	*B. subtilis, B. licheni-formis, Bacillus* spp., *Geobacillus* spp. *Paeni-bacillus* spp.	Endoglucanases	Acharya et al. (2012)
Garden compost	*B. subtilis, Bacillus* spp.	Cellulases	Mayende et al. (2006)
Poultry manure compost	*Paenibacillus cookie, Paenibacillus* spp.	Cellulase (Avicelase)	Wang et al. (2008)
Rice straw with pig manure	*Geobacillus* spp.	Cellulase	Ng et al. (2009)
Thermal compost	*B. subtilis*	Cellulase	Yang et al. (2010)

Table 6.1 (continued)

Compost type	Major bacterial groups	Enzyme(s)	Reference
Lignocellulosic substrates (corn stover and prairie grass)	Thermophilic microbial consortium (TMC)	Cellulase, xylanase	Zambare et al. (2011)
Saw dust and coffee residue compost	*Paenibacillus* spp. (*P. woosongensis*) *Cohnella* sp., *Streptomyces* spp.	Cellulase, xylanase, β-glucosidase, mannanase	Eida et al. (2012)
Hot compost	*B. subtilis*	CMCase, Avicelase and β-glucosidase	Kim et al. (2002)
Rotting rice straw	*B. subtilis*	CMCase, cellobiase	Shu-bin et al. (2012)
Lignocellulosic biomass	*Bacillus* sp.	Endoglucanase	Nizamudeen and Bajaj (2009)

counts for 1.5×10^9 t year^{-1} of dry biomass for the conversion to bioethanol, thus projecting the total potential bioethanol production from crop residues and waste crops to approximately 491.0 GL year^{-1}).

Sugarcane is among the principal agricultural crops cultivated in tropical countries. Bagasse is the residue obtained from the sugarcane after the processing of sugarcane juice and is used as a source of sugar and ethanol production. It has an advantage over other crop residues as it is a by-product, and does not require special and costly collection and transportation measures. The annual world production of sugarcane is 1.6 billion t, and it generates 279 million t of biomass residues (bagasse and leaves). For the past three decades, bagasse and leaves have been explored for use in lignocellulosic bioconversion (Beukes and Pletschke 2011; Chandel et al. 2007) (Fig. 6.2).

As with other raw materials, when discussing the use of crop residues as raw material for biofuel, their alternative uses should be considered. Of special importance is their use as soil conditioner and for increasing the levels of soil organic matter, with important effects on soil structure, preventing erosion, the supply of nutrients, acidification, and water holding capacity of soils, all affecting soil fertility and health (Lal 2004).

The contribution of the agricultural sector to emissions of climate change gases is becoming better understood. At the same time, the potential role of the sector as a means through which to tackle climate change, widely neglected in the past, is becoming more widely acknowledged. The absorption potential of agricultural soils could contribute significantly to constraining increases in greenhouse gas emission levels, while also contributing to improvements in soil quality in some areas. In addition to the measures listed above, other benefits of compost application may have some relevance. Some of these measures include the replacement of chemical fertilizers (implying avoidance of greenhouse gases related to their production),

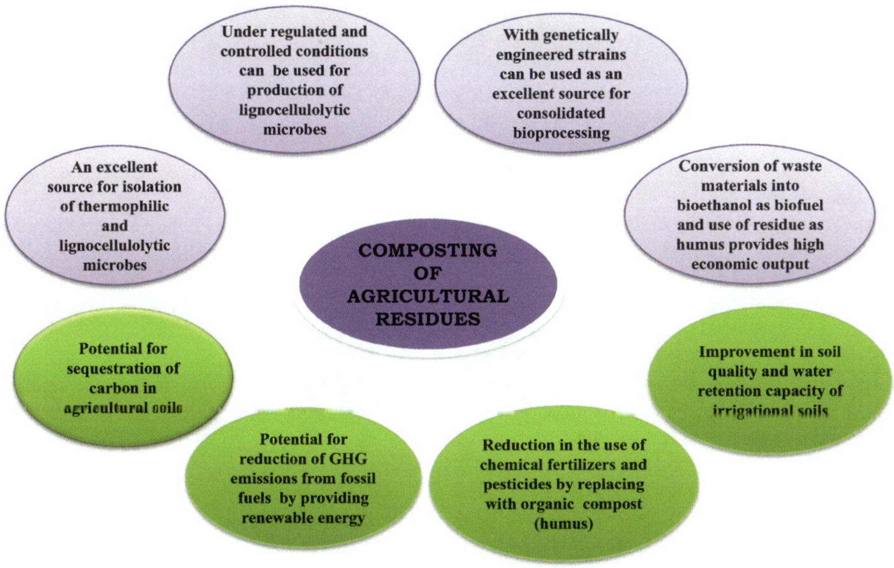

Fig. 6.2 Economic and ecological advantages of composting of agricultural residues

reduced use of pesticides (avoiding emissions associated with their production) and improved tilth and workability (less consumption of fuels).

Acknowledgements The authors would like to express their gratitude to Rhodes University for granting A.B. a Rhodes University Postdoctoral Fellowship.

References

Acharya A, Joshi DR, Shrestha K, Bhatta DR (2012) Isolation and screening of thermophilic cellulolytic bacteria from compost piles. Scientific World 10(10):43–46

Alfreider A, Peters S, Tebbe C, Rangger A, Insam H (2002) Microbial community dynamics during composting of organic matter as determined by 16S ribosomal DNA analysis. Sci Util 10:303–312

Amore A, Pepe O, Ventorino V, Birolo L, Giangrande C, Faraco V (2013) Industrial waste based compost as a source of novel cellulolytic strains and enzymes. FEMS Microbiol Lett 339:93–101

Beeby M, O'Connor BD, Ryttersgaard C, Boutz DR, Perry LJ, Yeates TO (2005) The genomics of disulfide bonding and protein stabilization in thermophiles. PLoS Biol 3(9):e309

Beffa T, Blanc M, Aragno M (1996) Obligately and facultatively autotrophic, sulfur-and hydrogen-oxidizing thermophilic bacteria isolated from hot composts. Arch Microbiol 165:34–40

Beukes N, Pletsche BI (2011) Effect of alkaline pre-treatment on enzyme synergy for efficient hemicellulose hydrolysis in sugarcane bagasse. Bioresour Technol 102:5207–5213

Blanc M, Marilley L, Beffa T, Aragno M (1997) Rapid identification of heterotrophic, thermophilic spore forming bacteria isolated from hot composts. Int J Bacteriol 47:1246–1248

Brock TD (1978) Introduction. In: Brock TD (ed) Thermophilic microorganisms and life at high temperatures. Springer, New York, pp 1–16

Caporaso JG, Lauber CL, Walters WA, Berg-Lyons D, Huntley J et al (2012) Ultra-high throughput microbial community analysis on the IlluminaHiSeq and MiSeq platforms. ISME J 6:1621–1624
Castaldi P, Garau G, Melis P (2008) Maturity assessment of compost from municipal waste through the study of enzyme activities and water soluble fractions. Waste Manage 28:534–540
Cayuela ML, Mondini C, Sanchez-Monedero MA, Roig A (2008) Chemical properties and hydrolytic enzyme activities for the characterization of two phase olive mill wastes composting. Bioresour Technol 99:4255–4262
Chandel AK, Chan ES, Rudravaram R, Narasu ML, Rao VL, Ravindra P (2007) Economics and environmental impact of bioethanol production technologies: an appraisal. Biotech Mol Biol Rev 2(1):14–32
Charbonneau DM, Meddeb-Mouelhi F, Boissiont M, Sirois M, Beauregard M (2012) Identification of thermophilic bacterial strains producing thermotolerant hydrolytic enzymes from manure compost. Indian J Microbiol 52(1):41–47
Cicicopol C, Peters J, Kellermann J, Baumeister W (1994) Primary structure of multimeric protein, homologous to the PEP-utilizing enzyme family and isolated from a hyperthermophilic archaebacterium. FEBS Lett 356:345–350
Cihan AZ, Tekin N, Ozcan B, Cokmus C (2012) The genetic diversity of genus *Bacillus* and the related genera revealed by 16S rRNA gene sequences and ARDRA analyses isolated from geothermal regions of Turkey. Brazilian J Microbiol 12:309–324
Das R, Gerstein M (2000) The stability of thermophilic proteins: a study based on comprehensive genome comparison. Funct Integr Genomics 1:76–88
Dees PM, Ghiorse WC (2001) Microbial diversity in hot synthetic compost as revealed by PCR-amplified rRNA sequences from cultivated isolates and extracted DNA. FEMS Microbiol Ecol 35:207–216
Derry AM, Staddon WJ, Trevors JT (1998) Functional diversity and community structure of microorganisms in uncontaminated and creosote-contaminated soils as determined by sole carbon-source-utilization. World J Microbiol Biotechnol 14:571–578
Eida MF, Nagaoka T, Wasaki J, Kouno K (2012) isolation and characterization of cellulose-decomposing bacteria inhabiting sawdust and coffee residue composts. Microbes Environ 27(3):226–233
Eiland F, Klamer M, Lind A, Leth M, Baath E (2001) Influence of initial C/N ratio on chemical and microbial composition during long term composting of straw Microb Ecol 41:272–280
Fakruddin M, Mannann KSB (2013) Methods for analysing diversity of microbial communities in natural environments. Ceylon J Sci (Bio Sci) 42(1):19–33
Fakruddin M, Mazumdar RM, Chowdhury A, Hossain MN, Mannan KSB (2012) Pyrosequencing-Prospects and Applications. Int J Life Sci Pharma Res 2(2):65–76
Finstein MS, Morris ML (1975) Microbiology of municipal solid waste composting. Adv Appl Microbiol 19:113–151
Garcia C, Moreno JL, Hernandez T, Costa F, Polo A (1995) Effect of composting on sewage sludges contaminated with heavy metals. Bioresour Technol 53(1):13–19
Garland JL, Mills AL (1991) Classification and characterization of heterotrophic microbial communities on the basis of patterns of community level sole carbon source utilization. App Environ Microbiol 57:2351–2359
Goris J, Konstantinidis KT, Klappenbach JA, Coenye T, Vandamme P, Tiedje JM (2007) DNA–DNA hybridization values and their relationship to whole-genome sequence similarities. Int J System Evol Microbiol 57:81–91
Goyal S, Dhull SK, Kapoor KK (2005) Chemical and biological changes during composting of different organic wastes and assessment of compost maturity. Bioresour Technol 96:1584–1591
Hassen A, Belguith K, Jedids N, Cherif A, Cherif M, Boudabous A (2001) Microbial characterization during composting of municipal solid waste. Bioresour Technol 80:217–225
He Y, Xie K, Xu P, Huang X, Gu W, Zhang F, Tang S(2013) Evolution of microbial community diversity and enzymatic activity during composting. Res Microbiol 164:189–198

Ishii K, Fukui M, Takii S (2000) Microbial succession during a composting process as evaluated by denaturing gradient gel electrophoresis analysis. J Appl Microbiol 89:768–777
Jaenicke R, Bohm G (1998) The stability of proteins in extreme environments. Curr Opin Struct Biol 8(6):738–748
Kim KY, Lee SC, Cho YY, Oh HJ, Ko YH (2012) Isolation of cellulolytic *Bacillus subtilis* strains from agricultural environments. ISRN Microbiology doi:10.5402/2012/650563
Kim S, Dale EB (2004) Global potential bioethanol production from wasted crops and crop residues. Biomass Bioenergy 26:361–375
Kim YK, Bae JH, Oh BK, Lee WH, Choi JW (2002) Enhancement of proteolytic enzyme activity excreted from *Bacillus stearothermophillus* for a thermophilic aerobic digestion process. Bioresour Technol 82:157–164
Ladenstein R, Antranikian G (1998) Proteins from hyperthermophiles: stability and enzymatic catalysis close to the boiling point of water. Adv Biochem Eng Biotech 61:38–85
Lal R (2004) Soil carbon sequestration impacts on global climate change and food security. Sci 304:1623–1627
Lauberet CL, Hamady M, Knight R, Fierer N (2009) Pyrosequencing-based assessment of soil pH as a predictor of soil bacterial community structure at the continental scale. App Environ Microbiol 75:5111–5120
Lieph R, Veloso FA, Holmes DS (2006) Thermophiles like hot T. Trends Microbiol 14(10):423–426
Liu D, Zhang R, Wu H, Xu D, Tang Z et al (2011) Changes in biochemical and microbiological parameters during the period of rapid composting of daily manure with rice chaff. Bioresour Technol 102(19):9040–9049
Liu WT, Marsh TL, Cheng H, Forney LJ (1997) Characterization of microbial diversity by determining terminal restriction fragment length polymorphisms of genes encoding 16S rRNA. App Environ Microbiol 63:4516–4522
Logan NA, Berge O, Bishop AH, Busse HJ, de Vos P, Fritze D et al (2009) Proposed minimal standards for describing new taxa of aerobic, endospore-forming bacteria. Int J Syst Evol Microbiol 59:2114–2121
Ludwig W, Schleifer KH, Whitman WB (2007) Revised road map to the phylum Firmicutes. Retrieved October 1, 2013 from Bergey's Manual Trust website. http://www.bergeys.org/outlines/Bergeys, Vol 3, Outline
Mayende L, Wilhelmi BS, Pletschke BI (2006) Cellulases (CMCases) and polyphenol oxidases from thermophilic *Bacillus* spp. isolated from compost. Soil Biol Biochem 38:2963–2966
Moyer CL, Tiedje JM, Dobbs FC, Karl DM (1996) A computer-simulated restriction fragment length polymorphism analysis of Bacterial Small-subunit rRNA genes: efficacy of selected tetraneric restriction enzymes for studies of microbial diversity in Nature. App Environ Microbiol 62:2501–2507
Mühling M, Woolven-Allen J, Murrell JC, Joint I (2008) Improved group-specific PCR primers for denaturing gradient gel electrophoresis analysis of the genetic diversity of complex microbial communities. ISME J 2:379–392
Muyzer G (1999) DGGE/TGGE a method for identifying genes from natural ecosystems. Curr Opin Microbiol 2:317–322
Ng SI, Li CW, Yeh YF, Chen PT et al (2009) A novel endo-glucanse from thermophilic bacterium *Geobacillus* sp. 70PC53 with high activity and stability overa broad range of temperatures. Extrem 13:425–435
Nizamudeen S, Bajaj BK (2009) A novel thermo-alkalitolerant endoglucanase production using cost-effective agricultural residues as substrates by a newly isolated *Bacillus* sp. NZ. Food Technol Biotechnol 47(4):435–440
Ntougias S, Zervakis GI, Kavroulakis N, Ehaliotis C, Papadopoulou KK (2004) Bacterial diversity in spent mushroom compost assessed by amplified rDNA restriction analysis and sequencing of cultivated isolates. Syst Appl Microbiol 27:746–754
Odlare M (2005) Organic residues: a resource for arable soils. PhD thesis, Swedish University of Agricultural Sciences, Uppsala

Partanen P, Hultman J, Paulin L, Auvinen P, Romantschuk M (2010) Bacterial diversity at different stages of the composting process BMC Microbiol 10:94

Peters S, Koschinsky S, Schwieger F, Tebbe CC (2000) Succession of microbial communities during hot composting as detected by PCR-single-strand conformation polymorphism-based genetic profiles of small subunit rRNA genes. App Environ Microbiol 66:930–936

Raut MP, William PSPM, Bhattacharya JK, Chakrabarti T, Devotta S (2008) Microbial dynamics and enzyme activities during rapid composting of municipal solid waste-Acompost maturity analysis perspective. Bioresour Technol 99:6512–6519

Rawat S, Agarwal PK, Chaudhary DK, Johri BN (2005) Microbial diversity and community dynamics of mushroom compost ecosystem. In Satyanarayana T and Johri BN (eds) Microbial diversity: Current perspectives and applications. International Book Distributing Co, Lucknow, pp 181–206

Rawat S, Johri BN (2013) Role of thermophilic microflora in composting. In: Satyanarayana T, Littlechild J, Kawarabayasi Y (eds) Thermophilic microbes in environmental and industrial biotechnology, 2nd edn. Biotechnology of thermophiles, Springer, pp 137–170

Ros M, Garcia C, Hernandez T (2006) A full-scale study of treatment of pig slurry by composting: kinetic changes in chemical and microbial properties. Waste Management 26:1108–1118

Ryckeboer J, Mergaert J, Coosemans J, Deprins K, Swings J (2003) Microbiological aspects of biowaste during composting in a monitored compost bin. J Appl Microbiol 94(1):127–137

Saiki T, Beppu T, Arima K, Izawa T, Morimoto K, Misaizu Y, Higaki S (1978) Changes in microbial flora, including thermophiles, during composting of animal manure In: Friedman SM (eds) Biochemistry of thermophily. Academic Press, New York, pp 103–115

Schloss PD, Hay AG, Wilson DB, Walker LP (2003) Tracking temporal changes of bacterial community fingerprints during the initial stages of composting FEMS Microbiol Ecol 46:1–9

Shu-bin L, Reh-chao Z, Xia L, Chu-yi C, Ai-lin Y (2012) Solid-state fermentation with okara for production of cellobiase rich cellulases preparation by a selected *Bacillus subtilis* pa5. African J Biotechnol 11(11):2720–2730

Theron J, Cloete TE (2000) Molecular techniques for determining microbial diversity and community structure in natural environments. Critical Rev Microbiol 26:37–57

Thompson MJ, Eisenberg D (1999) Transproteomic evidence of a loop-deletion mechanism for enhancing protein thermostability.J Mol Biol 290:595–604

Tiago I, Texieria I, Silva S, Chung P, Verissimo A, Manaia M (2004) Metabolic and genetic diversity of mesophilic and thermophilic bacteria isolated from composted municipal sludge on poly-ε-caprolactones. Curr Microbiol 49:407–414

Tiedje JM, Asuming-Brempong S, Nusslein K, Marsh TL, Flynn SJ (1999) Opening the black box of soil microbial diversity. Appl Soil Ecol 13:109–122

Tiquia SM, Wan JHC, Tam NFY (2002) Extracellular enzyme profiles during co-*composting* of poultry manure and yard. Compost Sci Util 10:150–161

Torsvik V, Sorheim R, Goksoyr J (1996) Total bacterial diversity in soil and sediment communities—a review. J Ind Microbiol 17:170–178

Wang CM, Shyu CL, Ho SP, Chiou SH (2008) Characterization of novel thermophilic, cellulose-degrading bacterium *Paenibacillus* sp. strain B39. Lett Appl Microbiol 47:46–53

Wei H, Tucker PM, Baker JO, Harris M, Luo Y, Xu Q et al (2012) Tracking dynamics of plant biomass composting by changes in substrate structure, microbial community and enzyme activity. Biotechnol Biofuels 5:20

Yang D, Weng H, Wang M, Xu W, Li Y, Yang H (2010) Cloning and expression of a novel thermostable cellulase from newly isolated *Bacillus subtilis* strain 115. Mol Biol Rep 37:1923–1929

Zambare V, Zambare A, Muthukumarappan K, Christopher LP (2011) Biochemical characterization of thermophilic lignocellulose degradating and their potential biomass for bioprocessing. Intl J Energy Environment 2(1):99–112

Chapter 7
Agronomic, Soil Quality and Environmental Consequences of Using Compost in Vegetable Production

Simon M. Eldridge, K. Yin Chan and Nerida J. Donovan

Abstract This chapter summarises many of the findings from a long term compost vegetable field experiment at Camden in south western Sydney, Australia. Large applications of garden organics compost resulted in significant improvements to soil quality (physical, chemical and biological) compared to farmer practice. These included soil structural stability, soil carbon, cation exchange capacity, pH and microbial biomass carbon. However, conventional tillage with the rotary hoe eroded away these improvements over time by accelerating the loss of soil carbon and pulverising the soil structure. The compost treatment matched the farmer practice treatment in terms of crop yield for all crops, and exceeded it for some crops. The compost treatment was found to be an economic alternative to farmer practice in the Sydney basin, with additional environmental benefits. Targeted applications of compost and minimum tillage may help optimise benefits. A repeat application of compost resulted in a more significant and sustained response in the soil biology.

Keywords Soil quality · Soil health · Food security

7.1 Introduction

Reports of the beneficial effects of composts on crop growth go back as far as 800 BC in the Mediterranean (Semple 1928). But the use of compost and other organic amendments went out of favour in the 1960's and 1970's during a period commonly referred to as the "green revolution" where there was a widespread adoption

S. M. Eldridge (✉)
Wollongbar Primary Industries Institute, NSW Department of Primary Industries, 1243 Bruxner Highway, Wollongbar, NSW 2477, Australia
e-mail: simon.eldridge@dpi.nsw.gov.au

K. Y. Chan
Formerly NSW Department of Primary Industries, Richmond, NSW 2753, Australia

N. J. Donovan
Elizabeth Macarthur Agricultural Institute, NSW Department of Primary Industries, Menangle, NSW 2568, Australia

D. K. Maheshwari (ed.), *Composting for Sustainable Agriculture*, Sustainable Development and Biodiversity 3, DOI 10.1007/978-3-319-08004-8_7

of soluble inorganic NPK fertilizers and chemical pesticides/herbicides by farmers (Lal 2010a). A three fold increase in food production was attributed to this change (Childers et al. 2011). However, since that time there have been many research publications highlighting the importance of organic C to soil quality and function including soil structure, water-holding capacity, drainage, aeration, cation exchange capacity and biological activity (Feller et al. 2010).

Food security is now a major challenge for agriculture in the twenty first century, with there being a need to increase food production by more than 60 % over the next 50 years (Bruinsma 2009) in order to be able to feed a projected world population of 9.2 billion people (UNESA 2008). Improving soil quality by increasing soil organic C levels is seen as one potential way of improving food security (Lal 2004, 2010a). Increasing soil carbon levels in intensive agricultural systems has proven difficult (Heenan et al. 2004; Chan et al. 2011a), but composts have been identified as a potential source of stable organic carbon for this purpose (Gibson et al. 2002). Recent research has also reported reductions in greenhouse gas emissions in association with the application of green waste compost (Dalal et al. 2009, 2010; Vaughan et al. 2011). Composts have potential to improve soil biological function (e.g., nutrient cycling) and have also been found to suppress some soil borne pathogens for vegetables and other crops (Termorshuizen et al. 2006; van der Gaag et al. 2007; Pane et al. 2013; Suárez-Estrella et al. 2013).

During the year 2005 a long term compost vegetable field trial was started at the Centre for Recycled Organics in Agriculture (CROA) at Camden in south western Sydney to evaluate the benefits and risks associated with using compost in vegetable production systems. Data from this field trial provides a valuable case study for the use of compost in intensive horticulture and this will form the basis for much of the following discussion in this chapter. This field trial was commissioned for a number of reasons. Firstly, a survey of the soils of thc vegetable farms of the Sydney Basin found that many of these soils had severely depleted levels of soil carbon, degraded soil structure and very high levels of available phosphorus compared to adjacent non-farmed soils (Chan et al. 2007a). As such, it was apparent that there was a need for organic C inputs and improved nutrient management in these systems. However, there was little information available on this particularly for the longer term. Secondly, the successful diversion of garden organics waste from its previous destination of landfill to composted garden organics (cGO) via government legislation and strategies, was starting to generate large quantities of cGO (i.e., ~0.3 mt/year) in the Sydney basin (Figs. 7.1 and 7.2) and these quantities are predicted to increase (Chan et al. 2007b, 2008). Around 87 % of the recycled organics generated in the area was utilised in the urban amenity market segments which included landscaping and domestic gardens, but this was thought to be approaching saturation (DEC 2004). In contrast, only ~4 % of the cGO was being used in agriculture, and as such it was thought that there was great potential for its use in agriculture around the Sydney Basin. This was the other driving force behind setting up the compost vegetable field trial.

The following sections are on the results of the long term compost vegetable field experiment at CROA and their implications for the beneficial use of compost in vegetable production systems.

Fig. 7.1 Source separated garden organics (i.e., grass clippings and shrub pruning's) ready for composting at a commercial composting facility, collected from suburban 'green lid bin' kerbside collections

Fig. 7.2 Source separated green-waste compost ready for distribution to farmers at an urban composting facility in south western Sydney

7.2 CROA Compost Field Experiment Design

The field trial was located at the NSW Department of Primary Industries 'Centre for Recycled Organics in Agriculture' (CROA) near Camden (150°42′32″E, 34°05′45.6″S), NSW, Australia, at a site with a long history of intensive cropping prior to the experiment. The soil at the site was a Chromosol/Dermosol inter-grade (Isbell 1996) [Lixisol (FAO 2006)] with topsoil which was hard-setting with low organic C levels and a silt-clay-loam texture and chemistry as presented in Table 7.1.

The field trial design is outlined in detail in Chan et al. (2008). Briefly, it consisted of seven treatments in a randomised complete block design with 4 replicates of each treatment. The treatments were; T1 = high soil P and conventional farmer practice (half poultry manure and half chemical fertilizer); T2 = high soil P and full compost; T3 = high soil P and compost and chemical fertilizer (half:half); T4 = low soil P and conventional farmer practice (half poultry manure and half chemical fertilizer); T5 = low soil P and full compost; T6 = low soil P and compost and chemical fertilizer (half:half); T7 = control (nil inputs).

Table 7.1 Properties of the soil (T=0), poultry manure and compost used in the compost vegetable field trial

Treatment	pHCa[a]	EC[b]	TOC	TN	C/N	Colwell P	Exchangeable cations [cmol (+) kg^{-1}]			
		dS m^{-1}	g 100 g^{-1}	g 100 g^{-1}		mg kg^{-1}	Na	K	Ca	Mg
Field trial soil (0–10 cm)	5.2	0.13	1.1	0.11	10	29	0.12	0.29	5.35	1.25
	pHw[b]	EC[b]	TOC	TN	C/N	Colwell P	TP	N:P ratio		
		dS m^{-1}	g 100 g^{-1}	g 100 g^{-1}		mg kg^{-1}	g 100 g^{-1}			
Compost no. 1 (crop 1)	5.6	3.14	21	1.1	19.1	1200	0.38	2.9		
Poultry manure (crops 1–10)	8.1	9.20	32	3.1	10.3	7500	2.60	1.2		
Compost no.2 (Crop 6)	6.9	5.3	30	1.6	18.8	2200	0.72	2.2		

[a] pH in 1:5 soil/0.01 M $CaCl_2$
[b] Electrical conductivity and pHw in 1:5 soil: water extract
TOC total organic carbon, *TP* total P

Fig. 7.3 The garden organics compost (cGO) being weighed out for distribution to compost treatment plots for the second application of compost prior to planting crop 6—capsicum at the CROA field experiment site

High and low initial levels of soil extractable P was included as a factor in the experimental design because high P levels were found to be typical of vegetable farm soils in the Sydney basin (Chan et al. 2007a, b) and as such it was considered important to assess the impact on vegetable production (Chan et al. 2008). For the high P treatments (T1, T2 and T3), triple superphosphate was applied to each plot at a rate equivalent to 680 kg P ha^{-1} and incorporated to 0.10 m, to raise the soil extractable P concentrations to levels similar to those observed in vegetable farm soils (~250 mg/kg in 0.10 m, Chan et al. 2007a, 2008) prior to the commencement of the field experiment. The site soil had a low initial concentration of bicarbonate extractable P (29 mg/kg) and as such ensured the other treatments (T4, T5, T6, T7) were representative of new vegetable farms with no prior history of high fertilizer inputs.

The compost used was derived from source separated garden organics blended with 10 % poultry (laying chickens) manure that was composted according to the Australian Standard AS 4454-2003. The properties of the compost and poultry manure used in this field trial are presented in Table 7.1. The compost used in the experiment is shown in Fig. 7.3 along with its application to the compost treatment plots prior to incorporation in Fig. 7.4.

In the case of these treatment descriptions, half refers to half the recommended dosage of total chemical NPK fertilizer application rates which was based on crop specific industry expert recommendations for each crop (NSW DPI Agfact/Primefact series (Agfact/Primefact Series 2013) and district horticulturalist advice). For the two organic amendments, poultry manure and compost, the rate was based on their total nitrogen (N) and an assumed availability index of 0.60 and 0.10 for poultry manure and compost respectively (i.e., it was assumed that, 60 % of the poultry manure total N and 10 % of the compost total N would be available to the crop) according to Evanylo and Sherony (2002). For the chemical fertilizers, phosphorus was applied as Triple superphosphate and incorporated into the soil to a depth of 0.10 m prior to crop seedling planting, whilst potassium and nitrogen were applied as muriate of potash and urea respectively, in split surface applications over the

Fig. 7.4 The garden organics compost (cGO) being spread out on the compost treatment plots prior to incorporation into the soil with a rotary hoe, prior to the planting of crop 6—capsicum, at the CROA field experiment site

duration of the crop. Poultry manure was applied with P for treatments T1 and T4 and incorporated into the soil prior to the planting of each crop.

The compost was applied to the compost treatments in a single application prior to the first crop. The full compost application rate for treatments T2 and T5 was determined to be 125 dry t ha^{-1} based on the recommended agronomic rate for N for the first crop (broccoli), the total N content of the compost, and the availability index of 0.10 for compost (Evanylo and Sherony 2002). The half compost rate for treatments T3 and T6, was, therefore, 62.5 dry t ha^{-1}. For the full compost treatments (T2 and T5), urea was only applied when petiole sap test results confirmed crop observations of low nitrate levels compared to the farmer practice treatments (T1 and T4). Applications of inorganic N fertilizer were not required for the first two crops of the full compost treatments (T2 and T5), but were required for crops 3 to 5. The full compost treatments received no muriate of potash over the course of the trial. The inorganic chemical and organic fertilizer inputs for each treatment for each of the first five crops is summarised in Table 7.2. Following the first 5 vegetable crops, a repeat application of compost was applied to the compost treatments, and a further 5 crops were grown. For the purpose of the following discussions, treatments T1 and T4, T2 and T5, T3 and T6, and T7 will be referred to collectively as 'farmer practice', 'compost', 'mixed' and 'control' treatments respectively.

The cropping sequence for this experiment was; 1. Broccoli (*Brassica oleracea* var. *botrytis* L.), 2. Egg-plant (*Solanum melongena* L.), 3. Cabbage (*Brassica oleracea* L.), 4. Capsicum or bell pepper (*Capsicum annuum* L.), 5.Leek (*Allium ampeloprasum* var. *porrum* L.), 6. Capsicum or bell pepper ('*Capsicum annuum* L.'), 7. Broccoli (*Brassica oleracea* var. *botrytis* L.), 8. lettuce (Lactuca sativa var. *capitata*), 9. Cabbage (*Brassica oleracea* L.) (Figs. 7.5 and 7.6), and 10. Sweet corn (*Zea mays*) (Fig. 7.7). After the harvesting of each crop, all of the non-harvestable crop residues on each plot were incorporated into the soil by rotary hoeing. Crops were managed following recommendations from the NSW Department of Primary Industries (Agfact/Primefact Series 2013) and an industry handbook (Salvestrin 1998).

Fig. 7.5 Cabbage seedlings (crop 9), shortly after being transplanted into the experiment plot beds

Fig. 7.6 Harvesting the cabbages at the end of crop 9 at the CROA long term compost-vegetable field trial site

The crops were drip irrigated with irrigation scheduling based on gypsum blocks (G bug) soil moisture monitoring of plots. More details of the field trial management are provided in Chan et al. (2008) and Chan et al. (2010).

7.3 Impacts of Compost on Intensive Vegetable Production Systems

Intensive vegetable production systems can degrade soil quality and function, and as a consequence lead to a decline in crop yields over time. Inputs of compost have been found to improve a number of measures of soil quality, sometimes resulting in crop yield benefits. Some of these impacts are outlined in the following sections.

Fig. 7.7 Sweet Corn (crop 10), just prior to harvest at the CROA long term field experiment

7.3.1 Agronomic and Economic Impacts

The marketable yield data (fresh weight in t ha^{-1}) from the first five crops of the field trial revealed that the large one off application of 125 dry t ha^{-1} of compost associated with the full compost treatments (T2 and T5) induced a crop yield response which matched the farmers practice treatments (T1 and T4) for four of the first five crops and exceeded it for one crop (Chan et al. 2008, 2011a, b). No significant difference ($p<0.05$) was found between the mean yields of farmer practice and full compost treatments for crops namely broccoli, eggplant, cabbage and leek whilst the yields of the full compost treatments for capsicum or bell pepper and was found to be 22% higher than that of the comparable farmer practice treatments (Chan et al. 2008, 2011a, b). Over the period of the first five crops, the compost treatment also resulted in significant savings from reduced fertilizer use compared to farmer practice (Table 7.2), with a 36% saving in urea as well as a 100% saving for K and P fertilizers (Chan et al. 2011b). The economic analysis of the yield and inputs for the first five crops determined that the full compost treatment had a benefit cost ratio (BCR) of 1 compared to farmer practice, indicating that this compost practice was very close to breaking even for the first five crops (Chan et al. 2011b). Although, a BCR of 1 on its own may not seem that encouraging for those considering practice change, it was thought at the time, that given the additional benefits measured for soil quality (Chan et al. 2008) and the environment with reduced water quality risk (Chan et al. 2010), that this was a fairly encouraging result. In contrast, the mixed compost treatment (½(half compost; half chemical fertilizer) with a one off application of 62.5 dry t ha^{-1} of cGO compost at the start of the field trial, although matching the yields of the farmer practice treatment for the first four crops, had a significantly lower ($p<0.05$) yield for the leek crop which was 64% of that of farmers practice (Chan et al. 2008, 2011b). This resulted in a negative BCR of −1.15 for the economic analysis of the mixed compost treatment versus farmer practice,

Table 7.2 Organic and inorganic fertilizer inputs applied to each treatment for each of the first 5 crops grown at the compost-vegetable field trial (Comp = garden organic compost; Mix = garden organic compost and inorganic fertilizer ½:½; FP = farmers practice (1/2 inorganic fertilizer and ½ poultry manure). (Adapted from Chan et al. 2011b)

Input[1]	1. Broccoli				2. Eggplant				3. Cabbage				4. Capsicum				5. Leek			
	Comp	Mix	FP	Cont	Comp	Mix	FP	Cont	Comp	Mix	FP	Cont	Comp	Mix	FP	Cont	Comp	Mix	FP	Cont
Compost (kg ha^{-1})	125	62.5	–	–	–	–	–	–	–	–	–	–	–	–	–	–	–	–	–	–
Poultry manure (kg ha^{-1})	–		4.03	–	–	–	3.24	–	–	–	4.3	–	–	–	3.25	–	–	–	3.62	–
Urea (kg ha^{-1})	–	163	163	–	–	130	130	–	133	200	200	–	200	265	266	–	425	425	425	–
Triple P (kg ha^{-1})	–	143	143	–	–	200	200	–	–	190	190	–	–	119	119	–	–	72	72	–
Muriate of Potash (kg ha^{-1})	–		–	–	–	47	47	–	–	57	57	–	–	43	43	–	–	–	58	–

[1] Note high P and low P treatments received the same fertilizer inputs after the initial adjustment of P levels at the start of the trial

which suggested that the lower compost application rate was not as economic as the larger application rate over a time frame of five crops, at least (Chan et al. 2011b).

In 2008, prior to the growing capsicum, the compost applications were repeated for the compost treatments (T2 and T5) and the mixed compost treatments (T3 and T6). A further five vegetable crops were grown in the field trial following the experiment protocols of the first five crops. The detailed results for these crops (currently unpublished), again found that the marketable yields for the full compost treatment matched or exceeded the yields of the farmer practice for crops capsicum, broccoli, lettuce, cabbage and sweet corn. The most extraordinary result was for the crop 6 (capsicum) which was the first crop grown following the second compost application. For this crop, the full compost treatment achieved yields which were almost double the farmers practice ($p<0.05$), whilst the mixed compost treatment (62.5 dry t cGO ha^{-1}) attained a mean yield which was more than 50% higher than the farmer practice yield (Fig. 7.8). To put the extent of the response of the capsicum crop to the compost treatment (i.e., a yield of ~60 t ha^{-1}) into context, the farmer practice mean yield of around 32 t ha^{-1} for this experiment was only just below the perceived potential yield for capsicums of 40 t ha^{-1} (Bartha 1983). The compost treatments, therefore, helped the capsicum crop achieve its optimal level of production. The only other crop where the full compost treatment had a significantly higher mean yield than the farmer practice treatment was crop 8 (lettuce) where the compost yield was ~22% higher than farmer practice. The benefit cost ratios (BCR) from the financial analysis of this experiment over ten crops with two applications of compost for the full compost and mixed compost treatments (currently unpublished) using the same methodology and assumptions outlined in Chan et al. (2011b), were well over 1 for both the full compost and mixed compost treatments. This was largely due to the fact that capsicum or bell pepper (i.e., the crop which had a significant yield response to the compost treatment) was a high value crop (Dorahy et al. 2013). These results demonstrate that such high compost input systems can be economical for vegetable growers over the 10 crop cycle, provided crops that are responsive to improvements in soil quality are selected for planting early in the cropping sequence following the application of the compost.

The yield results from the ten crops grown in this compost vegetable trial demonstrated that large applications (62.5 and 125 dry t ha^{-1}) of a blended garden organics green-waste compost product (80–90% garden organics composted with 10–20% chicken manure) supplemented with inorganic N fertilizer was able to match the current farmers practice (half inorganic fertilizer; half poultry manure) for the Sydney Basin region based on vegetable crop marketable yields. The results also demonstrated that some crops, in terms of their marketable yield, are more responsive to soil quality improvements than others. This experiment revealed that capsicum was one such crop. It is, therefore, important to evaluate local crops to establish which crops are more responsive to soil quality improvements, and ensure that they are planted as the first crops following the application of compost. If these are also high value crops like capsicum, then the chance of maximising economic return may also be increased.

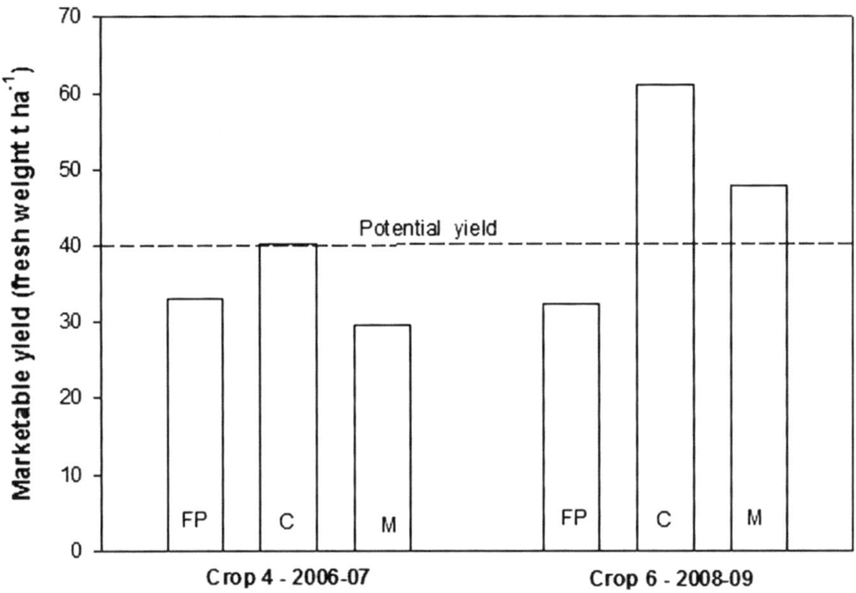

Fig. 7.8 Capsicum yields for the two capsicum crops grown at the field trial for the farmers practice (FP), compost (C), and mixed (M) treatments. Crop 4 was the fourth crop following the first compost application, while crop 6 was the first crop following the repeat compost application. The perceived maximum crop yield of 40 t ha^{-1} (Bartha 1983) is presented as a dashed line

In summary, the lessons gained from the yield and economic data on these crops were; (a) that certain crops are more responsive than others to improvements in soil quality, (b) it is important to work out which crops are responsive and which are not, (c) that it is important to plant responsive crops early in the cropping sequence following compost amendments in order to maximise yield benefits, especially if the responsive crops are high value crops, (d) that yield responses from responsive crops may be greater following repeat applications of compost than they were initially following the first application, possibly due to enhanced response by soil microbiology, as observed in the CROA experiment, and (e) that alternative vegetable production systems based on significant inputs of compost can be an economical alternative to conventional systems based on chicken manure and synthetic inorganic fertilizers.

7.3.2 Impact of Composts on Soil Quality

Soil quality includes the physical, chemical and biological properties of the soil which together influence soil function which is vital for sustainable agriculture. Improving soil quality is very important for maintaining and improving food production from agricultural land.

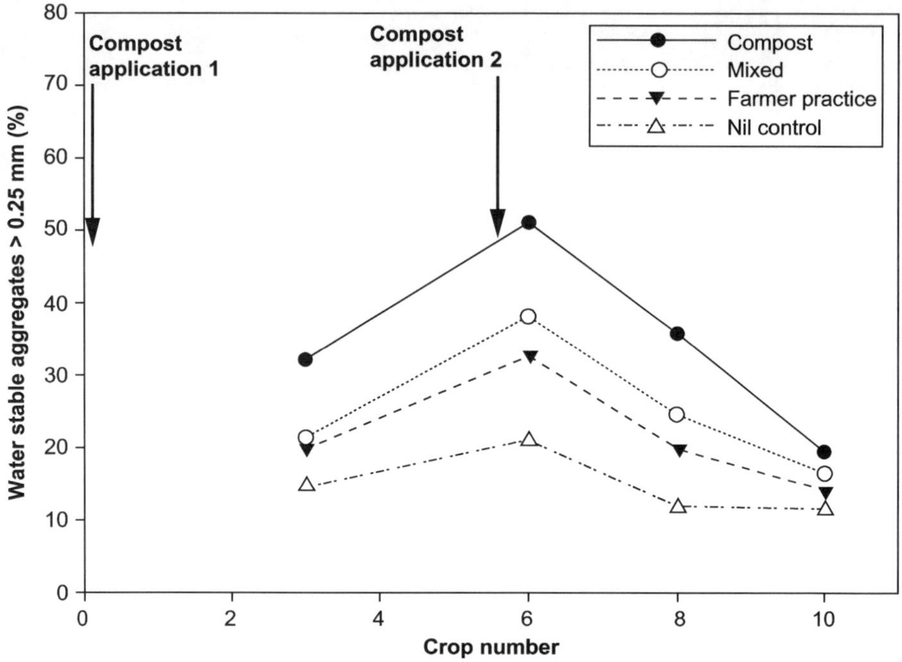

Fig. 7.9 The mean proportion of the soil as water stable aggregates >0.25 mm for each treatment over the cropping sequence at the CROA vegetable-compost field trial

7.3.2.1 Physical Soil Quality

Soil structural stability is important for maintaining pore spaces in the soil to help with soil drainage, aeration and crop root growth. The application of compost (and its associated organic matter) at a high rate was found to greatly improve the soil physical structure in the compost treatments of the field trial, resulting in a significantly ($p<0.05$) higher proportion of the soil as water stable aggregates which was still evident in the third crop following the first application of compost (Chan et al. 2008). The second application of compost similarly improved the soil structure in the compost treatment soils relative to the farmer practice treatment soils (Figs. 7.9 and 7.10) where the compost treatment had more than 50% of its mass present as water stable aggregates >0.25 mm. Kremer and Hezel (2013) in another field trial in Missouri also found that organic inputs including composted vegetative residues had similar benefits for soil structural stability, increasing soil water stable aggregates by up to 72% compared to conventional farming systems.

However, the benefit of investigating soil structural stability over successive crops can be seen in Figs. 7.9 and 7.10 where it is apparent that the structural stability of the soil is gradually being degraded over time across all treatments including the compost treatment. By the tenth crop there was little difference in the structural stability of the soils across the treatments, and the structure of the soil across all

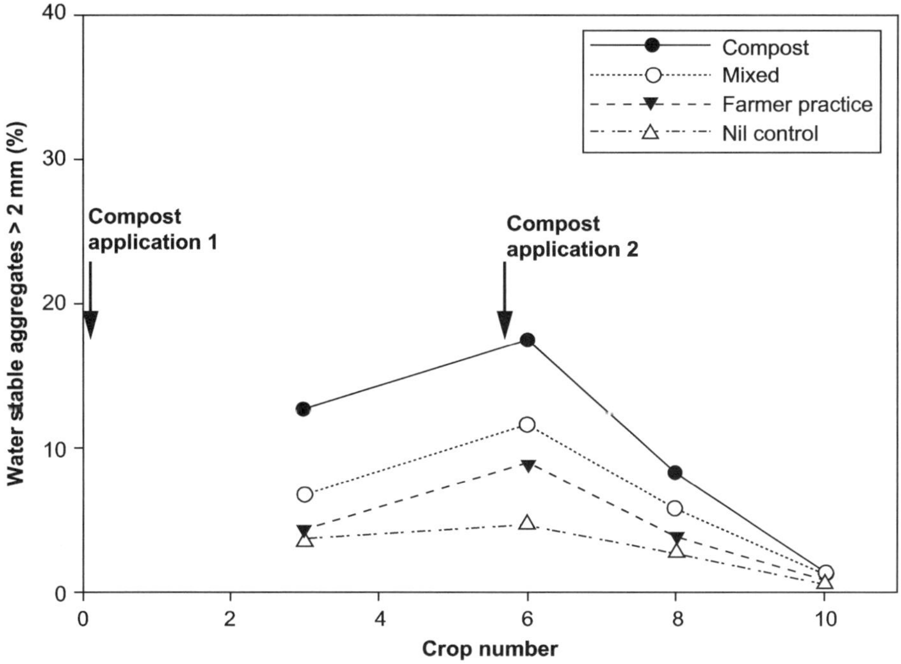

Fig. 7.10 The mean proportion of the soil as water stable aggregates >2 mm for each treatment over the cropping sequence at the CROA vegetable-compost field trial

treatments was degraded with less than 25 % of the soil as water stable aggregates >0.25 mm and less than 2 % of the soil as water stable aggregates >2 mm. This result largely reflects the influence of tillage on soil structure. The tillage in this study was done by rotary hoe which is a fairly aggressive tillage implement and is the standard practice for vegetable production in the Sydney basin. The results illustrate very well how the longevity of the benefits of the compost application to soil structure can be undermined by aggressive tillage which accelerates the breakdown of organic C and also physically pulverises the soil (Chan et al. 2007a, 2011a). It thus seems from these results that a minimum tillage regime may help extend the longevity of the benefits to soil quality that come from compost application.

Another soil physical property that is important to crop growth is soil strength or resistance to penetration, as it affects the ease with which roots can grow and explore the soil for nutrients and water. This property was measured to a depth of 45 cm in the experiment plots after the harvest of capsicum using a penetrometer (Rimik®) as described in Chan et al. (2006). The mean penetration resistance for each treatment in the experiment presented in Fig. 7.11 where the effect of the compost on soil strength is very apparent down the whole 45 cm profile for both the full compost (125 dry t ha^{-1}) and the mixed (62.5 dry t ha^{-1}) treatments. The compost applications reduced the penetration resistance of the soil, which has implications for effective root growth and crop access to nutrients and moisture in the soil. It is

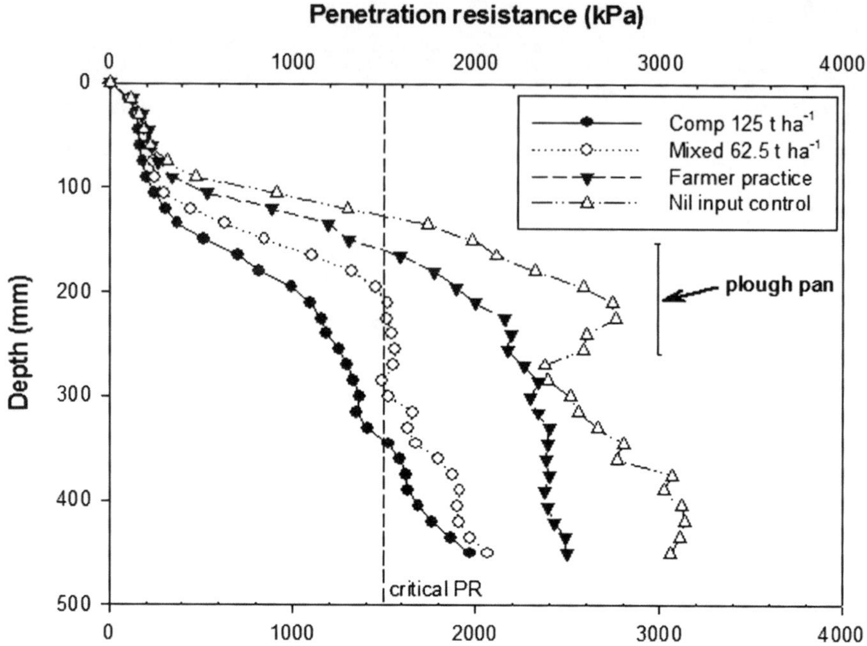

Fig. 7.11 Average penetration resistance (kPa) within the top 45 cm of the soil profile for each treatment

also apparent in, that a ‘plough pan’ or compaction layer is already present at 15 to 20 cm depth in the control treatment soil and appears to be starting to form in the farmers practice treatment soil at a similar depth. Absence of ‘plough pan’ formation in the compost treatment is thought to be a consequence of better drainage due to improvements in soil structure, which has resulted in a lower soil moisture content at plough depth, and as such less shearing stress and compaction of the soil at the plough depth during tillage operations.

7.3.2.2 Soil Chemistry

The application of blended cGO compost at high rates (125 dry t ha^{-1}) resulted in improvements to many key soil chemical properties generally associated with soil quality including cation exchange capacity (eCEC) important for nutrient storage, total organic C, plant available potassium (exch. K), exchangeable calcium (exch. Ca), plant available phosphorus, and pH, and many of these benefits persisted for several crops (Eldridge et al. 2014). The soil test results (0–15 cm depth) for the compost field trial treatments are presented in Table 7.3 for crops 1 and 4 and in Table 7.4 for crop 3. The main disadvantage of large applications of compost was a moderate increase in soil salinity in the short term immediately after application (Table 7.3). This means that crops sensitive to salt such as lettuce should be avoided

Table 7.3 Comparison of farming system means for key soil properties in crop 1 (broccoli) and crop 4 (capsicum). (Adapted from Eldridge et al. 2014)

Treatment	Total OC g 100 g^{-1}		eCEC[a] cmol (+) kg^{-1}		Exch. K cmol (+) kg^{-1}		Exch. Na cmol (+) kg^{-1}		pHca		EC dS m-1	
	Crop 1	Crop 4	Crop 1	Crop 4	Crop 1	Crop 4	Crop 1	Crop 4	Crop 1	Crop 4	Crop 1	Crop 4
Farmer practice —low P	1.23c	1.43b	8.26b	11.49a	0.45c	0.85a	0.181c	0.403a	5.187c	5.387b	0.185bc	0.384a
Compost (125 t ha^{-1}) —low P	2.10a1	2.00a	11.81a	12.30a	1.18a	0.62b	0.439a	0.271b	5.725a	5.663a	0.388a	0.298a
Mixed (62.5 t ha^{-1}) —low P	1.59b	1.59b	9.47b	10.03b	0.73b	0.49b	0.283b	0.253b	5.475b	5.300b	0.259b	0.296a
Nil input control	1.13c	1.18c	7.01c	7.96c	0.29d	0.24c	0.118c	0.218b	5.175c	5.300b	0.130c	0.158b
l.s.d. (p=0.05)	0.20	0.166	0.95	1.16	0.15	0.12	0.075	0.069	0.134	0.109	0.088	0.091

[a] Different lower case letters down each column indicate a significant difference (p=0.05) in soil property means between treatments

Table 7.4 Soil quality (chemical) under the different treatments at the transplanting of crop 3 (cabbage). (Adapted from Chan et al. 2008)

Treatment	EC	pH	N	C	C/N	Exchangeable cations [cmol (+) kg^{-1}]				
	dS m-1	$CaCl_2$	g 100 g^{-1}	g 100 g^{-1}		Ca	K	Mg	Na	eCEC[a]
Farmer practice—low P	0.31	5.43	0.16	1.35	8.4	6.13	0.64	1.73	0.41	8.96
Compost (125 t ha¯1)—low P	0.17	5.88	0.20	2.03	10.2	7.78	0.65	2.00	0.23	10.69
Mixed (62.5 t ha¯1)—low P	0.22	5.30	0.18	1.68	9.3	6.83	0.41	1.56	0.23	9.01
Nil input control	0.15	5.33	0.14	1.20	8.6	5.65	0.24	1.43	0.23	7.55
l.s.d. (p=0.05)	0.11	0.17	0.04	0.26	–	0.81	0.17	0.38	0.09	1.27

[a] Effective cation exchange capacity

early in the cropping sequence after compost application (Eldridge et al. 2014). The field trial results also showed some soil carbon sequestration benefits from the compost treatments relative to the farmer practice for the Sydney basin, being 38.5 % higher for the third crop (Chan et al. 2008). However, the benefits of compost in this regard, could be extended with the adoption of a minimum tillage approach, as rotary hoe tillage within the CROA experiment was relatively intensive.

7.3.2.3 Nutrient Cycling and the Environment

The experimental data from the CROA field trial also provides some valuable insights into the sustainability of both the current farmer practice and the other trial treatments in terms of the nutrient cycling for the important plant macro-nutrients NPK and associated environmental risk. A partial mass balance was determined for the phosphorus nutrient for each treatment in the experiment by Chan et al. (2010) and this is presented in Table 7.5. It is apparent in Table 7.5 that the farmer practice, compost and mixed treatments as farming systems, are all loading the soil up with P in excess to crop requirements, with only 6, 9, and 9 % respectively being removed from each treatment in harvested crop produce. None of these systems are sustainable in the long term without adjustment, which in part reflects the problem of applying organic amendments at N fertilizer rates when they have N:P content ratios (see Table 7.1) that are much lower than the crop uptake N:P ratio. As such P can accumulate in the soil over time, eventually posing a risk to water quality and the environment. In some situations it might, therefore, be more appropriate to use crop P requirements and soil available P levels as the criteria for determining compost application rates.

An assessment of the relative environmental risk posed by the farmer practice and compost treatments was done by Chan et al. (2010) at the end of the first five vegetable crops by a simulated rainfall study and analyses of soil P and runoff water samples. It was found that the compost and mixed compost treatment soils had significantly lower levels of available P (Colwell and $CaCl_2$) and total P than the

Table 7.5 Phosphorus inputs (kg ha^{-1}) from inorganic fertilizer, poultry litter, and compost and removal of P (kg ha^{-1}) by the harvesting of vegetable crops for the different treatments for the first 5 vegetable crops in the field trial. Numbers in brackets are the removal of P by harvesting expressed as a % of total P inputs. (Adapted from Chan et al. 2010)

Treatment	Inorganic fertilizer	Poultry litter	Compost	Total inputs	Total removal by crops	Partial balance
kg ha^{-1}						
Farmers practice —low P	151.9	493.4	0.0	645.3	38.4 (6 %)	607
Compost (125 t ha^{-1})—low P	0.0	0.0	487.5	487.5	43.4 (9 %)	441
Mixed (62.5 t ha^{-1}) —low P	151.9	0.0	243.8	395.7	36.3 (9 %)	359
Nil input control	0.0	0.0	0.0	0.0	11.9	−11.9

Table 7.6 Phosphorus levels in the soil and runoff water (rainfall simulations) from field trial treatment plots after the fifth vegetable crop. Within columns values followed by the same letter are not significantly different at $p=0.05$; $^{**}p<0.01$: $^{***}p<0.001$. (Adapted from Chan et al. 2010)

Treatment	Soil Colwell P	Soil $CaCl_2$ P	Soil total P	Runoff total P	Runoff soluble P
	mg kg^{-1}			mg L^{-1}	
Farmers practice —low P	235b	17.6ab	800b	14.5ab	2.88c
Compost (125 t ha^{-1}) —low P	99c	3.2cd	600c	4.6c	0.53d
Mixed (62.5 t ha^{-1}) —low P	116c	3.5cd	500cd	10.4bc	0.82d
Nil input control	26d	0.5d	300d	4.1c	0.07e
Significance	***	***	***	**	***

farmer practice treatment, and that this translated into significantly lower levels of soluble P in the runoff water (Table 7.6). On this basis, Chan et al. (2010) concluded that replacing poultry manure with blended garden organics compost could pose less risk to water quality from vegetable production systems. But nevertheless, soil P levels can also build up over time with such large applications of blended garden organics compost (i.e., 125 dry t ha^{-1}), and this in the end becomes the environmental limit for this system when using low contaminant compost. It is also worth noting that the second compost applied in this experiment (Table 7.1) had almost double the amount of total and available P which was believed to be due to a higher proportion of poultry manure in the compost blend (i.e., around 20% instead of the 10% for compost 1), and this compost application resulted in a proportionally larger increase in soil P levels (Dougherty and Chan 2014). Thus, the soil P level results from the CROA experiment indicate that although such large applications of compost have great value in rejuvenating soil quality, environmental considerations suggest that their application at such rates should not be done on a continual basis, but rather as an occasional treatment to rejuvenate soil quality. Although, as with all farming systems, the soil nutrient levels and properties need to be monitored and adjustments made to inputs accordingly. The effect of a large initial application of compost followed by regular smaller targeted applications (e.g., along the plant lines in beds) should be explored further.

It is apparent from the level of exchangeable K in the compost treatment soil at the start of the first crop broccoli (Table 7.3) that the blended garden organics compost product supplied large reserves of available K to the soil raising it to 1.2 cmol (+) kg^{-1} soil from its original level of 0.29 cmol (+) kg^{-1} soil (Table 7.1). However, by the fourth crop of this experiment (Table 7.3), the substantial reserves of K that were present in the soil at the start of the experiment for the compost treatment were almost halved to 0.62 cmol (+) kg^{-1}. This loss of K reserves reflects the susceptibility of K to leaching when applied at large rates (Huang 2005). As such, it does raise the question of whether applying compost at such large rates is the most efficient way of using the nutrient reserves within the compost product, from a nutrient use efficiency perspective.

The field trial records suggest that the Evanylo and Sherony (2002) assumption of 10% of the total N in compost being available for the first crop, provided a reasonable estimate of the plant available N (PAN) supply for the blended garden organics compost (i.e., composted blend of 90% garden organics and 10% chicken manure) used in this field trial, as evident in the petiole sap test results and crop monitoring for crop 1. In addition, the full compost and the incorporated crop 1 residues provided sufficient PAN supply for crop 2 as well, for the full compost treatment. For the repeat application of 125 dry t ha^{-1} of compost after the harvesting of crop 5, sufficient PAN was supplied for the first crop (i.e., crop 6 capsicum) only, with supplementary inorganic N (urea) required for all subsequent crops. However, it is worth noting that the compost treatments only received half the available N fertiliser that was applied to the farmer practice treatments (i.e., only the same amount of urea as the farmer practice treatments, but with no poultry manure N) and this represented an N fertiliser use efficiency gain in the compost treatment relative to farmer practice. This may have been due to improved N cycling by the soil microorganisms in the compost treatment soils. However, the PAN supply for this compost product was dependent on prompt incorporation of the compost immediately following delivery and spreading, so as to minimise N loss to the atmosphere via ammonia volatilisation. At some of our demonstration sites where the same product was used but was not immediately incorporated into the soil, supplementary inorganic N fertilizer was required for the first crop. Other demonstration trials using composts of predominantly garden organics (i.e., >90% garden organics green-waste), revealed significant N immobilisation in the soil (i.e., 'N drawdown') throughout the early crop phases and a real need for inorganic N fertilizer. For these composts it is best to assume negligible PAN supply and a need for supplementary inorganic N fertilizer. In contrast, studies have found that composts derived from largely vegetable food waste (e.g. source separated municipal solid waste compost) or animal manures (composted broiler litter) generally have been found to yield >10% of their total N as PAN for the first crop (Pratt and Castellanos 1981; Sims and Stehouwer 2008). This is generally correlated with their total N content and their C/N ratio which usually reflects differences in their molecular composition which influences their decomposition and N mineralisation (Eldridge et al. 2013). Given the difficulty in predicting the mineral N or PAN supply from composts, sound advice for compost use in agriculture and horticulture, would be to always monitor crop condition carefully, and be ready to apply supplementary inorganic N fertilizer when required. The use of inorganic N fertilizer strips (i.e., applying inorganic N fertiliser at the recommended rate to a small area of the field) is also highly recommended as a strategy for the early detection of 'N drawdown' or inadequate N supply symptoms in crops in fields receiving compost. This practice can allow an early response to crop N deficiency with inorganic N fertilizer applications and as such minimise the risk of 'N drawdown' impacts on crop yield.

7.3.2.4 Soil Microbiology

Basic soil biological properties were measured for each crop in the CROA field trial to examine the effect of the compost treatments compared to farmer practice and these included soil respiration, microbial biomass carbon, and the hydrolysis of fluoroscein diacetate (FDA). The results for the first seven crops are presented in Table 7.7 and Fig. 7.12.

Soil respiration in the compost treatment (see Table 7.7) was found to be significantly higher ($p<0.05$) than the farmer practice treatment for broccoli, the first crop following the initial application of 125 dry t ha^{-1} compost, but there were no significant differences found between the treatments for the soils of the subsequent four crops (Donovan et al. 2014). The soil respiration of the compost treatments was again found to be significantly higher than that of farmers practice in the capsicum crop (crop 6) which followed the repeat application of 125 dry t ha^{-1} compost, and this significant difference was still observed in the soil of crop 7 (broccoli). The elevated soil respiration most likely reflects the substantial increase in available carbon substrate for the soil micro-organisms to utilise, which results from such a large application rate.

Donovan et al. (2014) found no significant differences between the soil microbial biomass carbon levels of the compost and farmer practice treatments for any of the five crops following the first application of compost at the CROA field trial (Fig. 7.12). However, the second application of 125 dry t ha^{-1} compost did result in significantly higher ($p<0.05$) microbial biomass C levels for the compost treatment compared to the farmer practice treatment for crop 6 (capsicum) and crop 7 (broccoli) (Donovan et al. 2014). In fact the microbial biomass C levels in the soil of the compost treatment was found to be up to 100 % higher than that of the comparable farmer practice soil (Fig. 7.12). The FDA results in contrast found few significant ($p<0.05$) differences between the compost and farmer practice treatment soils (Donovan et al. 2014). The microbial biomass results from the second application of compost suggest some benefit in repeat applications of compost. The initial application may have had a priming effect on the biological community allowing it to be more responsive to subsequent later additions of compost applications.. There may be merit in following up an initial large application of compost to rejuvenate soil quality, with smaller more frequent applications of compost to provide sustained potential benefits to soil biology. Other studies (Kremer and Hezel 2013; Reeve et al. 2010) found that organically managed agricultural systems with high organic inputs significantly increased soil microbial activity. Kremer and Hezel (2013) also found that organic systems with high inputs of composted vegetable residues significantly increased ($p<0.05$) soil enzyme activity and soil function. The impact of compost amendments on soil function and the transformation of organic matter and the cycling of nutrients and carbon is certainly an area of research which requires more attention.

Table 7.7 Effect of compost application on basal soil respiration in 7 vegetable crops in soil samples collected at time of crop planting or harvest in a long-term vegetable field trial (means followed by the same letter are not significantly different). (Adapted from Donovan et al. 2014)

Mean soil respiration rates (μg CO2-C/g o.d.e[a] soil/h)

Treatment	1. Broccoli (Planting)	2. Eggplant (Planting)	3. Cabbage (Planting)	4. Capsicum (Planting)	5. Leek (Planting)	5. Leek (Harvest)	6. Capsicum (Planting)	6. Capsicum (Harvest)	7. Broccoli (Planting)
Farmer practice—low P	2.154 de	1.632 b	0.715 a	0.573 a	0.689 ab	0.964 a	0.7299 b	0.4332 cd	0.6341 cd
Compost (125 t ha¯1)—low P	4.549 a	1.762 ab	0.720 a	0.574 a	0.959 a	0.866 a	1.6683 a	0.7692 a	1.2166 a
Mixed (62.5 t ha¯1)—low P	3.492 bc	1.512 b	0.712 a	0.497 a	0.750 ab	0.827 a	0.6808 b	0.5298 bc	0.598 cd
Nil input control	1.598 e	1.264 c	0.568 a	0.286 b	0.668 b	0.587 a	0.4133 b	0.2608 d	0.3947 d
l.s.d. ($p = 0.05$)	0.722	0.314	0.209	0.147	0.304	0.391	0.4881	0.2143	0.3210

[a] o.d.e = oven dry equivalent weight of fresh soil

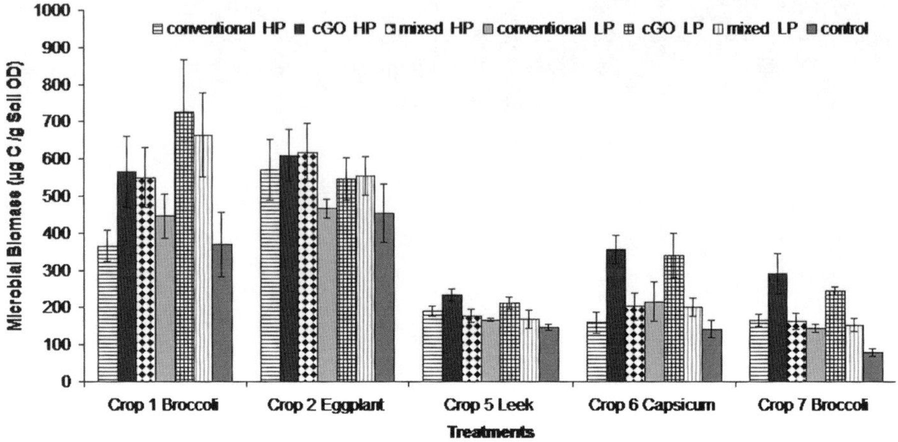

Fig. 7.12 Effect of compost application on microbial biomass carbon in soil samples collected at time of planting in a long-term vegetable field trial (data shown for 5 out of 7 crops grown, SE bars shown). (Adapted from Donovan et al. 2014). Conventional farmer practice treatment (inorganic fertilizer and poultry manure)

7.3.3 Contaminants Issue

The composts used in the long term compost–vegetable field trial at CROA were composts made from source separated garden organics waste (i.e., domestic grass clippings and pruning's) and chicken manure, and these composts met the Australian Standard (AS4454) for composts, which meant they were low contaminant composts. To protect food quality and human health, it is imperative that the composts that are applied to agricultural soils are low in chemical contaminants such as heavy metals and pesticides which can persist in the environment. Separating waste streams at the source (i.e., waste generation and collection point) and government regulation can help to achieve this. Composts made from non-source separated mixed municipal solid waste streams, should generally be avoided, as they often contain high levels of contaminants. However, the organic food scraps component of domestic municipal solid waste, if kept separate from other wastes at the household collection point and then all the way through to composting (i.e., to minimise contaminant levels), has great potential as an input for compost production.

7.4 Conclusions

It is apparent from the results of the field experiment at CROA, that the application of compost at large rates (i.e., >62 dry t ha^{-1}) can significantly improve a number of measures of soil quality (physical, chemical, and biological), and that such improvements can result in yield benefits for certain crops, with positive economic outcomes for farmers. Environmental benefits from incorporating compost inputs in vegetable productions systems were also noted. Vigorous tillage with rotary hoes was found to undermine some of these soil quality benefits by accelerating soil carbon losses.

The challenge is to further refine our use of composts in the farming systems to maximise the potential benefits. Combinations of compost applications with minimum tillage may help to further extend the soil quality benefits. More information on the yield response of the different crops in any given farming system to compost applications along with economic analyses, analysis can potentially help farmers to make decisions that will maximise financial outcomes and improve the environmental performance of their farms.

Acknowledgement The long term compost-vegetable field trial at the Centre for Recycled Organics in Agriculture (CROA) was a joint research and development project of NSW Department of Primary Industries (2005–2013), with collaboration and funding support from NSW Department of Environment and Climate Change (DECC) (2005–2008), the Australian Centre for International Agricultural Research (ACIAR) (2008–2012) and Horticulture Australia Ltd (HAL) (2010–2013). Technical assistance provided by Darren Fahey, Lynette Muirhead, Fadi Saleh, Ildiko Meszaros, and Brett Enman is gratefully acknowledged.

References

Agfact/Primefact Series (2013). www.dpi.nsw.gov.au/agriculture/horticulture/vegetables/commodity. Accessed 17 Jan 2013

Bartha L (1983) Yields of vegetable crops. Agnote, Department of Agriculture (Victoria). Order No. 2240/83, Agdex 207/01. ISSN 0155-0217, F.D. Atkinson, Government Printer, Melbourne, Australia

Bruinsma J (2009) The resource outlook to 2050: By how much do land, water, and crop yields need to increase by 2050? Expert meeting on how to feed the world in 2050, 24–26 June 2009. (Food and Agriculture Organisation of the United Nations, Rome)

Chan KY, Oates A, Swan AD, Hayes RC, Dear BS, Peoples MB (2006) Agronomic consequences of tractor wheel compaction on a clay soil. Soil Tillage Res 89:13–21

Chan KY, Dorahy CG, Tyler S, Wells AT, Milham PP, Barchia I (2007a) Phosphorus accumulation and other changes in soil properties as a consequence of vegetable production, Sydney region, Australia. Aust J Soil Res 45:139–146

Chan KY, Dorahy CG, Tyler S (2007b) Determining the agronomic value of compost produced from garden organics from metropolitan areas of New South Wales, Australia. Aust J Exp Agric 47:1377–1382

Chan KY, Dorahy C, Wells T, Fahey D, Donovan N, Saleh F, Barchia I (2008) Use of garden organic compost in vegetable production under contrasting soil P status. Aust J Agric Res 59:374–382

Chan KY, Wells T, Fahey D, Eldridge SM, Dorahy CG (2010) Assessing P fertilizer use in vegetable production: agronomic and environmental implications. Aust J Soil Res 48:674–681

Chan KY, Conyers MK, Li GD, Helyar KR, Poile G, Oates A, Barchia IM (2011a) Soil carbon dynamics under different cropping and pasture management in temperate Australia: results of three long-term experiments. Soil Res 49:320–328

Chan KY, Orr L, Fahey D, Dorahy CG (2011b) Agronomic and economic benefits of garden organics compost in vegetable production. Compost Sci Util 19:97–104

Childers DL, Corman J, Edwards M, Elser JJ (2011) Sustainability challenges of phosphorus and food: solutions from closing the human phosphorus cycle. Bioscience 61:117–124

Cordell D, Drangert JO, White S (2009) The story of phosphorus: global food security and food for thought. Global Environ Change 19:292–305
Dalal RC, Gibson IR, Menzies NW (2009) Nitrous oxide emission from feedlot manure and green waste compost applied to vertisols. Biol Fertil Soils 45:809–819
Dalal RC, Gibson IR, Allen DE, Menzies NW (2010) Green waste compost reduces nitrous oxide emissions from feedlot manure applied to soil. Agric Ecosyst Environ 136:273–281
DEC (2004) Draft aggregated analysis of the Department of Environment and Conservation's 2002/03 survey of organics processing in NSW. Department of Environment and Conservation (NSW), Parramatta, NSW, Australia
Donovan NJ, Saleh F, Chan KY, Eldridge SM, Fahey D, Muirhead L, Meszaros I (2014) Use of garden organic compost in a long-term vegetable field trial: biological soil health. Acta Hortic 1018:47–55
Dorahy C, Tulin A, Eldridge S, Mercado A, Salvani J, Lapoot C, Justo V, Duna L, Gonzaga N, Quinones CM, Rallos R, Rãnises M, Galambao M, Chan Y, Donovan N (2013) Enhancing profitability of selected vegetable value chains in the southern Philippines and Australia-Component 1-Integrated soil and crop nutrient management. (Project no. HORT/2007/066/1)-Final Report to the Australian Centre for International Agricultural Research (ACIAR), (unpublished), 79 p.
Dougherty WJ, Chan KY (2014) Soil properties and nutrient export of a duplex hard-setting soil amended with compost. Compost Sci Utili (In Press)
Eldridge SM, Chen CR, Xu ZH, Nelson PN, Boyd SE, Meszaros I, Chan KY (2013) Molecular composition of recycled organic wastes, as determined by solid-state 13C NMR and elemental analyses. Waste Manage 33:2157–2169
Eldridge SM, Chan KY, Donovan NJ, Saleh F, Fahey D, Meszaros I, Muirhead L, Barchia I (2014) Changes in soil quality over five consecutive vegetable crops following the application of garden organics compost. Acta Hortic 1018:57–71
Evanylo GK, Sherony CA (2002) Agronomic and environmental effects of compost use for sustainable vegetable production. In: International Compost Science and Utilisation Conference. 6-8 May, Columbus, OH, pp 730–740
Food and Agriculture Organization of the United Nations (FAO) (2006) World reference base for soil resources 2006: a framework for international classification, correlation, and communication. World Soils Reports No. 103, FAO, Rome
Feller C, Blanchart E, Bernoux M, Lal R, Manlay R, Ollivier T (2010) Organic matter knowledge and management in soils of the tropics related to ecosystem services. In: Lal R, Stewart BA (eds) Food security and soil quality. CRC, Boca Raton, pp 241–275
Gibson TS, Chan KY, Sharma G, Shearman R (2002) Soil carbon sequestration: utilising recycled organics: a review of the scientific literature. Organic waste recycling unit, NSW Agriculture.
Heenan DP, Chan KY, Knight PG (2004) Long term impact of rotation, tillage, and stubble management on the loss of soil organic carbon and nitrogen from a chromic luvisol. Soil Tillage Res 76:59–68
Huang PM (2005) Chemistry of potassium in soils. In: Tabatabai MA, Sparks DL (eds) Chemical processes in soils. SSSA Inc., Madison, pp 227–292
Isbell RF (1996) The Australian soil classification. CSIRO Australia, Collingwood, Vic., Australia
Kremer RJ, Hezel LF (2013) Soil quality improvement under an ecologically based farming system in northwest Missouri. Renewable Agric Food Syst 28:245–254
Lal R (2004) Soil carbon sequestration impacts on global climate change and food security. Science 304:1623–1627
Lal R (2010a) Managing soils to address global issues of the twenty-first century. In: Lal R, Stewart BA (eds) Food security and soil quality. CRC, Boca Raton, pp 5–22
Pane C, Piccolo A, Spaccini R, Celano G, Villecco D, Zaccardelli M (2013) Agricultural waste-based composts exhibiting suppressivity to disease caused by the phytopathogenic soil-borne fungi Rhizoctonia solani and Sclerotinia minor. Appl Soil Ecol 65:43–51
Pratt PF, Castellanos JZ (1981) Available nitrogen from animal manures. Calif Agric July-August:24

Reeve JR, Schadt CW, Carpenter-Boggs L, Kang S, Zhou J, Reganold JP (2010) Effects of soil type and farm management on soil ecological functional genes and microbial activities. ISME J 4:1099–1107

Salvestrin J (eds) (1998) Australian vegetable growing handbook. Scope publishing, Frankston, Vic., Australia

Semple EC (1928) Ancient Mediterranean agriculture: part II. Manuring and seed selection. Agric His 2:129–156

Sims JT, Stehouwer RC (2008) Recycling of nitrogen through land application of agricultural, municipal, and industrial by-products. In: Schepers JS, Raun WR (eds) Nitrogen in agricultural systems. Agronomy monograph 49. ASA, CSSA, SSSA, Madison. 759–821

Suárez-Estrella F, Jurado MM, Vargas-Garćia MC, López MJ, Moreno J (2013) Isolation of bio-protective microbial agents from eco-composts. Biol Control 67:66–74

Termorshuizen AJ, van Rijn E, van der Gaag DJ, Alabouvette C, Chen Y, Lagerlöf J, Malandrakis AA, Paplomatas EJ, Rämert B, Ryckeboer J, Steinberg C, Zmora-Nahum S (2006) Suppressiveness of 18 composts against 7 pathosystems: variability in pathogen response. Soil Biol Biochem 38:2461–2477

UN ESA (2008) Population data of the U.N. Economic and social affair.http://esa.un.org.undp/

van derGDJ, van Noort LHM, Stapel-Cuijpers C, de Kreij AJ, Termorshuizen AJ, van Rijn E, Zmora-Nahum S, Chen Y (2007) The use of green waste compost in peat-based potting mixtures: fertilization and suppressiveness against soilborne diseases. Sci Hortic 114:289–297

Vaughan SM, Dalal RC, Harper S, Menzies NW (2011) Effect of green waste compost on mineral nitrogen, nitrous oxide and carbon dioxide from a vertosol. Waste Manage 31:1720–1728

Chapter 8
Principles of Compost-based Plant Diseases Control and Innovative New Developments

Catello Pane and Massimo Zaccardelli

Abstract Compost is considered one of the different available sustainable approaches that may be used to prevent, mitigate or to control plant diseases. This is possible through the exploitation of the compost suppressive properties. This chapter aims to address the several aspects of compost suppressivity in order to derive the key principles of this phenomena. In particular, will be elucidate the concepts of compost suppressivity, the mechanisms governing it and the ecological aspects that lead it. Finally, will be reported how compost can be exploited in plant disease management both directly and as suitable source of plant protectants.

Keywords Biological control · Soil-borne pathogens · Disease ecology · Plant resistance inducer · Suppression mechanisms

8.1 Introduction

Plant pathogens and diseases can be well controlled with composts by exploiting their suppressive properties. This actual fact, perceived already in empirical ancient agriculture (Parr and Hornick 1992), has been rigorously studied by scientists since 1970s (Bonanomi et al. 2007). To date, the findings produced by research groups worldwide, have contributed to advance the knowledge on this topic and to encourage the compost applications in agriculture, also with aim of plant protection. However, it is needless to say, right from the beginning dissertation, that suppressive composts are not fungicides that able to give standard performances. But, their increasing utilization, as we will see later, will be able to drastically reduce the needs for external inputs in crop protection management. Therefore, although disease control effectiveness by compost can be variable (Termorshuizen et al. 2007), the economic and environmental benefits deriving from its use can win any form of distrust that could hover on operators.

C. Pane (✉) · M. Zaccardelli
Consiglio per la Ricerca e la Sperimentazione in Agricoltura, Centro di Ricerca per l'Orticoltura, via dei Cavalleggeri 25, 84098 Pontecagnano, SA, Italy
e-mail: catello.pane@entecra.it

D. K. Maheshwari (ed.), *Composting for Sustainable Agriculture,* Sustainable Development and Biodiversity 3, DOI 10.1007/978-3-319-08004-8_8

8.1.1 Plant Pathogens and Diseases

The plants in the agricultural ecosystems, as in the natural ones, are subjected to a wide range of pathogenic agents that cause the so called biotic stresses or, more simply, diseases. Plant diseases can be schematized in soil, air and seed-borne ones, depending by the type of host plant organ (hypogeal, epigean or seed, respectively) from which the infection began. Soil-borne phytopathogens are responsible for destructive diseases of roots, stems and vascular tissues, occurring in a broad-spectrum of susceptible hosts under favorable environmental conditions.

Plant pathogens belong to several different phyla. In the plethora of microorganisms, potential plant pathogens, filamentous fungi and bacteria, are among the most feared and dangerous. This because of their extraordinary ability to survive, conserve and diffusion themselves both in the space and in the time, even in extreme conditions; to breakdown plant genetic resistance mechanisms and, finally, to overcame toxicity of chemical compounds.

Disease occurrence generates physiological disorders in the plants that significantly reduce yield and quality until, eventually, to cause the complete crop losses. Therefore, for these reasons, in productive eco-systems cannot totally renounce to an efficacious control program. The methods of disease control are always finalized to the breaking of the disease cycle. This last is the succession of interlinked stages occurring in the complex interaction between the virulent pathogen and the susceptible host plant that establish the pathogenesis. The more sensible phases of this cycle to the action of the control tools include: pathogenic inoculum conservation and dissemination; penetration, infection and spreading in the host plant; evasion and production of propagules for future infections.

8.1.2 The Old Compost as Innovative Tool for Diseases Control

Since the agrarian crops are susceptible to many bacterial and fungal attacks, disease management is one of the most important components in the production system. Generally, disease control is entrusted to chemically synthesized fungicides and fumigants. When severe biotic epidemics occur, growers generally respond with applications of broad-spectrum chemicals on a calendar basis, with up to 12 to 16 applications per season (Reitz et al. 2008) that weights by 4.1 % on total costs of production (Engindeniz 2006). But, health reasons and environmental sensitivity are driving the market towards products obtained by sustainable cropping systems, stimulating the reduction of synthetic pesticides (banning of more dangerous ones) and the use of alternative and eco-compatible plant disease management (Wagner 1999). Moreover, this issue is of great relevance today in Europe, that is exploring new directions toward integrated pest management (IPM) in the view of implementing the Directive 2009/128/EC on the reducing pesticide use-and-risk in EU. In this context, the suppressive properties of compost represent a real opportunity for the development of a green agriculture, safe, competitive and sustainable in nature.

The compost is a complex organic matter commonly used as soil amender. This peculiarity has prompted its utilization in the area of the plant protection mainly in the context of diseases caused by telluric pathogens. At any rate, there are several interesting studies on the application of composts and their derivatives, to suppress aerial plant diseases. In the large landscape of the different available approaches that may be used to prevent, mitigate or control plant diseases, the use of suppressive composts may be referred in total as a biological control tool. It, in fact, fully corresponds to U.S. National Research Council definition of biological control: "the use of natural or modified organisms, genes, or gene products, to reduce the effects of undesirable organisms and to favor desirable organisms such as crops, beneficial insects, and microorganisms" (Pal and McSpadden 2006). Accordingly to that, compost suppressivity may be assured by microbial communities, that may play antagonistic functions but, also, by organic molecules that may be ascribed as microbe "gene products".

8.2 The Concept of Compost Suppressivity

The ability of compost to generate an environment partially or totally adverse to the development of plant disease(s), although a pathogen might be present and the host plant is susceptible to it, is defined suppressivity. Generally, to indicate this property, in the field of phytopathology, in addition, is also used variables of the term suppressivity bearing the same root (i.e., suppressiveness, suppressive, suppression, etc.), and/or its synonym such as repressiveness. The establishment of condition(s) exactly opposite to those indicated for suppressivity occurrence is named conductivity. Thus, conductive composts are those that favor the development of plant pathogen(s) and/or plant disease(s). The compost that does not affect any phytopathological parameter is reported to be null effect.

The theory of disease suppressivity related to the compost is borrowed from that of the natural suppressive soils (Haas and Défago 2005). Although, in this case, it were principally referred to the suppression of soil borne plant pathogens (Mazzola 2002), in the particular case of compost, it has also been extended to the aerial and seed-borne diseases control ability.

In the compost different components co-exists both biotic and abiotic ones that, potentially, might sustain in varying degree the suppressive functions. Based on how the different determinants concur to realize it, compost suppressivity has also been described according to a more informative classification that include biotic or abiotic, general or specific and, finally, direct or indirect suppressivity. Hence, this further separation, evidently, is closely related to the mechanisms of action of the compost. Therefore, hereinafter, we will report only the definitions, reserving the details in the specific paragraphs.

8.2.1 Biotic and Abiotic Suppressivity

This primary distinction is much important and emphasizes the contribution offered by living microorganisms against that, instead, provided by the organic and inorganic abiotic molecules. It responds to the question: what causes the suppressivity and in what proportions? There are a number of examples in which both biotic and abiotic elements of compost, have relevance in disease control. On the one part, the biological component, represented mainly by fungi, bacteria, actinomycetes, protozoa, nematodes and their products, play the crucial role in determining biotic suppressivity. On the other hand, compounds showing significant antifungal and disease suppressive bioactivity are indicated as determinants of abiotic suppressivity. In this category enters also physical compost features conditioning the pathogenesis. Obviously, the biological suppressivity is completely loss following compost sterilization by heating, fumigation or other non-thermic antibiotic treatments (Malandraki et al. 2008). Whereas, the suppressivity can be restored after cooling when the antagonistic micro-flora naturally recolonize the compost or by re-adding to the sterilized null-effective one, a minimal amount of the raw compost or its water extracts. The total or partial persistence of suppressive properties in biologically vacuum composts, highlights unequivocally the role of the abiotic component.

8.2.2 General and Specific Suppressivity

Based on the amount and diversity of the involved actors, the compost suppressivity can be divided into two major categories, general and specific. In particular, general suppressivity is that generated by the sum of the activities exerted by the overall potential agents of suppressivity. In this case, the responsibility of the effect is distributed on all elements that can be biotic and abiotic, including the groups of microbial compost community and/or the whole spectrum of the chemical and physical features. The specific suppression, instead, is considered to be generated through the actions of one or few particular determinants. Often, this effect is established by few specific microbial groups or single microbes exhibiting species-specific antagonistic activity against phytopathogens. But, also single abiotic character of compost can act according to a specific model. Apparently, this contraposition between the general and specific models may seem to be connected to the effectiveness of the system in diseases controlling. But it is not true. Actually, as we shall see later, the levels of suppressivity in compost depend from the challenge between the mode of action and the specific pathosystem physiology. In other terms, limited to the objectives of this section, it is sufficient to say that some pathogens and/or diseases are more manageable with general suppressivity model, while others are better controllable with specific ones.

8.2.3 Direct and Indirect Suppressivity

The suppression of diseases can be mediated by compost in both direct or indirect way. It is direct if exist a clear causal-effect linkage between the pathosystem and the compost suppressive determinants. Thus, disease suppression is closely related to action of these last. Almost all published papers on this topic are based on direct suppressivity observations. Nevertheless, compost may also indirectly counteract the development of disease, for example, by improving the vegetative status of plants or by stimulating the native soil inhabiting antagonists before cultivation.

8.2.4 *From Potential to Multiple Suppressivity*

The set of characteristics of compost that are relevant for plant disease control including, for example, In-vitro antagonism, fungitoxicity, organic matter pathogen availability, microbial activity indexes, etc., define the potential of suppressivity. Therefore, compost may be evaluated as potentially suppressive. But, only the challenge with a suitable set of pathosystems can ascertain it's real in host-plant effectiveness. In the majority of literature reports, the study of this functional property is focused only on one phytopathogen/host plant system. Less frequently, the composts are tested against two or more diseases, revealing their significant variability. Composts able to control at least two different diseases are defined multisuppressive.

8.3 The Mechanisms Governing Compost Suppressivity

Compost suppressivity is caused by the totality of physical, chemical, biochemical and microbiological dynamic interactions between all its parts and plant-pathogen system. This complexity expresses univocally the suppressive ecosystem function, so that externally appears as a super-organism (Fig. 8.1). Behind all of this, there are a number of precise mechanisms of action, whose full understanding has been and to date still remains the main challenge for researcher worldwide.

8.3.1 *Antagonistic Models*

The major role in compost suppressivity is played by microbes directly responsible for antagonistic interactions with pathogens. The category includes a variety of potential antagonistic microbial groups including various genera of bacteria, such as *Bacillus* and *Pseudomonas*, other than fungi, such as *Trichoderma*. These microbes exert their antagonism by microbiostasis, antibiosis and hyperparasitism (Fig. 8.2), and/or for the stimulation of systemic resistance in host plants (Zhang et al. 1996).

8.3.1.1 Microbiostasis

The close competition for nutrients induced by compost-inhabiting microbial populations, results in the suppression of pathogen spore germination, in decreased

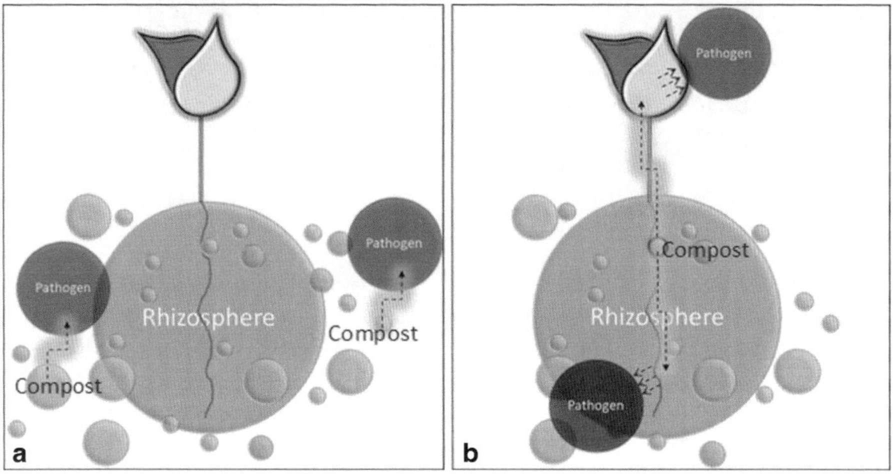

Fig. 8.1 Schematic representation of the two possible mode of action through which the compost can suppress disease development: toward the pathogen (**a**) or toward the plant stimulating its systemic reaction (**b**)

growth, activity and fecundity: this phenomenon is called microbiostasis (Hoitink and Fahy 1986; Lockwood 1990). The nutrient competition for a pathogen often means contend space and infection sites too.

8.3.1.2 Antibiosis

Antibiotic-mediate suppression involves those microbes that produce and secrete one or more compounds with detrimental activity for various plant pathogens; a process known as antibiosis (Hoitink et al. 1996). This antagonistic path in addition to the antibiotics can also include lytic enzymes, and other products of microbial metabolism. Bacteria belonging to the thermophylic *Bacillus* group, for example, have received greater attention because they produce antimicrobial compounds associated with bio-control with a broad-spectrum activity. These include antibiotics, such as bacillomycin, iturin A, mycosubtilin and Zwittermycin A (Leifert et al. 1995), lipopeptides (Ahimoua et al. 2000; Pal Bais et al. 2004) and antifungal proteins (Liu et al. 2007). Moreover, are well known the 2,4-diacetylphloroglucinol, phenazines, pyoluteorin and pyrrolnitrins produced by *Pseudomonas fluorescens*, the pseudanes (quinolinones) from *Burkholderia cepacea* or the gliotoxin secreted by fungi belonging to *Trichoderma* groups. An important role in this suppressive path was also played by lytic enzymes, such as various chitinases, glucanases and proteases. Singh et al. (2010) has reported *Pseudomonas aeruginosa* PN1 produces citinase and β-1,3-glucanase caused 69 % colony growth inhibition of fungal phytopathogen *Macrophomina phaseolina* and exhibited a strong bio-control agent. In this series presence of protease, chitinase, lipase and β-1,3 glucanase (lysogenic enzymes) in the compost, indicates a possible role in fungal degradation (El-Masry et al. 2002).

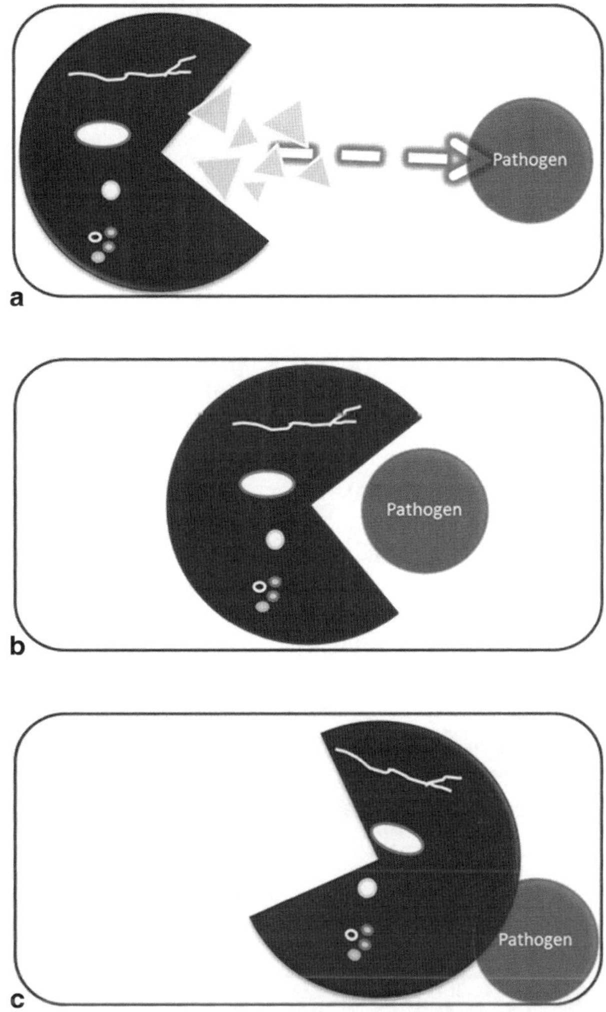

Fig. 8.2 Hypothetical mechanisms of antagonism that can be established by useful compost microflora (*black sphere*) toward the pathogen (*grew sphere*): antibiosis (**a**), hyperparasitism (**b**) and microbiostasis (**c**)

8.3.1.3 Hyperparasitism

Compost can be populated by predators that act directly on the pathogen by kill it or its propagules. The parasite of a pathogen is called hyperparasite and, subsequently, the trophic relation is called hyperparasitism. This antagonistic path is the most specific and not very frequent, although it is very effective when occurs. The most popular microbial hyperparasite belong to *Trichoderma* group. However, this antagonistic mechanism in composts is rather rare, because it is linked to the presence of specific pathogen predators.

Ascomycetes communities carried by biosolid compost suppressing *Sclerotium rolfsii* on sclerotia surface, were identified by a metagenomic approach, as mycoparasites belonging to *Chaetomium*, *Geomyces*, *Penicillium*, *Thielavia*, *Petriella* and *Trichoderma* species (Danon et al. 2010).

8.3.1.4 Activation of Plant Disease-Resistance

A further suppressive mechanism, moderately introduced in the recent times, concerns the induction of systemic resistance in the plants mediated by compost. It has been stated that reduction by composts of foliar diseases, as well as all diseases in which the site of treatment and the site of infection are spatially separated, is mediated by induced resistance. A side-grafted split-root system was used to demonstrate the occurrence of induced resistance phenomena in compost suppressing wilt disease caused by the soil-borne pathogens *Fusarium oxysporum* on melon (Yogev et al. 2010). While, Paplomatas et al. (2005) used an incised plexiglas chamber, that allow to separate the root system in two parts, to assess induced resistance triggered by compost in *Arabidopsis* against *Verticillium dahliae*. In response to specific stimuli, plants are able to enhance their innate defensive capacity against a wide spectrum of pathogens. Two distinct pathways that typically are triggered by non-pathogenic and pathogenic microbes lead to induced systemic resistance (ISR) and systemic acquired resistance (SAR), respectively. ISR develops systematically via jasmonic acid and ethylene, whereas SAR signaling moves through salicylic acid (SA) and the induction of pathogenesis related (PR) proteins, which occur specifically during pathological or related situations. As ISR is generally driven by useful microorganisms, in compost-grown plants it is triggered by colonization of plant roots by plant growth-promoting rhizobacteria (PGPR) and fungi (PGPF) inhabiting organic matter. However, many studies reported the compost-induced suppressivity driven preferably by enhanced expression of the PR and defense related genes, which share a SAR-like model. Enhanced peroxidase activity in cucumber plants treated with pine bark compost indicated the putative role of SAR in anthracnose suppression (Zhang et al. 1996). Induced expression of two SA-dependent pathogenesis related genes, PR-1 and β-1,3-glucanase in *Arabidopsis* grown in compost amended soils in pathogen-free condition, conferred resistance to *Pseudomonas syringae*, whereas SAR deficient mutants were susceptible (Vallad et al. 2003). Similarly, Segarra et al. (2013a) demonstrated with the aid of *Arabidopsis thaliana* jasmonate insensitive and salicylic acid-deficient mutants, the need for SA signaling in compost-induced resistance against *Botrytis cinerea*. Successively, a microarray analyses revealed that a compost from olive marc and olive tree leaves interfere in *B. cinerea*/*Arabidopsis* system by triggering SAR and ABA-dependent/independent abiotic stress responses (Segarra et al. 2013b). On the basis of several receptor-like kinase upregulated genes in compost-grown plants, has been moreover hypothesized that compost might be recognized as PAMPs or MAMPs. These are pathogen or microbe associated molecular patterns involved in pathogen recognizing by receptors located in the plasma membrane of the plant cell as stated by Cassells and Rafferty-McArdle (2012).

This model is believed to be also implied in the development of structural barriers in the plants against pathogenic infections. Compost water extracts applied to roots, elicited lignin accumulation in pepper stems against aerial infections of *Phytophthora capsici* or *Colletotrichum coccodes* (Sang et al. 2010). While, a compost derived from agricultural residues promoted the pathogen exclusion by inducing callose deposition in tomato root cells evidenced with a GUS expressing strain of *Fusarium oxysporum* f. sp. *radicis-lycopersici* (Kavroulakis et al. 2005). SAR-like defense response induced by compost can evolve in a primed status (priming) of plant that allow it to react immediately to future pathogen attacks. Sang et al. (2010) showed as root treatments by compost water extracts, induced priming state against *C. coccodes* on pepper and *C. orbiculare* on cucumber, by enhancing PR proteins, defense related enzyme production and H_2O_2 generation under pathogen-inoculated conditions. SAR strategy driven by compost is effective for a variety of plant species, while the ability of PGPR and PGPF to promote ISR is specific to certain plant species and genotypes (Van der Ent et al. 2009).

8.3.2 Role of Chemicals in the Compost Suppressivity

Compost provides a number of elements, ions, organic and inorganic molecules, originated from the abiotic matrix that may potentially affect the pathogenesis. These factors can have a direct toxicity effect on plant pathogen growth and survival, or may indirectly contribute to diseases suppression by strengthening plant stand.

Composts undergoing degradation, form and release a variety of substances toxic to pathogens. Ethanol, methanol, formaldehyde, ammonia and ethyl esters have all been identified as such substances (Bollen 1993). Volatiles formed during decomposition of crop residues exhibited suppressive effects on germination of *Verticillium longisporum* sclerotia, as well as inhibition of sporulation of other parasitic and saprophytic fungi (Bollen 1993). Extracts of tree bark compost have been found to release compounds toxic to fungi, i.e., inhibitors that lyse zoospores and sporangia of *Phytophthora* sp. (Hoitink et al. 1993). Ethyl esters of hydroxy-oleic acids were the most toxic compounds in 6-month old composted hardwood bark suppressive to *Phytophthora* sp. (Hoitink and Fahy 1986). Several organic chemicals present in composts or released by compost inhabiting microorganisms have been identified as providing disease suppressive effects, including phenolic compounds, volatile fatty acids and salicyclic acid (ROU 2006). Tenuta et al. (2002) found toxicity of volatile fatty acids (acetic, propionic, isobutyric, n-butyric, n-valeric, isovaleric, and n-caproic acids) in liquid swine manure to *Verticillium dahliae* microsclerotia. Olive mill waste composts showed the abiotic capability produced by chemical compounds, such as phenols, for suppression of *Phytophthora capsici* (Cayuela et al. 2008). Ammonia, nitrous acid or volatile fatty acids accumulated in organic amended soils with low rate of nitrification, are detrimental for *Verticillium dahliae* microsclerotia (Tenuta and Lazarovits 2002). This property can be implemented with the use of a matter with low organic carbon content (Tenuta and Lazarovits 2004). The effective NH_3 concentration was the major inhibitory component in compost extracts of *Sclerotium rolfsii* germination (Zmora-Nahum et al. 2008). A nutritional-driven mechanism lead by compost is also described. For example, N supply via compost, at agronomic levels, is beneficial to plants and functional to physiological and structural defenses, while excessive N loading have also been implicated in exacerbating numerous diseases. Composted biosolids applied to nutrient-deficient soils, improved turfgrass resistance against leaf rust severity caused by *Puccinia* sp. by enhanced nitrogen nutrition (Loschinkohl and Boehm 2001).

Minor specific elements can also play a role in increasing plant disease resistance. Enhanced foliar content of Mo, Ca and Si in compost amended cucumber plants was positively related to increased suppressivity against *Botrytis cinerea* (Segarra et al. 2007). Instead, Al has been indicated to be functional for suppressive media against damping-off of ornamental bedding plants caused by *Phytophthora parasitica* (Benson 1995). Hydrophobic humic molecules in compost are believed to directly or indirectly improve its disease control capability (Mathur 1991). Pascual et al. (2002) verified this property by separately assaying the various fractions of municipal compost in a suppressive assay with *Pythium ultimum* on pea seedling. The mechanisms by which composts and their humic fractions inhibit plant pathogens are not fully understood, although recent studies showed a significant relationship between the chemical and functional properties of humic substances and their capacity to suppress soil pathogenic fungi (Loffredo et al. 2007, 2008).

8.3.3 Physical Aspects of Composts Suppressivity

The compost is able to modify the physical environment in which disease develops by affecting moisture, free water containment, pH, electrical conductivity, temperature, water-holding capacity and bulk density (Chen et al. 1988; Hoitink et al. 2001). Of course, all the changes determining conditions that are unfavorable to pathogen growth and pathogenesis activation, lead to suppressivity.

8.4 Ecological Aspects Related to Compost Suppressivity

As seen, the microbial community in compost was considered to be the most significant factor of plant pathogens inhibition in function of its suppressive-like structure. Ecological relationships among microbes that achieve over time within the compost nutritional niches in the time, move a sort of dynamic evolution process that involve both biological and non-living factors (Fig. 8.3). Changes in microbial communities induced by the carbon food competition have relevant implications on functionalities, including those linked to suppressivity (Pane et al. 2011). Quality and bioavailability of organic matter has been linked to compost ability to provide biotic suppressivity (Veeken et al. 2005; Termorshuizen et al. 2007). Boehm et al. (1997) demonstrated the crucial role played by NMR-assessed carbohydrates to sustain peat microflora antagonism against *Pythium ultimum* on pea. While, shifts in microbial communities favored by phenolic and methoxyl C fractions, were considered to be decisive for the reduction of *Rhizoctonia* and *Sclerotinia* damping-off of *Lepidium sativum* (Pane et al. 2013). Similarly, (Castaño et al. 2011) evidenced the relevance of hydrophylic carbon moieties in compost-based bio-control of tomato Fusarium wilt disease. The carbon types distribution among several molecular classes incite changes in microbial structure, that can be evidenced by specific metabolic profiles used to distinguish non-suppressive from suppressive situations (Itoh et al. 2002). For example, Borrero et al. (2006) reported that Biolog culturable community, extracted from Fusarium wilt-suppressive compost, prefer not easily biodegradable compounds, such as carboxylic acids, aminoacids, amines, phenolic compounds and polymers, than non-suppressive ones. Similarly, Pane et al. (2013) pointed out the relation between the ability of suppressive community to metabolize complex carbon sources and the relative abundance of hydrophobic C in the compost. The compost is a highly complex and dynamic environment that sustain a great diversity of microbes that are involved directly and indirectly in the provision of a wide range of ecosystem services. The microbial structure more than the size of diversity in the compost is believed, generally, directly related to disease suppressivity (Xu et al. 2012). Thus, it is imperative to increase the level of understanding of compost microbial ecology and population dynamics. Changes in compost microbial community structures have been extensively explored with holistic approach by using various tools ranging from traditional plate counting and biochemical-based cultural methods until to innovative molecular-based techniques and "omic" strategies, including metagenomic analysis. Newer researches to measure compost microbial biodiversity, in fact, are focusing on genetic diversity viewed as the amount and distribution of genomic information within the generality of microbial living species. Metagenomic strategy analyze the set of the total microbiota genomes, termed metagenome, extracted from indigenous community in a given substrate (Rondon et al. 2000). This approach have a lot of credit respect to traditional methods that provide very limited information, because only a small percentage of total soil microbiota, estimated between 0.1 and 10 %, is cultivable In-vitro (Torsvik and Øvreås 2002). Hadar (2011), reviewing on the role of compost microbial ecology in suppressivity, stressed the opportunities coming from the application in this field of the next generation sequencing technologies including metagenomics, metatranscriptomics and bioinformatics. Studies on microbiota diversity in suppressive composts by molecular approach are increasing in this decade and concur to distinctly clarify implied groups. For example, Pérez-Piqueres et al. (2006) described a clear influence of different types of compost on bacterial and fungal communities composition in enhanced suppressive soils by using terminal restriction length polymorphism (T-RFLP). While a PCR-DGGE methodology has used to evidence the positive influence of compost amendment on rhizospheric streptomycetes (Inbar et al. 2005) and fungal (Kowalchuk et al. 2003) community composition. PCR amplified and cloned metagenomic sequences revealed that compost inoculation of the pathogenic oomycetes *Pythium ultimum* induces distinct shifts in microbial community favouring α-Proteobacteria suppressive groups (Hagn et al. 2008). Obviously, differential changes in microbial structure can be associated to modification of compost bioactivities. van Rijn et al. (2007) found that only significant shifts in 16S-rDNA DGGE banding patterns of the composts, occurred during storage, are related to significant reduction of suppression of *Fusarium oxysporum* f. sp. *lini*.

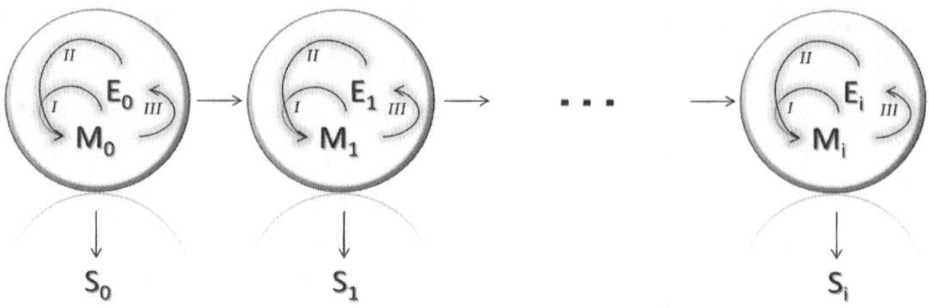

$$Suppression = \sum_{t=0}^{i} S$$

Fig. 8.3 Schematization of the ecological theory of compost suppressivity based on the interactive coevolution between the microbial community (*M*) and their trophic niche or environment (*E*) that occur dynamically in a determined period of time. The suppressivity results from the sum of the effects of simple and complex microbe-to-microbe (*I*), microbes-to-environment (*II*) and environment-to-microbes (*III*) interactions, in each timeframe (*0–i*)

8.5 Compost Suppressivity in Integrated Pest Management

Thus, compost-based suppressivity can be considered a successful tool for disease management only through a new ecological and global vision of the agriculture. In order to potentiate the effectiveness of compost phytoiatric applications, the integration with other control methods is desirable. Additional suppressive factors is often indispensable because they improve efficiency of compost, extending the spectrum of controlled pathogens and prolonging the time of protection. For example, the compost could be actually fortified by combining it with the incorporation of various microbial antagonists, which contribute with their specific mechanisms (Siddiqui et al. 2008). Trillas et al. (2006) used cork compost enriched with the Spanish strain T-34 of the bio-control agent *Trichoderma asperellum* as plant growth media for protecting cucumber seedling from Rhizoctonia disease. Similarly, the addition of a compost-derived isolate of *Trichoderma harzianum* significantly increased the ability of rice straw and empty fruit bunch null-effective compost extracts in controlling *Choanephora* wet rot in okra production (Siddiqui et al. 2008). Same bio-control formulations prepared by amending a paddy straw compost whit singles antagonistic cyanobacteria isolates, such as *Anabena oscillarioides* (C12) and *Bacillus subtilis* (B5), proved more effective in reduction of tomato diseases caused by a plant pathogenic fungal consortium including *Fusarium oxysporum*, *Pythium debaryum*, *P. aphanidermatum* and *Rhizoctonia solani* (Dukare et al. 2011). Fortification of composts with antagonistic microbes seems to be a suitable strategy of integrated biological control. However, the effectiveness levels depend by the compost type and the plant pathosystem, as demonstrated by using antagonistic isolates of *Verticillium biguttatum* and non-pathogenic *F. oxysporum* in various bioassays (Postma et al. 2003).

In this general framework, also the integrated use of compost combined with solarization could be very interesting. Gamliel and Stapleton (1993) showed the positive effects gained by integration of chicken compost amendment and soil solarization on lettuce growth and suppressed infections by *Pythium ultimun*. Similarly, Choi et al. (2007) showed the potential of spent mushroom sawdust compost used in combination with calcium cyanamide and solarization, in the control of Fusarium basal stem rot of the cactus *Hylocereus trigonus*. Compost complements solarization efficacy by preservation of soil quality from heat damage and inducing significant changes in microbial community structure. This was seen with DGGE profiles that showed how mature household compost incited shift in solarized soils towards a community with enhanced antagonism against *R. solanacearum* (Schönfeld et al. 2003).

8.6 Compost as Source of Plant Protectants

Compost suppressivity can be fully exploited as source of suppressive biotic or abiotic agents that can be isolated from composts and used directly in protecting plants (Fig. 8.4). In this way, the composts became a suitable source of new products that, after extraction from the original matrix, will follow a completely independent path of development and application. Essentially, from suppressive composts can be derivate of two types of plant protectants: biological control agents and natural compounds.

8.6.1 Microbial Antagonists

Compost is a suitable source of microbial antagonists which can be used singularly or in consortia to implement plant disease management (Fig. 8.5). The research in the field of biological control of plant pathogens, then, take advantage of compost suppressive properties that origin from the biotic components. A number of examples shows how the application of a screening strategy to a suppressive compost-derived micro-flora, can lead to the isolation of the best individuals exhibiting the desired characteristics. Composted urban organic and yard wastes, for example, has been used as source for antagonists belonging to fungi, bacteria and oomycetes characterized against different pathogen species (Pugliese et al. 2010). In particular, in this last situation, strains of *Trichoderma* and of non-pathogenic *Fusarium* sp. showing antagonistic activity against some soil borne pathogens, such as *R. solani* and *F. oxysporum* f. sp. *basilici*, respectively, were remarked (Pugliese et al. 2008). Compost water extracts were found to contain *Bacillus* sp., *Micrococcus* sp., *Staphylococcus* sp. and *Corynebacterium* sp. and the fungi *Aspergillus* sp., *Rhizopus* sp., *Drechslera* sp. and various species actinomycetes responsible for antagonistic activity (El-Masry et al. 2002).

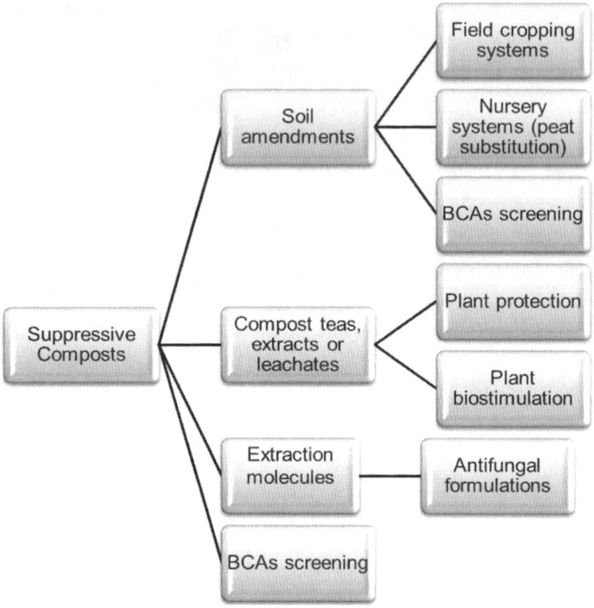

Fig. 8.4 Tree of resource that can be derived from suppressive composts exploitation and their applications

Useful bio-control agents can be isolated from rhizospheric soil treated whit suppressive composts. It is the case, for example, of some root-associated microbes of the genera *Bacillus*, *Lysinibacillus*, *Enterobacter* and *Serratia*, selected for their suppressivity to tomato crown and root rot disease caused by *F. oxysporum* f. sp. *radicis-lycopersici* (Kavroulakis et al. 2010). While, Pane et al. (2012a) showed a selective stepwise procedure that allowed to get *Bacillus methylotrophicus* and *Bacillus amyloliquefaciens* strains able to control *S. minor*, *R. solani* and *F. solani* from composts and compost amended soils.

8.6.2 Compost-derived Natural Substances

Nutrients and organic molecules, such as humic substances and phytochemicals extracted from composts, can prove valuable in limiting the detrimental effects of pathogens on plants. Organic substrates deriving humic acid fractions (HAs) has been characterized for the In-vitro suppression of fungal pathogens and the In-vivo protection of China aster infected with *F. oxysporum* f. sp. *callistephi* and impatiens infected with *P. ultimum* (Loffredo and Senesi 2009). In this case, however, the mechanisms of action are not completely understood. Nevertheless, recent studies showed the significant correlation between chemical and structural properties of the functional groups of these substances, such as total acidity, COOH group content and elemental composition, and their capability to inhibit fungal phytopathogens (Moliszewska and Pisarek 1996; Loffredo and Senesi 2009). Humic compounds derived from a prolonged biological degradation of organic matter are organized

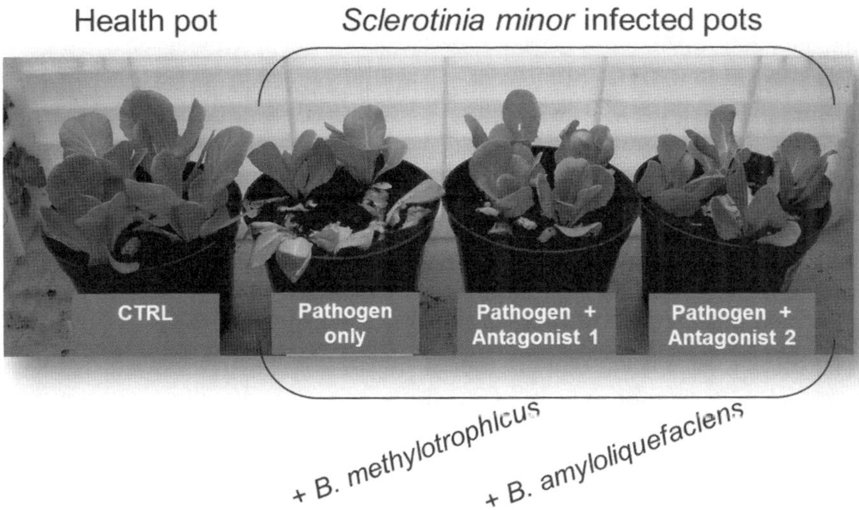

Fig. 8.5 Biocontrol assay of *Sclerotinia minor* on lettuce by two bacteria antagonistic strains belonging to *Bacillus methylotrophicus* and *Bacillus amyloliquefaciens* species isolated from compost amended soils

in more stable and complex structures chemically fractionable in humic and fulvic acids and humates. Siddiqui et al. (2009) concluded that the efficiency of palm oil compost derived humic acids, studied in controlling *Choanephora cucurbitarum* growth, was related to their concentration and carboxylic group content evidenced by FTIR spectra. Humic substances were assessed for In-vitro inhibition of two strains of *Fusarium oxysporum* (Loffredo et al. 2007). While, Loffredo et al. (2008) found that humic acid fractions (HAs) isolated from a mixture of an olive-oil-mill wastewater sludge after composting, caused the mycelial growth inhibition of *Sclerotinia sclerotiorum* related to some chemical and functional properties of HS. Instead, the same HS treatments, generally did not inhibit the growth of the two antagonistic strains of *Trichoderma viride* and *T. harzianum*. Phenolic phytochemicals with high antioxidant activity was extracted from black tea compost extracts (Nor Qhairul Izzreen and Mohd Fadzelly 2013). Pane et al. (2013) reported the implication of phenolic C compost fraction in suppressivity. A previous study showed the high potential of naturally occurring chemicals present in two-phase olive mill waste composts in suppression of a set of phytopathogenic fungi (Cayuela et al. 2008).

8.7 Compost Extracts, Leachate and Teas

Compost derived liquid products including aqueous extracts and aerated or non-aerated brewed teas, have a promising role in plant disease control. These organic formulates, reported in the majority of the cases simply as compost teas, are obtained by fermentation of compost carried out on liquid phase, in commercial or rudimental brewers, for a few days or up to 2 weeks, with or without active aeration, so as to produce aerated (ACTs) or non-areated (NCTs) compost teas, respectively (Ingham 1999; Lanthier 2007). Usually, also the compost extracts obtained by filtration of an aqueous suspension, after just a few hours of mixing, are referred as teas (Joshi et al. 2009). There are number of reports that demonstrate the ability of compost teas to suppress a wide range of plant pathogens, when applied as foliar spray and soil drenching (Scheuerell and Mahaffee 2002, 2004). Compost teas may, in fact, contain organic and inorganic soluble molecules, such as humic substances, phytochemicals and a large number of useful organisms, including bacteria, fungi, protozoa and nematodes, which may provide benefits for the vigor and health of the plants (Scheuerell and Mahaffee 2002). Composting pile-percolating leachates has been also proposed to reach the described scope (ROU 2006); but the risks deriving from a possible contamination by kinds of anaerobic human pathogens, gave restrictions to the use. A number of factors, which are involved in the compost extraction process, influencing the microbiological and chemical components of compost teas, are responsible for their efficacy in plant disease suppression. Firstly, a quality compost with good suppressive properties has the potential to make a high-quality compost tea, if made with proper preparation methodologies. The hypothesized mechanisms to explain compost teas ability in controlling plant diseases are similar to those viewed above for suppressive composts. In addition, compost teas contribute indirectly to the protection by inciting an improved physiological and vegetative status in the plants, making their less susceptible to pathogen attacks. Several nutritional and hormone-like substances suspended in the teas are believed to act in this direction. Compost tea suppressivity is ascribed, essentially, to its biotic component that has an antagonistic activity against pathogens (Noble and Coventry 2005; Pane et al. 2011, 2012b) and/or incite systemic resistance response in treated plants (Siddiqui et al. 2009). For these reasons, compost tea production techniques are based on the regulation of some parameters, such as ratio of compost to water, oxygenation levels, duration and temperature of fermentation, to generate conditions more favorable to the presence and development of beneficial organisms in the final product. In this context, the amendment of aqueous extractant with fermentation nutrients, such as food industry waste (such as molasses, borland, casein, whey, etc.), was also proposed in order to condition the microbiological characteristics of the teas. The introduced nutrients place selective pressure on the microbial community can potentially increase biological control agents that inhabit the compost tea.

8.8 Conclusion

Suppressivity should be viewed as an added value with a significant impact on crop management, because the use of disease suppressive compost will reduce crop losses caused by diseases. Such property made possible a global sustainable agriculture based on ecological, biological and natural principles. In the majority of circumstances the suppressivity occurs in the composts spontaneously and its intensity, unfortunately, is random. An important future challenge could be those to systematically produce suppressive composts with high standard quality. Using a sartorial approach, based on principles described above, perhaps, in a not too distant future, will be possible to see these new products make their way in innovative suppressive version. Until then, however, the agricultural world will still be able to take advantage from compost phytosanitary applications, providing the ecological implications of suppressivity.

Acknowledgement Authors thank the public institutions (European Community, Italian Ministry of Agricultural, Food and Forestry Policies and Campania Region) for supporting their works by financed research programs.

References

Ahimoua F, Jacques P, Deleua M (2000) Surfactin and Iturin A effects on *Bacillus subtilis* surface hydrophobicity. Enzyme Microb Tech 27:749–754

Benson DM (1995) Aluminium amendment of potting mixes for control of *Phytophthora* damping off in bedding plants. Hortic Sci 30:1413–1416

Boehm MJ, Wu T, Stone AG, Kraakman B, Iannotti DA (1997) Crosspolarized magic-angle spinning ^{13}C nuclear magnetic resonance spectroscopic characterization of soil organic matter relative to culturable bacterial species composition and sustained biological control of Pythium root rot. Appl Environ Microbiol 63:162–168

Bollen GJ (1993) Factors involved in inactivation of plant pathogens during composting of plant residues. In: Hoitink HAJ, Keener HM (eds) Science and engineering of composting. Renaissance Publications, Ohio, pp 301–318, 728 pp

Bonanomi G, Antignani V, Pane C, Scala F (2007) Suppression of soilborne fungal diseases with organic amendments. J Plant Pathol 89:311–324

Borrero C, Ordovás J, Trillas MI, Avilés M (2006) Tomato Fusarium wilt suppressiveness. The relationship between the organic plant growth media and their microbial communities as characterized by Biolog®. Soil Biol Biochem 38:1631–1637

Cassells AC, Rafferty-McArdle SM (2012) Priming of plant defences by PGPR against fungal and bacterial plant foliar pathogens. In: Maheshwari DK (ed) Bacteria in agrobiology: stress management. Springer, Berlin, pp 1–26

Castaño R, Borrero C, Avilés M (2011) Organic matter fractions by SP-MAS ^{13}C NMR and microbial communities involved in the suppression of Fusarium wilt in organic growth media. Biol Cont 58:286–293

Cayuela ML, Millner PD, Meyer SLF, Roig A (2008) Potential of olive mill waste and compost as biobased pesticides against weeds, fungi, and nematodes. Sci Total Environ 399:311–18

Chen W, Hoitink HAJ, Madden LV (1988) Microbial activity and biomass in container media predicting suppressiveness to damping-off caused by *Pythium ultimum*. Phytopathology 78:1447–1450

Choi HW, Chung M, Sin MH, Kim YS, Sim JB, Kim JW, Kim KD, Chun SC (2007) The effect of spent mushroom sawdust compost mixes, calcium cyanamide and solarization on basal stem rot of the cactus *Hylocereus trigonus* caused by *Fusarium oxysporum*. Crop Prot 26:162–168

Danon M, Chen Y, Hadar Y (2010) Ascomycete communities associated with suppression of *Sclerotium rolfsii* in compost. Fungal Ecol 3:20–30

Dukare AS, Prasanna R, Dubey SC, Nain L, Chaudhary V, Singh R, Saxena AK (2011) Evaluating novel microbe amended composts as biocontrol agents in tomato. Crop Prot 30:436–442

El-Masry MH, Khalil AI, Hassouna MS, Ibrahim HAH (2002) *In situ* and *in vitro* suppressive effect of agricultural composts and their water extracts on some phytopathogenic fungi. World J Microbiol Biot 18:551–558

Engindeniz S (2006) Economic analysis of pesticide use on processing tomato growing: a case study for Turkey. Crop Prot 25:534–541

Gamliel A, Stapleton JJ (1993) Effect of chicken compost or ammonium phosphate and solarization on pathogen control, rhizosphere microorganisms, and lettuce growth. Plant Dis 77:886–891

Haas D, Défago G (2005) Biological control of soil-borne pathogens by fluorescent pseudomonads. Nat Rev Microbiol 3:307–319

Hadar Y (2011) Suppressive compost: when plant pathology met microbial ecology. Phytoparasitica 39:311–314

Hagn A, Engel M, Kleikamp B, Munch JC, Schloter M, Bruns C (2008) Microbial community shifts in *Pythium ultimum*-inoculated suppressive substrates. Biol Fert Soils 44:481–490

Hoitink HAJ, Fahy PC (1986) Basis for the control of soilborne plant pathogens with composts. Ann Rev Phytopathol 24:93–114

Hoitink H, Boehm M, Hadar Y (1993) Mechanisms of suppression of soilborne plant pathogens in compost-amended substrates. In: Hoitink HAJ, Keener HM (eds) Science and engineering of composting. Renaissance Publications, Ohio, pp 601–621, 728 pp

Hoitink HAJ, Stone AG, Grebus ME (1996) Suppression of plant diseases by composts. In: De Bertoldi M, Sequi P, Lemmes B, Papi T (eds) The science of composting, vol 1. Blackie Academic & Professional, Glasgow, pp 373–381

Hoitink HAJ, Stone AG, Han DY (1997) Suppression of plant diseases by composts. HortScience 32:184–187

Hoitink HAJ, Krause MS, Han DY (2001) Spectrum and mechanisms of plant disease control with composts. In: Stofella PJ, Kahn BA (eds) Compost utilization in horticultural cropping systems. Lewis Publishers, Boca Raton, pp 263–274

Inbar E, Green SJ, Hadar Y, Minz D (2005) Competing factors of compost concentration and proximity to root affect the distribution of Streptomyces. Microb Ecol 50:73–81

Ingham ER (1999) What is compost tea? Part1. BioCycle 40:74–75

Itoh K, Goto D, Sueyasu K, Suyama K, Yamamoto H (2002) Effect of chloropicrin fumigation on microbial communities evaluated by community-level physiological profile and the resistance against Fusarium wilt. Soil Sci Plant Nutr 48:333–339

Joshi D, Hooda KS, Bhatt JC, Mina BL, Gupta HS (2009) Suppressive effects of composts on soil-borne and foliar diseases of French bean in the field in the western Indian Himalayas. Crop Prot 28:608–615

Kavroulakis N, Ehaliotis C, Ntougias S, Zervakis GI, Papadopoulou KK (2005) Local and systemic resistance against fungal pathogens of tomato plants elicited by a compost derived from agricultural residues. Physiol Mol Plant Pathol 66:163–174

Kavroulakis N, Ntougias S, Besi MI, Katsou P, Damaskinou A, Ehaliotis C, Zervakis GI, Papadopoulou KK (2010) Antagonistic bacteria of composted agro-industrial residues exhibit antibiosis against soil-borne fungal plant pathogens and protection of tomato plants from *Fusarium oxysporum* f. sp. *radicis-lycopersici*. Plant Soil 333:233–247

Kowalchuk GA, Os GJ, Aartrijk J, Veen JA (2003) Microbial community responses to disease management soil treatments used in flower bulb cultivation. Biol Fert Soils 37:55–63

Lanthier M (2007) Compost tea and its impact on plant diseases. BC Org Grower 10:7–11

Leifert C, Li H, Chidburee S, Hampson S, Workman S, Sigee D, Epton HAS, Harbour A (1995) Antibiotic production and biocontrol activity by *Bacillus subtilis* CL27 and *Bacillus pumilus* CL45. J Appl Bacteriol 78:97–108

Liu YF, Chen ZY, Ng TB, Zhang J, Zhou MG, Song FP, Lu F, Liu YZ (2007) Bacisubin, an antifungal protein with ribonuclease and hemagglutinating activities from *Bacillus subtilis* strain B-916. Peptides 28:553–559

Lockwood JL (1990) Relation of energy stress to behaviour of soilborne plant pathogens and to disease development. In: Hornby D (ed) Biological control of soilborne plant pathogens. CAB International, Wallingford, pp 197–214

Loffredo E, Senesi N (2009) *In vitro* and *in vivo* assessment of the potential of compost and its humic acid fraction to protect ornamental plants from soilborne pathogenic fungi. Sci Hort 122:432–439

Loffredo E, Berloco M, Casulli F, Senesi N (2007) *In vitro* assessment of the inhibition of humic substances on the growth of two strains of *Fusarium oxysporum*. Biol Fert Soils 43:759–769

Loffredo E, Berloco M, Senesi N (2008) The role of humic fractions from soil and compost in controlling the growth *in vitro* of phytopathogenic and antagonistic soil-borne fungi. Ecotox Environ Safe 69:350–357

Loschinkohl C, Boehm MJ (2001) Composted biosolids incorporation improves turfgrass establishment on disturbed urban soil and reduces leaf rust severity. HortScience 36:790–794

Malandraki I, Tjamos SE, Pantelides IS, Paplomatas EJ (2008) Thermal inactivation of compost suppressiveness implicates possible biological factors in disease management. Biol Cont 44:180–187

Mathur SP (1991) Composting process. In: Martin AM (ed) Bioconversion of waste materials to industrial products. Elsevier, New York, pp 147–186

Mazzola M (2002) Mechanisms of natural soil suppressiveness to soilborne diseases. Antonie van Leeuwenhock 81:557–564

Moliszewska E, Pisarek I (1996) Influence of humic substances on the growth of two phytopathogenic soil fungi. Environ Int 22:579–584

Noble R, Coventry E (2005) Suppression of soil-borne plant diseases with composts: a review. Biocontrol Sci Technol 15:3–20

Nor Qhairul Izzreen MN, Mohd Fadzelly AB (2013) Phytochemicals and antioxidant properties of different parts of *Camellia sinensis* leaves from Sabah tea plantation in Sabah, Malaysia. Int Food Res J 20:307–312

Pal KK, McSpadden GB (2006) Biological control of plant pathogens. The plant health instructor. APSnet, pp 1–25

Pal Bais H, Fall R, Vivanco JM (2004) Biocontrol of *Bacillus subtilis* against infection of Arabidopsis roots by *Pseudomonas syringae* is facilitated by biofilm formation and surfactin production. Plant Physiol 134:307–319

Pane C, Spaccini R, Piccolo A, Scala F, Bonanomi G (2011) Compost amendments enhance peat suppressiveness to *Pythium ultimum*, *Rhizoctonia solani* and *Sclerotinia minor*. Biol Cont 56:115–124

Pane C, Villecco D, Campanile F, Zaccardelli M (2012a) Novel strains of *Bacillus*, isolated from compost and compost-amended soils, as biological control agents against soil-borne phytopathogenic fungi. Biocontrol Sci Tech 22:1373–1388

Pane C, Celano G, Villecco D, Zaccardelli M (2012b) Control of *Botrytis cinerea*, *Alternaria alternata* and *Pyrenochaeta lycopersici* on tomato with whey compost-tea applications. Crop Prot 38:80–86

Pane C, Piccolo A, Spaccini R, Celano G, Villecco D, Zaccardelli M (2013) Agricultural waste-based composts exhibiting suppressivity to diseases caused by the phytopathogenic soil-borne fungi *Rhizoctonia solani* and *Sclerotinia* minor. Appl Soil Ecol 65:43–51

Paplomatas EJ, Tjamos SE, Malandrakis AA, Kafka AL, Zouvelou SV (2005) Evaluation of compost amendments for suppressiveness against Verticillium wilt of eggplant and study of mode of action using a novel Arabidopsis pathosystem. Eur J Plant Pathol 112:183–189

Parr JF, Hornick SB (1992) Agricultural use of organic amendments: a historical perspective. Am J Alt Agr 7:181–189

Pascual JA, Garcia C, Hernandez T, Lerma S, Lynch JM (2002) Effectiveness of municipal waste compost and its humic fraction in suppressing *Pythium ultimum*. Microb Ecol 44:59–68

Pérez-Piqueres A, Edel-Hermann V, Alabouvette C, Steinberg C (2006) Response of soil microbial communities to compost amendments. Soil Biol Biochem 38:460–470

Postma J, Montanari M, van den Boogert P (2003) Microbial enrichment to enhance the disease suppressive activity of compost. Eur J Soil Biol 39:157–163

Pugliese M, Liu BP, Gullino ML, Garibaldi A (2008) Selection of antagonists from compost to control soil-borne pathogens. J Plant Dis Prot 115:220–228

Pugliese M, Gullino ML, Garibaldi A (2010) Efficacy of microorganisms selected from compost to control soil-borne pathogens. Commun Agric Appl Biol Sci 75:665–669

Reitz SR, Maiorino G, Olson S, Sprenkel R, Crescenzi A, Momol MT (2008) Integrating plant essential oils and kaolin for the sustainable management of thrips and tomato spotted wilt on tomato. Plant Dis 92:878–886

Rondon MR, August PR, Bettermann AD, Brady SF, Grossman TH, Liles MR, Loiacono KA, Lynch BA, MacNeil IA, Minor C, Tiong CL, Gilman M, Osburne MS, Clardy J, Handelsman J, Goodman RM (2000) Cloning the soil metagenome: a strategy for accessing the genetic and functional diversity of uncultured microorganisms. Appl Environ Microbiol 66:2541–2547

ROU (2006) Compost use for pest and disease suppression in NSW. Recycled Organics Unit, The University of New South Wales, Sydney. www.recycledorganics.com

Sang MK, Kim JG, Kim KD (2010) Biocontrol activity and induction of systemic resistance in pepper by compost water extracts against *Phytophthora capsici*. Phytopathology 100:774–783

Scheuerell SJ, Mahaffee WF (2002) Compost tea: principles and prospects for plant disease control. Compost Sci Util 10:313–338

Scheuerell SJ, Mahaffee WF (2004) Compost tea as container medium drench for suppressing seedling damping-off caused by *Pythium ultimum*. Phytopathology 94:1156–1163

Schönfeld J, Gelsomino A, van Overbeek LS, Gorissen A, Smalla K, van Elsas JD (2003) Effects of compost addition and simulated solarisation on the fate of *Ralstonia solanacearum* biovar 2 and indigenous bacteria in soil. FEMS Microbiol Ecol 43:63–74

Segarra G, Casanova E, Borrero C, Avilés M, Trillas I (2007) The suppressive effects of composts used as growth media against *Botrytis cinerea* in cucumber plants. Eur J Plant Pathol 117:393–402

Segarra G, Elena G, Trillas I (2013a) Systemic resistance against *Botrytis cinerea* in Arabidopsis triggered by an olive marc compost substrate requires functional SA signaling. Physiol Mol Plant Pathol 82:46–50

Segarra G, Santpere G, Elena G, Trillas I (2013b) Enhanced *Botrytis cinerea* resistance of Arabidopsis plants grown in compost may be explained by increased expression of defense-related genes, as revealed by microarray analysis. PLoS One 8:e56075. doi:10.1371/journal.pone.0056075

Siddiqui Y, Meon S, Ismail MR, Ali A (2008) *Trichoderma*-fortified compost extracts for the control of Choanephora wet rot in okra production. Crop Prot 27:385–390

Siddiqui Y, Meon S, Ismail R, Rahmani M (2009) Bio-potential of compost tea from agro-waste to suppress *Choanephora cucurbitarum* L. The causal pathogen of wet rot of okra. Biol Cont 49:38–44

Singh N, Kumar S, Bajpai VK, Dubey RC, Maheshwari DK, Kang SC (2010) Biological control of *Macrophomina phaseolina* by chemotactic fluorescent *Pseudomonas aeruginosa* PN1 and its plant growth promotory activity in chir-pine. Crop Prot 29:1142–1147

Tenuta M, Lazarovits G (2002) Ammonia and nitrous acid from nitrogenous amendments kill the microsclerotia of *Verticillium dahliae*. Phytopathology 92:255–264

Tenuta M, Lazarovits G (2004) Soil properties associated with the variable effectiveness of meat and bone meal to kill microsclerotia of *Verticillium dahliae*. App Soil Ecol 25:219–236

Tenuta M, Conn KL, Lazarovits G (2002) Volatile fatty acids in liquid swine manure can kill microsclerotia of *Verticillium dahaliae*. Phytopathology 92:548–552

Termorshuizen AJ, van Rijn E, van der Gaag DJ, Alabouvette C, Chen Y, Lagerlöf J, Malandrakis AA, Paplomatas EJ, Rämert B, Ryckeboer J, Steinberg C, Zmora-Nahum S (2007) Suppressiveness of 18 composts against 7 pathosystems: variability in pathogen response. Soil Biol Biochem 38:2461–2477

Torsvik V, Øvreås L (2002) Microbial diversity and function in soil: from genes to ecosystems. Curr Opin Microbiol 5:240–245

Trillas MI, Casanova E, Cotxarrera L, Ordovás J, Borrero C, Avilés M (2006) Composts from agricultural waste and the *Trichoderma asperellum* strain T-34 suppress *Rhizoctonia solani* in cucumber seedlings. Biol Cont 39:32–38

Vallad GE, Cooperband L, Goodman RM (2003) Plant foliar disease suppression mediated by composted forms of paper mill residuals exhibits molecular features of induced resistance. Physiol Mol Plant Pathol 63:65–77

Van der Ent S, Van Wees SCM, Pieterse CMJ (2009) Jasmonate signaling in plant interactions with resistance-inducing beneficial microbes. Phytochemistry 70:1581–1588

van Rijn E, Termorshuizen AJ, van Bruggen AHC (2007) Storage method affects disease suppression of flax wilt induced by composts. Soil Biol Biochem 39:2743–2749

Veeken AHM, Blok WJ, Curci F, Coenen GCM, Temorshuizen AJ, Hamelers HVM (2005) Improving quality of composted biowaste to enhance disease suppressiveness of compost-amended, peat based potting mixes. Soil Biol Biochem 37:2131–2140

Wagner WC (1999) Sustainable agriculture: how to sustain a production system in a changing environment. Int J Parasitol 29:1–5

Xu D, Raza W, Yu G, Zhao Q, Shen Q, Huang Q (2012) Phytotoxicity analysis of extracts from compost and their ability to inhibit soil-borne pathogenic fungi and reduce root-knot nematodes. World J Microbiol Biot 28:1193–1201

Yogev A, Raviv A, Hadar Y, Cohen R, Wolf S, Gil L, Katan J (2010) Induced resistance as a putative component of compost suppressiveness. Biol Cont 54:46–51

Zhang W, Dick WA, Hoitink HAJ (1996) Compost-induced systemic acquired resistance in cucumber to Pythium root rot and anthracnose. Phytopathology 86:1066–1070

Zmora-Nahum S, Danon M, Hadar Y, Chen Y (2008) Chemical properties of compost extracts inhibitory to germination of *Sclerotium rolfsii*. Soil Biol Biochem 40:2523–2529

Chapter 9
Integrating Compost Teas in the Management of Fruit and Foliar Diseases for Sustainable Crop Yield and Quality

Katherine J. Evans and Alice K. Percy

Abstract Crop protectants are applied to crops to prevent loss of yield and pre-harvest spoilage by plant pathogens. Contemporary disease management focuses on the integration of cultural and biological controls to reduce or eliminate the need for synthetic chemicals. Compost tea is a watery extract of microorganisms and nutrients from compost for application to the soil or crop canopy. It is a type of biological control that has potential to suppress a broad range of plant pathogens. This review provides a framework for evaluating the efficacy and safety of compost teas for the management of fruit and foliar diseases. Mechanisms for integrated disease management are discussed in the context of mode of action, batch-to-batch variation in tea quality, spray timing and technique, and variation in disease suppression among sites and growing seasons. Future research is proposed to further identify the role of compost teas in sustaining crop yields, produce quality and rural livelihoods.

Keywords Crop protection · Food safety · Horticulture · Organic · Pathology

9.1 Introduction

Composting is a dynamic process involving complex interactions between physical, chemical and biological factors that lead to temporal changes in the relative abundance and diversity of different types of microorganisms (Epstein 1997). Compost tea is a watery extract of microorganisms and nutrients from compost. Farmers adopting organic and biodynamic practices have used compost teas to boost plant health since the 1920s (Brinton 1995); however, the subsequent Green Revolution

K. J. Evans (✉)
Perennial Horticulture Centre, Tasmanian Institute of Agriculture, University of Tasmania, 13 St Johns Avenue, New Town, Tasmania 7008, Australia
e-mail: Katherine.Evans@utas.edu.au

A. K. Percy
Sprout Tasmania, PO Box 2073, Howrah, Tasmania 7018, Australia

D. K. Maheshwari (ed.), *Composting for Sustainable Agriculture,* Sustainable Development and Biodiversity 3, DOI 10.1007/978-3-319-08004-8_9

ensured wide adoption of synthetic fertilizers and highly effective fungicides (Gaud 1968).

Today, fungicides and bactericides are applied to many food and ornamental crops to prevent loss of yield and pre-harvest spoilage by fungal and bacterial plant pathogens. Nevertheless, there has been resurgence of organic and biodynamic agriculture, perhaps catalysed by the publication of Rachel Carson's 'Silent Spring' (Carson 1962). Retailers and consumers continue to demand reduced inputs of synthetic fungicides because of concerns about the safety of pesticide residues in food and the potential for negative effects on non-target organisms (Dagostin et al. 2011). Development of pathogen resistance to fungicides (Walker et al. 2013) also provides incentive to manage chemical inputs judiciously and to seek alternative materials. Importantly, farmers in developing and developed countries seek to reduce the cost of crop inputs and manage land for both economic and environmental sustainability.

A common objective of contemporary plant disease management is the integration of cultural and biological controls to reduce or eliminate the need for synthetic chemicals. When applied to the soil or crop canopy, compost tea can be considered a type of biological control that has potential to suppress a broad range of plant pathogens (Scheuerell and Mahaffee 2002). Maheshwari (2013) also reported the application and advantages of biocontrol in disease management. Disease suppression is defined here as a statistically significant reduction in plant disease incidence and/or severity relative to a non-treated or water-treated control. There are now at least 30 refereed scientific publications that have examined the effectiveness of foliar applications of compost teas in detached-leaf assays, glasshouse and/or field trials.

This review provides a description of the characters of compost tea, methods of production using open-windrow compost and variation in disease suppression observed. A framework is proposed for evaluating the efficacy and safety of compost teas when applied to a crop canopy. This framework draws on results of specific investigations that demonstrate a relevant principle relating to the theory or practice of this type of biological control. Mechanisms for integrating compost tea in the management of fruit and foliar diseases in horticultural crops are then discussed, including the practicalities of crop protection and factors affecting adoption by farmers.

9.2 Aerated and Non-Aerated Compost Tea

Scheuerell and Mahaffee (2002) describe two dominant methods for producing compost teas. Aerated compost tea (ACT) results from any method in which water is actively aerated during contact with the compost. Non-aerated compost tea (NCT) results from any method that rarely disturbs the compost and water after initial mixing. Both methods require a vessel, compost, water and filtration to ensure large particles do not block equipment used for spraying or fertigation. Whey, rather than water, has also been used to prepare a pathogen-suppressive compost tea (Pane et al. 2012).

ACT is produced within 24–72 h using a custom-built vessel containing equipment for continuous aeration of dechlorinated water to maintain the activity of aerobic microorganisms (Ingham 2005; Palmer et al. 2010b). Filtration is achieved through use of a porous bag or screen. Practitioners often amend ACT with nutrients or adjuvants during or after preparation (Scheuerell and Mahaffee 2004). The perceived benefits, efficacy and safety of these practices are explored in later sections.

NCT takes anywhere between 3 and 14 days to produce and may be filtered after the extraction period (Koné et al. 2010; Scheuerell and Mahaffee 2002). NCT is associated with lower production costs and energy input, and is more likely to be used in regions where there is no electricity or the cost of energy is prohibitive. The generally faster production times for ACT have seen a plethora of commercial businesses establish to provide specialised equipment and services to high-input agriculture and horticulture in the USA, Europe and Australia (Anonymous 2006; Diver 2002).

9.3 Compost Teas as a Form of Biological Control

In order to understand potential levels of disease suppression with compost teas, it is worth considering the effectiveness of biological control more generally. A number of single species tested for biological control have been reported to be as effective as programs that use synthetic chemicals when applied in controlled environments such as greenhouses (Dik and Elad 1999; Elad et al. 1993; Guetsky et al. 2001). Indeed, compost is sometimes used as a substrate to propagate fungal biocontrol agents (Metcalf 2002; Ramona and Line 2002). In contrast, results under variable field conditions have been inconsistent and associated with disease severity observed in non-treated areas (Bisiach et al. 1985; Gullino and Garibaldi 1988). Single species may be constrained by mechanism of action (see Sect. 9.5), and/or environmental conditions such as temperature and surface-wetness duration that can affect spore germination and growth rates relative to the plant pathogen. One strategy has been to incorporate two or more biological agents to increase disease suppression (Guetsky et al. 2001; Stewart 2001).

In what appears to be the first scientific report about compost teas, Hunt et al. (1973) described how watery extracts of composted municipal refuse induced immobility in sting nematodes. Another foundational study used NCTs to suppress grapevine downy mildew, caused by the biotroph *Plasmopara viticola*, whereby NCT was applied to detached leaves from glasshouse-grown grapevines prior to inoculation (Weltzien and Ketterer 1986). Since then, disease suppression by ACTs and NCTs has been reviewed by several authors (Litterick et al. 2004; Scheuerell and Mahaffee 2002; Weltzien 1990). There are also a number of reports that conclude some teas are ineffective in terms of disease control (Al-Mughrabi 2006; Scheuerell and Mahaffee 2004; Sturz et al. 2006; Welke 2004). Where disease suppression has been demonstrated, the fungal pathogens causing disease were either biotrophic or necrotrophic in mode of nutrition (Elad and Shtienberg 1994), or, less frequently, bacterial plant pathogens (Al-Dahmani et al. 2003; Zhang et al. 1998).

In systematic comparisons of NCTs and ACTs, aerating compost tea did not significantly increase disease suppression when compared with results for NCTs

(Marín et al. 2013; Scheuerell and Mahaffee 2006). Litterick et al. (2004) list examples of disease or pathogen suppression from tests involving NCTs. Examples of variation in the level of disease suppression after foliar application of ACTs are provided in Table 9.1.

9.4 Production Variables

Methods of compost tea production have been variable and manipulated in an effort to increase disease suppression (Scheuerell and Mahaffee 2002). Variables include compost used, compost age or maturity, compost to water ratio during extraction, duration of compost extraction, addition of nutrients or adjuvants during or after production, and storage and dilution prior to application (Ketterer et al. 1992; Scheuerell 2003; Weltzien 1990).

9.4.1 Compost Raw Materials

Clearly the characters of a compost tea are related to those of the compost used in its production. Compost raw materials may be of animal and/or plant origin, such as animal manure, green and woody plant material. Materials are usually combined to obtain an initial carbon to nitrogen (C:N) ratio of 25–30:1 for rapid composting leading to a finished product with a C:N ratio of 10:1 (Golueke 1992; Tuomela et al. 2000). An excess of nitrogen at compost initiation can lead to volatile ammonia (Pagans et al. 2006), unpleasant odours (Finstein and Morris 1975) and/or reduced water quality from the leaching of excess nitrogen (Anonymous 1999). In contrast, excess carbon can lead to reduced microbial activity and slower degradation of compost and reduced nitrate, nitrate and ammonium in the resulting tea applied to crops (Boulter et al. 2000).

Given that the initial C:N ratio is an important determinant of the metabolic pathways of composting (de Bertoldi et al. 1996), this variable should be documented during any scientific investigation of compost tea. Preliminary evidence suggests that if the initial carbon to nitrogen (C:N) ratio of the compost windrow is standardised to 25–30:1, then a disease-suppressive ACT can be produced regardless of the sources of plant and animal material used to generate the compost (Palmer et al. 2010b).

Quality standards have been published for mature compost in various jurisdictions (Table 9.2), based on the pH, electrical conductivity and concentrations of various elements or nutrients. Adhering to specified standards for compost is one mechanism for achieving some consistency in compost tea production. A close-to-neutral pH of mature compost of between 5 and 7.5 (Anonymous 2012) is appropriate for most applications of compost tea to the soil or crop canopy. This pH range corresponds to the optimum pH for growth, which for most bacterial species is between 6.5 and 8, while for fungi, the optimum pH is between 2 and 6.5 (Matthies et al. 1997).

Table 9.1 Examples of variation in the level of disease suppression after foliar application of aerated compost teas (ACTs)

Pathogen, Disease, Host	ACT treatment	Disease suppression	Reference
Botrytis cinerea, grey mould, strawberry	Bi-weekly applications in the field	Small but significant reduction in disease incidence for fruit scored in the highest category for disease severity (51–95 % surface area infected), relative to water or non-treated control treatments	Welke (2004)
Botrytis cinerea, grey mould, geranium	Single foliar application to greenhouse-grown seedlings subsequently inoculated with the pathogen	Only one of six ACTs reduced disease severity slightly when compared with non-treated, inoculated control plants	Scheuerell and Mahaffee (2006)
Botrytis cinerea, grey mould, geranium	As above but ACTs amended with a mixture of kelp, rock dust and humic acids	Three of six batches of amended ACT reduced disease severity slightly when compared with non-treated, inoculated control plants	Scheuerell and Mahaffee (2006)
Alternaria solani, early blight, tomato; *Alternaria porri*, early blight, onion	Foliage sprayed in greenhouse and field trials	Reduced disease incidence to the same degree as conventional treatments	Haggag and Saber (2007)
Erysiphe necator, powdery mildew, grapevine	ACT prepared from immature compost; multiple applications at two vineyards in different growing seasons	Suppressed disease to <1 % mean severity on Chardonnay leaves (non-treated 79 % severity) and bunches (non-treated 77 % severity), and on Riesling leaves (non-treated 24 % severity)	Evans et al. (2013)
Botrytis cinerea, bunch rot, grapevine	As above	Reduced the incidence of Chardonnay bunches with latent *B. cinerea* and Riesling bunches with sporulating *B. cinerea*	Evans et al. (2013)

The electrical conductivity (EC) of compost is determined by the initial ingredients and compost maturity. Early secondary mesophilic compost has a much higher conductivity than late secondary mesophilic compost and the maturation process reduces the level of conductivity through leaching of salts and/or through microbial osmotic activity (Lau and Wong 2001). Electrical conductivity (EC) in the resulting compost tea should be tested prior to application to the crop canopy so that any risk of phytotoxicity due to excess salt is minimised. Ideally, the EC of compost tea should be as low as the level tolerated by the target crop plant and/or crop variety. Salts generally found in compost include potassium chloride, sodium chloride, various nitrates, compounds involving sulfates, calcium, magnesium, and potassium

Table 9.2 Typical ranges of characters of compost prepared according to the Australian Standard 4454. Composts, soil conditioners and mulches. (Anonymous 2012)

Variable	Typical range
pH	5–7.5
Electrical conductivity	No limit, although high EC can be detrimental to crop production
Soluble phosphorus (mg/L)	≤5 for phosphorus sensitive plants
Total phosphorus (% dry mass)	≤ 0.1 for phosphorous sensitive plants
Ammonium (mg/L)	<300
Ammonium + nitrate (mg/L)	≥100 if an input to plant nutrition is required
Nitrogen (% dry matter)	≥0.8 if an input to plant nutrition is required
Organic matter content (% dry matter)	≥25
Boron (mg/kg)	<200
Sodium (% dry mass)	<1
Moisture content (%)	25–40 (maximum dependent on % organic matter)

carbonates (Watson 2004). Compost prepared with animal manure, will lead to a compost with a higher EC than those prepared with plant materials (Watson 2004).

9.4.2 Compost Maturity and Microbial Community

The diversity, abundance and activity of microorganisms in compost vary according to compost source and maturity (Hoitink and Boehm 1999). Composts produced in open windrows undergo a predictable series of biological, physical and chemical processes (Fig. 9.1), including a microbial community that follows a predictable successional pattern (Herrmann and Shann 1997).

The internal windrow temperature of aerobic compost is the primary regulator of microbial diversity (Ishii et al. 2000; Peters et al. 2000; Tiquia 2005), activity (Ryckeboer et al. 2003a) and population structure (Herrmann and Shann 1997; Ryckeboer et al. 2003b). In the first stage of composting, known as the primary mesophilic stage, the fresh compost contains readily degradable compounds that can be utilised by mesophilic microorganisms resulting in a rapid increase in temperature from ambient temperature to around 60 °C. As the process continues, temperature stabilises at approximately 55–70 °C, humic substances and thermophilic bacterial numbers increase, while human, plant and animal pathogens, weed seeds and other microorganisms are destroyed.

At temperatures between 40 and 50 °C in the early secondary mesophilic phase of composting, thermophiles and mesophiles co-exist (Ryckeboer et al. 2003b). The following conditions are observed in compost at this time: (a) the optimum temperature for thermophilic fungi that moderate levels of nitrogen (Finstein and Morris 1975) (b) a diverse range of bacteria that use numerous enzymes to degrade organic material and transfer soluble materials into bacterial cells (Ryckeboer et al. 2003b), and (c) a great diversity and abundance of actinobacteria (Amner et al. 1988; Finstein and Morris 1975).

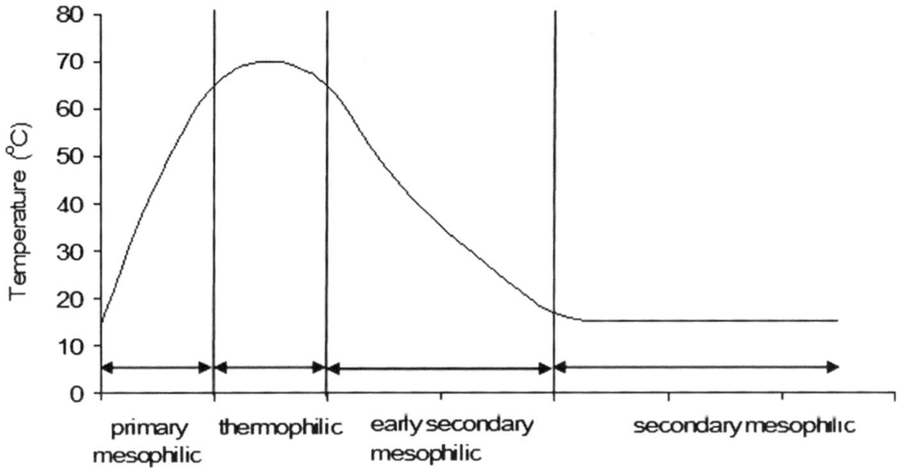

Fig. 9.1 Schematic representation of the temporal variation in compost internal windrow temperature during aerobic composting. Adapted from Epstein (1997) and reproduced from Palmer (2009).

In the final stage of composting, known as the secondary mesophilic stage, the temperature falls gradually from 55–70 °C to ambient temperature and remains at this temperature for several weeks. Mesophilic microorganisms, which are often different from the primary mesophilic microorganisms, recolonise the compost including those with the potential to suppress plant disease, such as actinobacteria, and species of *Trichoderma*, *Ulocladium*, *Penicillium* and *Cladosporium* (Ryckeboer et al. 2003b). In this late phase, there is a reduction in the amount of degradable products available and consequently a decline in microbial activity (Hoitink and Boehm 1999; Ryckeboer et al. 2003a; Tuomela et al. 2000).

Most studies evaluating compost tea have used mature compost. Immature composts have been tested occasionally; for example, Palmer et al. (2010b) found that ACTs produced from open-windrow composts sampled in the early secondary mesophilic stage of composting, when the internal compost temperature was ≤ 51 °C, inhibited colonisation of detached bean leaflets by the fungal pathogen *Botrytis cinerea* to a greater extent than ACTs produced from compost sampled in the later mesophilic stages. The diversity of bacteria and fungi in a tea from immature compost, determined by culture-independent analysis, was higher 28 days after windrow initiation than at later stages of composting, suggesting that microbial diversity should be explored further as a factor contributing to the suppression of *B. cinerea* (Palmer et al. 2010b). Analysis of terminal restriction fragments (T-RFs) (Osborn et al. 2000) in the same tea revealed 102 bacterial T-RFs and 23 fungal T-RFs when three restriction enzymes were used for each domain of microorganisms. A single T-RF can represent several species or taxonomic units. Knowledge of the microbial community present in immature or mature compost, such as the presence of actinobacteria and fungal genera that have shown potential as single biological control agents, might provide clues as to what components might be associated with pathogen suppression by the resulting compost tea.

There are numerous techniques for characterising the biological characters of compost teas, such as microbial biomass, activity, and/or diversity of compost and compost teas (Bandick and Dick 1999; Palmer et al. 2010b; Ryckeboer et al. 2003b); however, there is a lack of standard procedures for the purpose of quality control (Anonymous 2006). This situation is analogous to on-going attempts to develop practical indicators of the biological properties of soil (Bezuidenhout et al. 2012; Knight et al. 2013; Ugarte et al. 2013).

9.4.3 Extraction Time and Compost to Water Ratio

It has been suggested that the efficacy of ACT depends on the duration of extraction (Scheuerell 2003). Palmer et al. (2010b) steeped compost in in aerated water for 24, 48 or 72 h before application to detached bean leaflets and single-point inoculations of the fungal pathogen *Botrytis cinerea*. Relative to water, all batches of ACT, regardless of extraction time, reduced lesion development to a similar extent.

It has also been postulated that the quantity of compost in water influences biological and chemical characters of ACT and potential to control plant pathogens. According to Weltzien (1990), NCT controlled the plant pathogen *Phytophthora infestans* on detached potato leaflets when the compost to water ratio ranged between 1:3 and 1:10 but not when the compost to water ratio was 1:50. A more detailed in vitro experiment by Cronin et al. (1996) with *Venturia inaequalis*, the cause of apple scab or black spot, determined the EC_{50} (the concentration that inhibited germination of 50% of *V. inaequalis* conidia) of non-aerobic slurries of compost. There was a linear relationship between the $\log_{10}$cfu (colony forming units) of the extract concentration and spore germination inhibition. The extracts were mixed with sterile deionised water in ratios ranging from 1:3.16 to 1:100. In contrast to these studies, the quantity of compost in water did not influence grey mould suppression on field grown strawberries (Welke 2004). Similarly, compost weight to water volume ratios of 1:3, 1:10 or 1:30 tested by Palmer et al. (2010b) reduced the growth of *Botrytis cinerea* on bean leaflets to a similar degree.

9.4.4 Amendments and Adjuvants

Nutrient availability is critical for microbial metabolism and growth (Egli and Zinn 2003). It has been postulated that nutrients added to compost tea during or after its preparation increase the activity and abundance of microorganisms, and/or microbial composition, including materials rich in protein and amino acids, like fish emulsion or fish hydrolysate (El-Tarabily et al. 2003). Practitioners attempt to promote microbial growth by amendment of ACT with nutrients such as yeasts, sugar, molasses, kelp extract, fish hydrolysate, rock dust and/or humic acid (Ingham 2005). The effect of amendments on pH should be checked and aeration levels during preparation of ACT need to be monitored regularly to ensure any increase in microbial activity does not consume oxygen faster than it is introduced by aeration.

The benefit of using compost tea amended with nutrients appears to depend on the pathosystem. The most consistent ACT for suppression of damping-off on field-grown cucumbers caused by *Pythium ultimum* was ACT amended with kelp, humic acids and rock dust (Scheuerell and Mahaffee 2004). In another pathosystem, Scheuerell and Mahaffee (2006) illustrated no significant difference between ACT and ACT amended with nutrients for grey mould suppression on geranium, regardless of an increase in microbial numbers with additional nutrients.

Adjuvants are mixed with synthetic fungicides to reduce the surface tension of the spray droplets, helping the fungicide to adhere and increase its surface area on the leaf or fruit surface and/or penetration of the leaf cuticle (Steurbaut 1993; Zabkiewicz 2007). Materials compatible with organic agriculture include fish hydrolysate (de Ong 1927), vegetable oils and polysaccharides, although the benefit of each as an adjuvant needs to be determined empirically. Spray adjuvants, including Karaya gum and Yucca extract, have been shown to increase the effectiveness of ACTs in reducing *B. cinerea* development on glasshouse grown geranium (Scheuerell and Mahaffee 2006). In grapevine, however, adjuvants used as wetter-spreaders and penetrants have been shown to disintegrate the epicuticular waxes on berry surfaces, causing greater susceptibility to *B. cinerea* infection and depletion of the number of microorganisms on the fruit surface (Marois et al. 1987; Rogiers et al. 2005).

9.4.5 Storage

Practitioners expect to use compost tea soon after preparation meaning that product 'shelf life' is not a major concern; however, there may be occasions where a tea needs to be stored and/or transported to another location. Research is needed to determine the viability of microorganisms in compost tea during storage and transport, as well as changes in biological, chemical and physical characters resulting from microbial competition, parasitism and/or antibiosis. Foregoing commercialisation of bio-fungicides, experiments are required to determine the formulation that results in long-term survival or integrity of the active component/s. Abadias et al. (2001), for example, found that that the greatest survival rate and viability of *Candida sake* occurred when the biocontrol agent was freeze-dried and stored in lactose and skim milk. Compost tea, in contrast, comprises not only component microorganisms but the medium in which these microorganisms are sustained. The potential to concentrate and store compost tea produced in large quantities to maintain its disease-suppressive qualities is unknown.

9.4.6 Dilution Prior to Application

Some practitioners may seek to dilute a batch of compost tea with water to allow the same volume of tea to be applied over a greater area of a horticultural crop. Palmer et al. (2010b) evaluated the mean germination proportion of *Botrytis cinerea* conidia in vitro when different volumes of ACT prepared from immature compost were diluted in sterile water by adding 25, 15, 5, 1.67, 0.5 or 0 μl tea to a total volume of 50 μl. Diluted ACT was then mixed with 50 μl of 1×10^5 *B. cinerea* conidia ml^{-1} sterile water. After incubation, the mean germination proportion of conidia was less than 0.1 at all dilutions of ACT and 0.97 in water. However, the effect of diluting compost tea for crop protection needs to be tested under field conditions.

9.5 Mechanisms of Action

Disease suppression by biological control agents has been explained by mainly biotic mechanisms; including, direct antagonism through production of antibiotic or lytic compounds (Cronin et al. 1996; Fogliano et al. 2002), parasitism (Stewart 2001), competition for nutrients or niche (Kredics et al. 2004), siderophore-mediated competition for iron (Diánez et al. 2006) and inducible plant defence mechanisms (Droby et al. 2002; Reglinski et al. 2005; van Loon et al. 1998). Mechanisms of action for species of *Trichoderma*, for example, include parasitism of pathogenic fungal mycelia, antibiosis and/or production of cell wall degrading enzymes such as chitinases and β–1,3–glucanases (Elad 1994; Metcalf 2002; Stewart 2001). Myriad mechanisms of action might be associated with any particular compost tea, with the relative contribution of each mechanism varying depending on environmental conditions, including the physiological state of the host plant.

The presence of microorganisms in compost, and hence teas, has been demonstrated to be a contributing factor to the level of pathogen or disease suppression observed (Haggag and Saber 2007; Koné et al. 2010; McQuilken et al. 1994; Palmer et al. 2010b; Pane et al. 2012; Scheuerell 2004). Some authors postulate that the total bacterial count is associated with the efficacy of NCT (Al-Mughrabi 2006; Cronin et al. 1996; Diánez et al. 2006); however, a number of reports do not support this hypothesis. Assuming there is some minimum number of culturable microorganisms needed for disease suppression, the absolute number does not appear to be related to the level of disease suppression (Palmer et al. 2010b; Scheuerell and Mahaffee 2004, 2006). An alternative hypothesis is that specific microorganisms are important for disease suppression, rather than microbial abundance per se. If so, then toxic metabolites might be induced by specific microbial taxa in the presence of the plant pathogen (Martinez et al. 2006).

Inhibition of pathogen spore germination by compost tea has been reported previously; for example, NCT prepared from spent mushroom compost reduced the germination percentage of conidia of *Venturia inaequalis*, the cause of apple scab (Cronin et al. 1996). Palmer et al. (2010b) found that ACT prepared from immature compost inhibited the germination of *Botrytis cinerea* conidia to a greater extent than ACT in which microorganisms were removed by filtration, even when ACT was diluted 1 part in 10 with water. The large difference in activity between unfiltered and filtered ACT suggested that maximum inhibition of *B. cinerea* required close proximity between microorganisms in ACT and the pathogen. There was no evidence for the presence of water-soluble antibiotics in filtered ACT, even though it inhibited germination of *B. cinerea* to a significant degree. Other studies also found that filtration of ACT or NCT did not completely eliminate the ability of the tea to suppress various plant pathogens (Diánez et al. 2006; Marín et al. 2013). Diánez et al. (2006) provide some evidence to support the hypothesis that siderophores are released by microorganisms to capture iron in the media, so that this element is no longer available for other species.

The phyllosphere is a nutrient limiting environment (Lindow and Brandl 2003). Practitioners claim that nutrients in compost tea might help sustain microbial populations after application. When ACT was applied to grapevines in the field, the number of culturable microorganisms on the surface of Chardonnay leaves 10 days post application remained higher than observed on leaves 30 min pre-application (Evans et al. 2013). At a second vineyard site in the same study, the number of culturable bacteria and yeast declined to pre-application levels by 21 days after application of ACT to leaves of grapevine variety Riesling, whereas the number of culturable fungi remained higher at this time. These observations do not explain why microbial numbers were sustained for a period of time, but they do indicate that populations augmented by compost tea can decline with time and in a manner that is analogous to a fungicide decay curve.

There is limited evidence to suggest that compost teas induce host resistance to disease (Haggag and Saber 2007; Siddiqui et al. 2009; Zhang et al. 1998). Koné et al. (2010) treated tomato seedlings with various types of NCT, inoculated them with *Botrytis cinerea*, and observed control of grey mould in the greenhouse for up to 9 weeks. In another study involving preventative treatment, Evans et al. (2013) applied ACT to leaves of Cabernet Sauvignon grapevines in pots 7 days before inoculation with conidia of *Erysiphe necator*. This treatment reduced mean powdery mildew severity on the three youngest expanded leaves (at inoculation) to less than 1 %; mean severity on non-treated, inoculated leaves was 15 %. It was suggested that either the active components of the compost tea persisted on the treated plants and/or host resistance to a necrotrophic or biotrophic pathogen was induced. The latter mechanism would need to be investigated using appropriate methods, such as those described by Magnin-Robert et al. (2007). Such experiments must eliminate or isolate the possibility of direct toxicity by organic or other chemicals, such as phenolic compounds (Hoitink et al. 1997). The effect of nutrients in compost tea and changes in leaf nutrient status, especially calcium or silicon, or osmotic potential, could also be investigated in relation to levels of disease suppression observed (Segarra et al. 2007).

The effect of nutrients in compost tea on spore germination and growth of plant pathogens in the infection court remains unclear. Harper et al. (1981) found that nitrate or ammonium forms of nitrogen supported abundant growth of *Botrytis cinerea* in vitro. If these forms of nitrogen are present in compost tea and persist on plant surfaces, then there is potential for stimulation of saprophytic growth of *B. cinerea* on plant surfaces. It is postulated that a highly pathogen-suppressive compost tea might be associated with no or very low levels of nitrate in the tea. Absence of nitrate in compost tea, recorded for the disease suppressive ACTs as tested by Palmer et al. (2010b), suggests that nitrogen is present mostly in organic form, including microbial biomass. In the same study, ACT produced from compost windrows older than 42 days contained nitrate in agreement with the findings of other studies where maximum nitrification occurred in the secondary mesophilic stage of composting (Bishop and Godfrey 1983; Diaz et al. 1993). Again, the maturity of compost used to prepare tea may be a factor in its effectiveness, which may or may not be related to the form and availability of substrates in compost tea for pathogen metabolism.

9.6 Food Safety Issues

Some jurisdictions set standards that require commercially prepared composts to achieve conditions equivalent to pasteurisation for elimination of vegetative human and plant pathogens. In Australia, for example, application of the Australian Standard for production of compost, soil conditioners and mulches (Anonymous 2012) is required by law. Despite these standards, retailers of fresh food can be concerned about the risk of human pathogens in poor quality compost applied to horticultural crops. An additional concern is re-establishment and growth of human pathogenic bacteria in compost tea, especially when nutrients are added during extraction (Duffy et al. 2004).

There are few peer-reviewed reports on human pathogen presence and growth in compost tea prepared from mature, commercially available composts. Results from in vitro assays involving inoculation of ACT with *Escherichia coli* and/or *Salmonella enterica* revealed no significant increase in populations of these human pathogens (Duffy et al. 2004; Kannangara et al. 2006). However, such studies revealed a strong positive correlation between the concentration of molasses and kelp in ACT amended with these nutrients and *E. coli* numbers. Moreover, *E. coli* and *S. enterica* populations can proliferate in water-based solutions of fish hydrolysate, kelp, seaweed and/or humic acids, in the absence of ACT (Ingram and Millner 2007). To ascertain whether or not results from in vitro studies are relevant to commercial production conditions, Ingram and Millner (2007) inoculated commercially-available compost prepared and sold to produce ACT, or ACT amended with nutrients, with human pathogens. There was a significant increase in the numbers of *E. coli*, *S. enterica* and total fecal coliforms in ACT treatments supplemented with either a commercially available compost tea 'supplement' consisting of a mixture of molasses, bat guano, sea bird guano, powdered soluble kelp, citric acid, Epsom salts, "ancient seabed minerals" and calcium carbonate or any one of the following: powdered soluble kelp, liquid humic acids, and rock dust. These results suggest that practitioners adopting ACT need to be especially cautious when adding nutrients to compost extracts to prevent contamination of ACT with human pathogens during production, particularly if conditions during or after production enable pathogens to proliferate.

In general, practitioners have been using ACT on the assumption that human pathogenic bacteria, if present, do not proliferate in the presence of abundant microflora that outcompete human pathogenic bacteria for nutrients. Palmer et al. (2010a) prepared ACT using compost sampled from commercial open windrows during the cooling phase of composting when the internal windrow temperature was approximately 50 °C. *Escherichia coli*, *Listeria monocytogenes* and *Bacillus cereus* were not detected (< 1 cfu per 100 g) in this compost. ACT was inoculated with *E. coli* M23 strep^r at 1×10^7 cfu ml^{-1} at the beginning of extraction. This *E. coli* strain, although non-pathogenic, has growth characteristics similar to strains of *E. coli* pathogenic to humans (Brown et al. 1997; Salter et al. 1998). No significant change in the number of *E. coli* M23 strep^r was observed up to 72 h later. However, there was a significant increase in *E. coli* M23 strep^r numbers by 72 h when 0.8 % fish hydrolysate or 1 % molasses were introduced to ACT 24 h after extraction commenced. Introduction of 0.5–2 % liquid kelp or a mixture of 1.7 % liquid kelp and 0.8 % fish hydrolysate lead to a decline in the number of *E. coli* M23 strep^r. There was no relationship between the number of *E. coli* M23 strep^r and the abundance of culturable bacteria and fungi in ACTs amended with nutrients, although a low oxygen concentration, pH and high conductivity was associated with increased *E. coli* M23 strep^r numbers in an ACT amended with 1 % molasses. The results imply that methods should be identified and imposed to assure that human enteric pathogens do not contaminate amended ACTs, during or after preparation, and/or are prevented from attaining levels that pose a risk to humans consuming fruit and vegetables treated with them.

9.7 Methods for Evaluating Disease Suppression

Most studies for evaluating potential crop protectants, including compost tea, begin with in vitro or in planta tests to assess direct activity against the plant pathogen of interest. Commencing research with field trials is potentially costly and risky if the developmental product fails to control disease during commercial crop production. This phase of the work usually occurs after pathogen and/or disease suppression has been demonstrated in controlled-environment experiments.

9.7.1 Controlled-Environment Experiments

The detached-leaf bio-assay of Weltzien and Ketterer (1986) illustrates the type of rapid technique needed to reliably evaluate different teas prior to testing in the glasshouse or field. Other authors commence experimentation using growth media amended with different types or concentrations of compost tea to evaluate growth inhibition of target plant pathogens that can be cultured in the absence of the host plant (Koné et al. 2010). Growth media and controlled incubation conditions typically favour colonisation by the plant pathogen. It follows that the same media will favour growth and metabolic activity of a sub-set of culturable microorganisms within the compost tea if their growth rate and nutritional needs do not vary significantly from those of the plant pathogen. A large proportion of microbial taxa are non-culturable (Amann et al. 1995). Compost tea is likely to contain non-culturable

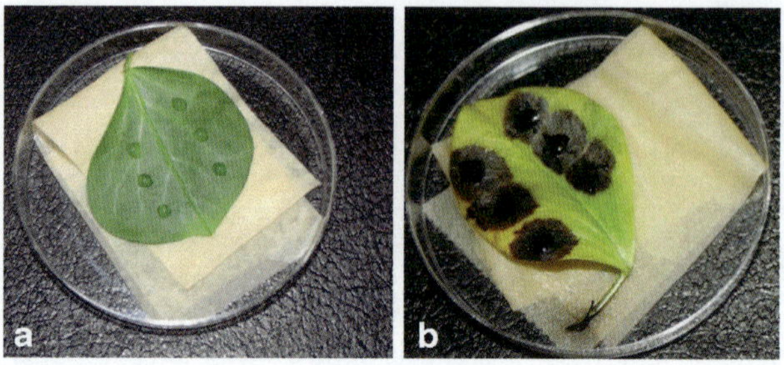

Fig. 9.2 Detached bean leaflets with (**a**) droplets of a suspension of *Botrytis cinerea* conidia prior to incubation, and (**b**) necrotic lesions after inoculation with *B. cinerea* and incubation of sealed plates for 5 days in darkness at 21 ± 4 °C. Reproduced from Palmer et al. (2010b)

microorganisms with unknown metabolic activity. In short, pathogen growth inhibition in vitro may be a consequence of the selection pressure exerted by the growth medium, and it may or may not be an indicator of the potential for disease suppression involving other substrates and environmental conditions.

When the goal is to investigate disease suppressiveness of a compost tea directly, then one approach is to select a model pathosystem amenable to rapid and reproducible bioassay. *B. cinerea* is a suitablc fungal pathogen for initial testing of compost tea because it is common in the environment and colonises a wide range of fruit, vegetable and ornamental crops in many different climates (Coley-Smith et al. 1980). The necrotrophic mode of nutrition of *B. cinerea* means that it generally colonises senescing plant tissues and can alternate between being a saprobe and plant pathogen. Important diseases of flowers and fruit caused by *B. cinerea* include bunch rot of wine and table grapes and grey mould of strawberries. Palmer et al. (2010b) described a method for single-point inoculation of detached bean leaflets (Fig. 9.2) whereby the area or dimensions of the resulting discrete lesions are measured readily by image analysis.

If the plant pathogen of interest is an obligate biotroph, such as a mildew or rust fungus, then glasshouse trials involving inoculation and treatment of whole plants can provide information under conditions that are ideal for both host and pathogen growth. These trials are also useful for studying whether or not a single application of compost tea can eradicate young mildew colonies or infections established within 24–72 h of inoculation. The effect of the period between treatment with compost tea and pathogen inoculation on the level of disease suppression can also be investigated, along with relevant methodology to investigate the potential of compost tea to induce disease resistance in the host plant.

9.7.2 Field Trials

The European and Mediterranean Plant Protection Organization (EPPO) publishes general and specific standards for plant pathogens or pathosystems with regard to the conduct of field trials to evaluate the efficacy of plant protection products (Anonymous 2013). Field trials are often conducted in replicated small plots using a randomised complete-block design amenable to analysis of variance. These designs should have sufficient residual degrees of freedom to separate treatment means statistically when the difference between means is also of practical significance.

As compost tea has potential to control more than one plant disease, all plant diseases that appear in experimental plots should be assessed. Arthropod pests should be controlled biologically or through use of sprays that are unlikely to interact with the effect of the compost tea and/or confound the results of disease assessment. Similarly, any foliar applications of nutrients and all other crop inputs should be recorded. Importantly, the trial should be surrounded by buffer plots or rows to ensure adjacent spray operations do not result in spray drifting onto the experimental treatments.

Applications of compost teas in the first field trials should commence at a time during the growing season before there is a risk of primary infection and continue at the minimum, practical spray interval. A small spray interval minimises the risk of undue crop loss and maximises potential disease suppression. If a weekly spray interval is selected, then sprays are likely to be applied every 6–10 days in practice, given advances or delays in spraying due to inclement weather. It would seem reasonable to check the viability of microorganisms before and after they have passed through the spray nozzle.

Given potential batch-to-batch variation in compost teas, each batch should be considered a different spray material for the purpose of data analyses. Care needs to be taken in comparing efficacy of a compost tea treatment relative to a spray program based on standard commercial practice in which materials and their time of application may be different to those used in the compost tea program. A treatment comprising multiple applications of compost teas represents one spray program; therefore, results should be presented as the effect of one spray program relative to another, such as a standard spray program or one where water is applied instead of compost tea. If the objective is to investigate the timing of a spray application, then the effect of omitting an application of a single batch of compost tea can be investigated as long as there is another treatment in which all other spray applications are identical.

Once initial field trials indicate a useful level of disease suppression, trials should be repeated over multiple growing seasons to understand variation in the extent of disease suppression across the range of environmental conditions likely to be encountered in a particular location. Whole-of-block experimentation can potentially shorten the period of field experimentation because disease suppression is quantified across fields where disease severity varies across the landscape (Bramley et al. 2011). This approach can also address the question of whether or not two treatments, such as a spray program based on compost tea or the current (standard, organic) spray program, result in statistically similar or different responses depending on the location in the field. To adopt this method, the co-operating grower must agree to accept some level of disease over a wider area of crop than might otherwise be tolerated in small-plot trials. The trade-off is a rich data set that can inform the grower of site-specific factors resulting in so-called disease 'hot spots' (Bramley et al. 2011). Cultural interventions to reduce disease risk in 'hot spots' can produce more desirable results with compost teas when overall disease risk in a field has been reduced.

9.8 Integrating Compost Teas in Disease Management

Adoption of compost tea as a means of biological control depends on strategies for integrating it into existing crop and disease management that allow the timing and number of applications to be optimised in relation to disease risk. How well the crop protectant is applied to the crop surface can also influence efficacy. Spray technique influences the distribution of spray droplet size and how well those droplets intercept and cover the target plant tissue (Landers 2010). As such, biological control is best implemented when there is a good understanding of disease epidemiology and spray technology, including the key pathogen, crop and environmental factors driving disease development. Disease management tactics are also dictated by the level of disease control considered acceptable by the farming business.

9.8.1 Seasonal Variation and Disease Thresholds

The extent of disease suppression using any form of crop protection is likely to vary from one growing season to the next according to how the crop protectant interacts with the prevailing environmental conditions, the arrival and concentration of virulent pathogen inoculum and host growth factors affecting the susceptibility of various plant tissues. In contrast, the level of acceptable disease control is defined by the farming business and is influenced by the farming philosophy, business-style and/or quality of produce demanded by customers and consumers. Evidence for disease suppression from small-plot field experiments does not imply that application of compost teas under commercial farming conditions would result in effective disease control, as defined by the farmer. Nil-tolcrancc for a particular disease may be necessary when customers reject a diseased crop on the basis of reduced produce quality. Otherwise, there is usually a pre-determined disease severity, below which there is no significant impact on the long-term sustainability of the farming business. A certain frequency of crop failure may also be tolerated by some farmers because high returns in abundant seasons maintain business viability.

9.8.2 Strategies for Implementation

The mechanism of action of compost tea will inform whether or not compost tea is applied as a protectant, to prevent infection by a plant pathogen, and/or as a post-infection eradicant or suppressant of pathogen colonisation. The first protective spray is usually timed in the weeks immediately preceding the expected arrival of pathogen inoculum for primary infection of the host plant. Evidence of environmental conditions favouring dispersal of primary inoculum and/or primary infection might also be used to time a spray application (Magarey et al. 2002). Monitoring for a low incidence of the first symptoms of disease can also be used; however, this method may be unsuitable for those diseases where incidence and severity increase rapidly under certain environmental conditions.

Spray materials should be applied during environmental conditions that maximise interception of spray droplets by the target plant. Whereas the potential for phytotoxicity by some chemicals is increased when there is slow drying of spray droplets, it is not known whether these same conditions would enhance or reduce the efficacy of compost tea. Indeed, environmental conditions during and immediately after application of compost tea, including the temperature, relative humidity, wind speed and UV index, could be investigated in relation to spatio-temporal changes in microbial populations on plant surfaces.

Once the first spray has been applied, the interval between spray applications will depend on crop phenology and disease susceptibility, plus environmental conditions affecting disease development. If the presence of live microorganisms in compost tea contributes to its efficacy, then environmental conditions are also likely to affect their survival on plant surfaces. In a field trial of ACT applied to grapevines, there was an increase in the number of bacteria on Riesling leaves between 5 and 13 days post application, which might have been promoted by rainfall 6 days following ACT application (Evans et al. 2013).

Plant growth also influences the spray interval. New leaf tissue emerging since the previous spray application may be unprotected from infection by those pathogens that colonise green, juvenile tissue (e.g., mildews). If a compost tea inhibits a plant pathogen only in the location in which it is deposited, like a 'contact' fungicide, then the interval between spray applications needs to be relatively short during periods of rapid shoot growth. Similarly, there may be key phenological stages when a crop is highly susceptible to infection by a plant pathogen. Again, the interval of any 'contact' spray material may need to be as short as possible during these periods.

The integration of biological control with other disease management practices is especially important when biocontrol is effective only when disease severity in the absence of crop protection is low to moderate. Farmers who understand crop cultural factors conducive to disease development can sometimes manipulate these to reduce disease development; for example, orchardists and vineyard managers can adopt measures to increase air circulation around fruit, which in turn reduces the duration of surface wetness and lowers relative humidity (Thomas et al. 1988). These measures might include selection of appropriate sites and row orientation at planting, and pruning, feeding and watering plants for a good balance between crop vegetative growth and fruit production (crop load). The extent of shading within a crop canopy can also be manipulated to increase exposure of foliage and fruit to UV light, which is known to reduce the severity of powdery mildew by reducing the colonisation of epiphytic fungal hyphae (Austin et al. 2011; Willocquet et al. 1996).

If there is a commitment to the adoption of organic practices, then farmers need to plan their disease management strategy in growing seasons that are highly conducive to disease development. Disease forecasting systems and long-range weather forecasts can provide alerts to looming risks. If significant crop loss or failure can be tolerated, then farming practices can proceed as before. Otherwise, every known cultural and biological control practice needs to be implemented when disease risk is very high, assuming multiple measures will be additive or synergistic in their effects.

9.8.2.1 Integration with Synthetic Chemicals

If the use of synthetic chemicals is tolerated, then these can be viewed as additional control measures in seasons of high disease risk. In glasshouse and field experiments with grapevines, for example, Reglinski et al. (2005) improved the efficacy of a single biological control agent, *Ulocladium oedemansii*, by applying the fungal antagonist in a program that included applications of a chemical elicitor of induced plant resistance [5-chloro salicylic acid (5-CSA)].

Integrating compost tea in a spray program with reduced inputs of synthetic fungicides would need to be done carefully if fungicide residues deplete surface microflora. Application of synthetic fungicides to Riesling grapevines resulted in lower yeast and fungal populations on leaves than those observed after treatment with water or ACT (Evans et al. 2013). These results were consistent with those of Sholberg et al. (2006) who found that a standard viticultural spray program, including fungicides such as captan, iprodione, mancozeb, myclobutanil and sulfur, reduced the number of fungal epiphytes on grape berries and leaves. Given this knowledge, it may still be possible to time applications of compost tea during periods of low disease risk. For example, the pre-flowering, flowering and fruitset stages of wine grapes represent periods of high berry susceptibility to grape powdery mildew (Gadoury et al. 2012). Powdery mildew development on leaves prior to these stages provides a source of inoculum for grape berries (Calonnec et al. 2006). Given this knowledge, regular applications of compost tea until the pre-flowering stage may prevent significant powdery mildew development on grape leaves. Once flowering is imminent, the most effective fungicides (synthetic or organic) could be used for the critical period of bunch susceptibility to disease. After flowering and fruit set, compost teas might then be used for late season disease management of powdery mildew on leaves. Again, the effect of fungicide residues from flowering and fruit-set applications on the efficacy of late-season applications of compost tea would need to be investigated.

9.8.3 Sustainable Practices and Produce Quality

Any developmental product or system of disease management needs to be tested over several growing seasons to understand any non-target impacts on crop yield and/or quality. This is especially important for perennial crops where actions in one growing season can affect the outcome in subsequent seasons. Duxbury et al. (2004), for example, demonstrated that regular application of the chemical elicitor of induced disease resistance, 5-CSA, to grapevines caused a reduction in berry weight, leaf chlorosis and unacceptable residues in wine produced from treated grapes. The direct impact of a biological treatment on produce quality can be difficult to discern from those caused by fungal infection, especially where the level of disease varies between the experimental and standard treatment; for example, a low level of powdery mildew on grape berries can have a significant effect on grape juice quality (Stummer et al. 2005). Another issue for winemaking, or any crop utilising microorganisms during post-harvest processing, is the potential for late-season applications of compost tea to alter microbial population (e.g., yeasts) on the fruit surface. These changes may not be an issue if potentially deleterious microorganisms do not survive post-harvest processes like fermentation or heat treatment.

Regional variation in produce quality is a critical feature of perceived product identity and consumer appreciation. Knowledge of the microbial diversity of crop plant surfaces over time and space (Peiffer et al. 2013) are likely to provide important baseline information for understanding the consequence of adding microorganisms to that environment through application of compost tea or other human-mediated actions. There is evidence to suggest that microbial populations associated with wine grape surfaces are non-randomly associated with regional, varietal and climatic factors (Bokulich et al. 2013). It is not known if regional differences in microbial communities modulate wine sensory qualities; however, further studies might improve understanding of how these factors interact with interventions such as biological control to influence produce quality.

Even though many practitioners are committed to the philosophy of organic agriculture, there are still a number of questions about whether or not some approaches to disease management are environmentally and/or economically sustainable. Every time a spray is applied using tractor-driven machinery it uses (fossil) fuel and labour, with multiple passes of a tractor potentially contributing to soil compaction. The latter can be mitigated to some extent using controlled-traffic farming (Gasso et al. 2013). It could be argued that in some situations a single application of a well-timed and highly effective synthetic fungicide might be a more sustainable practice than multiple applications of a less effective organic crop protectant. Conversely, a single application of an effective organic crop protectant could suffice if the decision to apply the material is informed by a useful system for disease-risk assessment (Gent et al. 2013) and, if necessary, integrated with suitable cultural controls. Compost teas might be particularly useful in fields where the local pathogen population has developed resistance to commonly used fungicides. Breeding, selecting and/or mixing crop cultivars with durable disease resistance is perhaps the most sustainable disease management practice of all.

9.9 Future Research

Despite a gradual increase in the number of refereed, scientific reports about compost teas, additional replicated trials over many sites and seasons are needed to support, or otherwise, the many anecdotal claims made by practitioners about disease suppression. Microbial diversity in compost tea and on plant surfaces before and after application to crops can be studied further by analyses of community DNA; for example, actinobacteria are known to be present in early secondary mesophilic compost (Herrmann and Shann 1997) and their diversity could be studied by application of appropriate analyses of community DNA (Conn and Franco 2004; Osborn et al. 2000).

The field of metagenomics is providing the tools to study not only what culturable and non-culturable organisms are present, but to also answer questions about the metabolic activity and function of different microbial groups (Simon and Daniel 2011). Metagenomics generates largely unbiased samples from the environment of all genes from all the members of the microbial communities present. This approach has potential to identify and predict factors associated with disease suppression. It cannot show which of these processes are active; however, the extraction and analysis of metagenomic mRNA (the metatranscriptome) can provide information on the regulation and expression profiles of complex communities (Simon and Daniel 2011).

The following represents a preliminary list of hypotheses to be addressed or to be tested further for additional host-pathogen interactions and/or for a wider range of production and application environments:

- A consistently pathogen-suppressive ACT can be produced using immature compost produced by a wide range of variable raw ingredients within a defined C:N ratio
- The addition of certain additives or organic adjuvants to compost tea can improve the extent of discasc supprcssion
- The presence of nitrate or other metabolic substrates in compost tea influences its capacity to suppress leaf and fruit pathogens by supporting saprophytic and/or pathogenic microorganisms on the leaf or fruit surface
- Components of compost tea can sometimes induce host resistance or influence host tissue physiology in a way that can be characterised and related to the extent of plant pathogen colonisation
- Microbial communities of immature or mature compost can be identified and components correlated to the disease-suppressive qualities of an associated compost tea
- Some microbial taxa in compost tea survive on plant surfaces under conditions in which other taxa would perish. If so, these taxa can be identified and their diversity and persistence under a range of conditions can be quantified in relation to the level of disease suppression observed
- There is a relationship between the functional and metabolic diversity of microbial communities in compost tea and disease suppression in the field
- Microbial communities present naturally on plant surfaces influence crop quality (fresh or processed) and these communities can be modulated through application of compost tea.

9.10 Conclusion

Compost teas are like any other form of biological control in that single or multiple applications rarely provide 100 % crop protection. Whereas the persistence and growth of single microbial species used for biological control can be constrained by sub-optimal conditions of UV radiation, temperature, humidity, rainfall, nutrient availability and mechanism of action, evidence is needed to support the claim that microbial diversity in compost teas provides some buffering capacity for efficacy over a wider range of environmental conditions.

Relative to homogenous biocontrol formulations, compost teas are likely to result in greater within and among-season variation in the level of disease suppression because of batch-to-batch variation in biological, chemical and physical characters. This variation results, in part, from variation in materials and conditions during open-windrow composting. These features of commercial compost production cannot be controlled in practice or experimentally.

Standards and standard procedures for characterising the chemical, physical and biological characters of compost and compost tea would at least provide customers and consumers with reliable information on product quality and safety. Without these, there is the risk that teas produced under certain conditions might allow human pathogens to proliferate. It is less clear whether or not some teas can stimulate the growth of the plant pathogens targeted for control. Data on consistency in crop protection over multiple sites and for more than one or two growing seasons are needed.

Despite what might be seen as limitations by those who have access to and value synthetic fungicides, compost teas have the potential to fulfill the needs of those farmers and food producers seeking lower-cost, environmentally-friendly options for disease management that allow utilisation of materials that might otherwise be considered to be organic waste. With additional testing and cost-benefit analyses, compost teas may well prove to be a valuable tactic for integration with other forms of biological and cultural controls to meet the practical needs of farmers, the demands of customers and consumers, and to sustain crop yields, produce quality and rural livelihoods.

References

Abadias M, Teixido N, Usall J, Benabarre A, Vinas I (2001) Viability, efficacy, and storage stability of freeze-dried biocontrol agent *Candida sake* using different protective and rehydration media. J Food Prot 64:856–861

Al-Dahmani JH, Abbasi PA, Miller SA, Hoitink HAJ (2003) Suppression of bacterial spot of tomato with foliar sprays of compost extracts under greenhouse and field conditions. Plant Dis 87:913–919

Al-Mughrabi KI (2006) Antibiosis ability of aerobic compost tea against foliar and tuber potato diseases. Biotechnol Adv 5:69–74

Amann RI, Ludwig W, Schleifer KH (1995) Phylogenetic identification and in-situ detection of individual microbial cells without cultivation. Microbiological Reviews 59:143–169

Amner W, McCarthy AJ, Edwards C (1988) Quantitative assessment of factors affecting the recovery of indigenous and released thermophilic bacteria from compost. Appl Environ Microbiol 54:3107–3112

Anonymous (1999) Environmental impacts of animal feeding operations. In: U.S Environmental Protection Agency Office of Water Standards and Applied Sciences Division. http://www.epa.gov/guide/feedlots/envimpct.pdf. Accessed 26 Oct 2013.

Anonymous (2006) Overview of compost tea use in NSW. In: Recycled Organics Unit. http://www.recycledorganics.com/publications/reports/komposttea/komposttea.pdf. Accessed 20 Oct 2013.

Anonymous (2012) Australian Standard 4454. Composts, soil conditioners and mulches. Standards Australia International Ltd, Sydney NSW, Australia. ISBN 978-1-74342-036-2.

Anonymous (2013) EPPO Standards (PP1). In: Efficacy evaluation of plant protection products. List of EPPO Standards, EPPO Cloud. http://pp1.eppo.int/list.php. Accessed 31 Oct 2013.

Austin CN, Grove GG, Meyers JM, Wilcox WF (2011) Powdery mildew severity as a function of canopy density: associated impacts on sunlight penetration and spray coverage. Am J Enol Vitic 62:23–31

Bandick AK, Dick RP (1999) Field management effects on soil enzyme activities. Soil Biol Biochem 31:1471–1479

Bezuidenhout CN, van Antwerpen R, Berry SD (2012) An application of principal component analyses and correlation graphs to assess multivariate soil health properties. Soil Sci 177:498–505

Bishop PL, Godfrey C (1983) Nitrogen transformations during sludge composting. Biocycle 24:34–39

Bisiach M, Minervini G, Vercesi A, Zerbetto F (1985) Six years of experimental trials on biological control against grapevine grey mould. In: Proceedings 8th Botrytis Symposium, vol 9. Quaderni di Viticoltura ed Enologia dell' Universita di Torino, Alba, Torino, Italy, pp 285–297

Bokulich NA, Thorngate JH, Richardson PM, Mills DA (2013) Microbial biogeography of wine grapes is conditioned by cultivar, vintage, and climate. Proc Natl Acad Sci U S A (in press). doi:10.1073/pnas.1317377110

Boulter JI, Boland GJ, Trevors JT (2000) Compost: A study of the development process and end-product potential for suppression of turfgrass disease. World J Microbiol Biotechnol 16:115–134

Bramley RGV, Evans KJ, Dunne KJ, Gobbett DL (2011) Spatial variation in response to 'reduced input' spray programs for powdery mildew and botrytis identified through whole-of-block experimentation. Aust J Grape Wine Res 17:341–350

Brinton W (1995) The control of plant pathogenic fungi by use of compost teas. Biodynamics 197:12–15

Brown JL, Ross T, McMeekin TA, Nichols PD (1997) Acid habituation of *Escherichia coli* and the potential role of cyclopropane fatty acids in low pH tolerance. Int J Food Microbiol 37:163–173

Calonnec A, Cartolaro P, Deliere L, Chadoeuf J (2006) Powdery mildew on grapevine: the date of primary contamination affects disease development on leaves and damage on grape. Bulletin OILB/SROP 29:67–73

Carson R (1962) Silent spring. Houghton Miffin, Boston, USA

Coley-Smith JR, Verhoeff K, Jarvis WR (1980) The biology of *Botrytis*. Academic Press, New York, USA

Conn VM, Franco CMM (2004) Analysis of the endophytic actinobacterial population in the roots of wheat (Triticum aestivum L.) by terminal restriction fragment length polymorphism and sequencing of 16S rRNA clones. Appl Environ Microbiol 70:1787–1794

Cronin MJ, Yohalem DS, Harris RF, Andrews JH (1996) Putative mechanism and dynamics of inhibition of the apple scab pathogen *Venturia inaequalis* by compost extracts. Soil Biol Biochem 28:1241–1249

Dagostin S, Scharer HJ, Pertot I, Tamm L (2011) Are there alternatives to copper for controlling grapevine downy mildew in organic viticulture? Crop Prot 30:776–788

de Bertoldi M, Sequi P, Lemmes B, Papi T (1996) The science of composting. Blackie Academic & Professional, Glasgow, UK

de Ong ER (1927) Fish oil as an adhesive in control of grape berry and codling moths. J Econ Entomol 20:121–125

Diánez F, Santos M, Boix A, de Cara M, Trillas I, Aviles M, Tello JC (2006) Grape marc compost tea suppressiveness to plant pathogenic fungi: role of siderophores. Compost Sci Util 14:48–53

Diaz LF, Savage GM, Eggerth LI, Golueke CG (1993) Composting and recycling municipal solid waste. Lewis Publishers, Boca Rato, Florida, USA

Dik AJ, Elad Y (1999) Comparison of antagonists of *Botrytis cinerea* in greenhouse-grown cucumber and tomato under different climatic conditions. Eur J Plant Pathol 105:123–137

Diver S (2002) Notes on compost teas: a supplement to the ATTRA publication "Compost teas for plant disease control". http://www.planetnatural.com/wp-content/uploads/compost-tea-notes.pdf. Accessed 31 Oct 2013.
Droby S, Vinokur V, Weiss B, Cohen L, Daus A, Goldschmidt EE, Porat R (2002) Induction of resistance to *Penicillium digitatum* in grapefruit by the yeast biocontrol agent *Candida oleophila*. Phytopathology 92:393–399
Duffy B, Sarreal C, Ravva S, Stanker L (2004) Effect of molasses on regrowth of *E. coli* O157:H7 and *Salmonella* in compost teas. Compost Sci Util 12:93–96
Duxbury M, Hotter G, Reglinski T, Sharpe N (2004) Effect of chitosan and 5-chlorosalicylic acid on total phenolic content of grapes and wine. Am J Enol Vitic 55:191–194
Egli T, Zinn M (2003) The concept of multiple-nutrient-limited growth of microorganisms and its application in biotechnological processes. Biotechnol Adv 22:35–43
El-Tarabily KA, Nassar AH, Hardy GES, Sivasithamparam K (2003) Fish emulsion as a food base for rhizobacteria promoting growth of radish (*Raphanus sativus* L. var. *sativus*) in a sandy soil. Plant Soil 252:397–411
Elad Y (1994) Biological control of grape gray mold by *Trichoderma harzianum*. Crop Prot 13:35–38
Elad Y, Shtienberg D (1994) Effect of compost water extracts on gray mold (*Botrytis cinerea*) Crop Prot 13:109–114
Elad Y, Zimand G, Zaqs Y, Zuriel S, Chet I (1993) Use of *Trichoderma harzianum* in combination or alternation with fungicides to control cucumber gray mold (*Botrytis cinerea*) under commercial greenhouse conditions. Plant Pathology 42:324–332
Epstein R (1997) The science of composting. Technomic Publishing Company, Lancaster, PA, USA
Evans KJ, Palmer AK, Metcalf DA (2013) Effect of aerated compost tea on grapevine powdery mildew, botrytis bunch rot and microbial abundance on leaves. Eur J Plant Pathol 135:661–673
Finstein MS, Morris ML (1975) Microbiology of municipal solid waste composting. Adv Appl Microbiol 19:113–151
Fogliano V, Ballio A, Gallo M, Woo S, Scala F, Lorito M (2002) *Pseudomonas lipodepsipeptides* and fungal cell wall-degrading enzymes act synergistically in biological control. Mol Plant-Microbe Interact 15:323–333
Gadoury DM, Cadle-Davidson L, Wilcox WF, Dry IB, Seem RC, Milgroom MG (2012) Grapevine powdery mildew (*Erysiphe necator*): a fascinating system for the study of the biology, ecology and epidemiology of an obligate biotroph. Mol Plant Pathol 13:1–16
Gasso V, Sorensen CAG, Oudshoorn FW, Green O (2013) Controlled traffic farming: a review of the environmental impacts. Eur J Agron 48:66–73
Gaud WS (1968) The green revolution: accomplishments and apprehensions. In: AgBioWorld http://www.agbioworld.org/biotech-info/topics/borlaug/borlaug-green.html. Accessed 20 Oct 2013.
Gent DH, Mahaffee WF, McRoberts N, Pfender WF (2013) The use and role of predictive systems in disease management. Annu Rev Phytopathol 51:267–289
Golueke CG (1992) Bacteriology of composting. Biocycle 33:55–57
Guetsky R, Shtienberg D, Elad Y, Dinoor A (2001) Combining biocontrol agents to reduce the variability of biological control. Phytopathology 91:621–627
Gullino ML, Garibaldi A (1988) Biological and integrated control of grey mould of grapevine: results in Italy. EPPO Bulletin 18:9–12
Haggag WM, Saber MSM (2007) Suppression of early blight on tomato and purple blight on onion by foliar sprays of aerated and non-aerated compost teas. J Food Agric Env 5:302–309
Harper AM, Strange RN, Langcake P (1981) Characterization of the nutrients required by *Botrytis cinerea* to infect broad bean leaves. Physiol Plant Pathol 19:153–167
Herrmann RF, Shann JF (1997) Microbial community changes during the composting of municipal solid waste. Microb Ecol 33:78–85
Hoitink HAJ, Boehm MJ (1999) Biocontrol within the context of soil microbial communities: a substrate-dependent phenomenon. Annu Rev Phytopathol 37:427–446

Hoitink HAJ, Stone AG, Han DY (1997) Suppression of plant diseases by composts. Hortscience 32:184–187

Hunt PG, Smart GC, Eno CF (1973) Sting nematode, *Belonolaimus longicaudatus*, immotility induced by extracts of composted municipal refuse. J Nematol 5:60–63

Ingham ER (2005) The compost tea brewing manual. Fifth Edition edn. Soil Foodweb Incorporated, Oregon, USA

Ingram DT, Millner PD (2007) Factors affecting compost tea as a potential source of *Escherichia coli* and *Salmonella* on fresh produce. J Food Prot 70:828–834

Ishii K, Fukui M, Takii S (2000) Microbial succession during a composting process as evaluated by denaturing gradient gel electrophoresis analysis. J Appl Microbiol 89:768–777

Kannangara T, Forge T, Dang B (2006) Effects of aeration, molasses, kelp, compost type, and carrot juice on the growth of *Escherichia coli* in compost teas. Compost Sci Util 14:40–47

Ketterer N, Fisher B, Weltzien HC (1992) Biological control of *Botrytis cinerea* on grapevine by compost extracts and their microorganisms in pure culture. Recent Advances in Botrytis Research. Pudoc, Wageningen

Knight A, Cheng Z, Grewal SS, Islam KR, Kleinhenz MD, Grewal PS (2013) Soil health as a predictor of lettuce productivity and quality: a case study of urban vacant lots. Urban Ecosyst 16:637–656

Koné SB, Dionne A, Tweddell RJ, Antoun H, Avis TJ (2010) Suppressive effect of non-aerated compost teas on foliar fungal pathogens of tomato. Biol Control 52:167–173

Kredics L, Manczinger L, Antal Z, Penzes Z, Szekeres A, Kevei F, Nagy E (2004) In vitro water activity and pH dependence of mycelial growth and extracellular enzyme activities of *Trichoderma* strains with biocontrol potential. J Appl Microbiol 96:491–498

Landers AJ (2010) Effective vineyard spraying: a practical guide for growers. Keuka Park, New York. http://www.effectivespraying.com. Accessed 14 Aug 2014

Lau SSS, Wong JWC (2001) Toxicity evaluation of weathered coal fly ash-amended manure compost. Water Air Soil Pollut 128:243–254

Lindow SE, Brandl MT (2003) Microbiology of the phyllosphere. Appl Environ Microbiol 69:1875–1883

Litterick AM, Harrier L, Wallace P, Watson CA, Wood M (2004) The role of uncomposted materials, composts, manures, and compost extracts in reducing pest and disease incidence and severity in sustainable temperate agricultural and horticultural crop production-a review. Crit Rev Plant Sci 23:453–479

Magarey RD, Travis JW, Russo JM, Seem RC, Magarey PA (2002) Decision support systems: quenching the thirst. Plant Dis 86:4–14

Magnin-Robert M, Trotel-Aziz P, Quantinet D, Biagianti S, Aziz A (2007) Biological control of *Botrytis cinerea* by selected grapevine-associated bacteria and stimulation of chitinase and beta-1,3 glucanase activities under field conditions. Eur J Plant Pathol 118:43–57

Maheshwari DK (2013) Bacteria in agrobiology: disease management. Springer-Verlag Berlin Heidelberg, Germany

Marín F, Santos M, Diánez F, Carretero F, Gea FJ, Yau JA, Navarro MJ (2013) Characters of compost teas from different sources and their suppressive effect on fungal phytopathogens. World J Microbiol Biotechnol 29:1371–1382

Marois JJ, Bledsoe AM, Bostock RM, Gubler WD (1987) Effects of spray adjuvants on development of *Botrytis cinerea* on *Vitis vinifera* berries. Phytopathology 77:1148–1152

Martinez C, Avis TJ, Simard J-N, Labonte J, Belanger RR, Tweddell RJ (2006) The role of antibiosis in the antagonism of different bacteria towards *Helminthosporium solani*, the causal agent of potato silver scurf. Phytoprotection 87:69–75

Matthies C, Erhard HP, Drake HL (1997) Effects of pH on the comparative culturability of fungi and bacteria from acidic and less acidic forest soils. J Basic Microbiol 37:335–343

McQuilken MP, Whipps JM, Lynch JM (1994) Effects of water extracts of a composted manure-straw mixture on the plant pathogen *Botrytis cinerea*. World J Microbiol Biotechnol 10:20–26

Metcalf DA (2002) Development of biological control agents which control onion white root rot under commercial field conditions. In: Proceedings Onions 2002 Conference. NSW Agriculture, Australia.
Osborn AM, Moore ERB, Timmis KN (2000) An evaluation of terminal-restriction fragment length polymorphism (T-RFLP) analysis for the study of microbial community structure and dynamics. Environmental Microbiology 2:39–50
Pagans E, Barrena R, Font X, Sanchez A (2006) Ammonia emissions from the composting of different organic wastes. Dependency on process temperature. Chemosphere 62:1534–1542
Palmer AK (2009) Standardised production of aerobic compost extract for disease management in sustainable viticulture. PhD thesis. University of Tasmania, Hobart, Tasmania, Australia
Palmer AK, Brown J, Ross T, Metcalf DA, Evans KJ (2010a) Potential for growth of *E. coli* in aerobic compost extract. Compost Sci Util 18:152–161
Palmer AK, Evans KJ, Metcalf DA (2010b) Characters of aerated compost tea from immature compost that limit colonization of bean leaflets by *Botrytis cinerea*. J Appl Microbiol 109:1619–1631
Pane C, Celano G, Villecco D, Zaccardelli M (2012) Control of *Botrytis cinerea*, *Alternaria alternata* and *Pyrenochaeta lycopersici* on tomato with whey compost-tea applications. Crop Prot 38:80–86
Peiffer JA, Spor A, Koren O, Jin Z, Tringe SG, Dangl JL, Buckler ES, Ley RE (2013) Diversity and heritability of the maize rhizosphere microbiome under field conditions. Proc Natl Acad Sci U S A 110:6548–6553
Peters S, Koschinsky S, Schwieger F, Tebbe CC (2000) Succession of microbial communities during hot composting as detected by PCR-single-strand-conformation polymorphism-based genetic profiles of small-subunit rRNA genes. Appl Environ Microbiol 66:930–936
Ramona Y, Line MA (2002) Potential for the large-scale production of a biocontrol fungus in raw and composted paper mill waste. Compost Sci Util 10:57–62
Reglinski T, Elmer PAG, Taylor JT, Parry FJ, Marsden R, Wood PN (2005) Suppression of *Botrytis* bunch rot in Chardonnay grapevines by induction of host resistance and fungal antagonism. Australas Plant Pathol 34:481–488
Rogiers SY, Whitelaw-Weckert M, Radovanovic-Tesic M, Greer LA, White RG, Steel CC (2005) Effects of spray adjuvants on grape (*Vitis vinifera*) berry microorganisms, epicuticular wax and susceptibility to infection by *Botrytis cinerea*. Australas Plant Pathol 34:221–228
Ryckeboer J, Mergaert J, Coosemans J, Deprins K, Swings J (2003a) Microbiological aspects of biowaste during composting in a monitored compost bin. J Appl Microbiol 94:127–137
Ryckeboer J, Mergaert J, Vaes K, Klammer S, De Clercq D, Coosemans J, Insam H, Swings J (2003b) A survey of bacteria and fungi occurring during composting and self-heating processes. Ann Microbiol 53:349–410
Salter MA, Ross T, McMeekin TA (1998) Applicability of a model for the non-pathogenic *Escherichia coli* for predicting the growth of pathogenic *Escherichia coli*. J Appl Microbiol 85:357–364
Scheuerell S (2003) Understanding how compost tea can control disease. Biocycle 44:20–25
Scheuerell S (2004) Going from compost to compost tea: weighing plant health benefits against human pathogen uncertainties. Phytopathology 94:S127–S127
Scheuerell S, Mahaffee W (2002) Compost tea: principles and prospects for plant disease control. Compost Sci Util 10:313–338
Scheuerell SJ, Mahaffee WF (2004) Compost tea as a container medium drench for suppressing seedling damping-off caused by *Pythium ultimum*. Phytopathology 94:1156–1163
Scheuerell SJ, Mahaffee WF (2006) Variability associated with suppression of gray mold (*Botrytis cinerea*) on geranium by foliar applications of nonaerated and aerated compost teas. Plant Dis 90:1201–1208
Segarra G, Casanova E, Borrero C, Aviles M, Trillas I (2007) The suppressive effects of composts used as growth media against *Botrytis cinerea* in cucumber plants. Eur J Plant Pathol 117:393–402
Sholberg P, Harlton C, Boule J, Haag P (2006) Fungicide and clay treatments for control of powdery mildew influence wine grape microflora. Hortscience 41:176–182

Siddiqui Y, Meon S, Ismail R, Rahmani M (2009) Bio-potential of compost tea from agro-waste to suppress *Choanephora cucurbitarum* L. the causal pathogen of wet rot of okra. Biol Control 49:38–44

Simon C, Daniel R (2011) Metagenomic analyses: past and future trends. Appl Environ Microbiol 77:1153–1161

Steurbaut W (1993) Adjuvants for use with foliar fungicides. Pestic Sci 38:85–91

Stewart A (2001) Commercial biocontrol-reality or fantasy? Australas Plant Pathol 30:127–131

Stummer BE, Francis IL, Zanker T, Lattey KA, Scott ES (2005) Effects of powdery mildew on the sensory properties and composition of Chardonnay juice and wine when grape sugar ripeness is standardised. Aust J Grape Wine Res 11:66–76

Sturz AV, Lynch DH, Martin RC, Driscoll AM (2006) Influence of compost tea, powdered kelp, and Manzate (R) 75 on bacterial-community composition, and antibiosis against *Phytophthora infestans* in the potato phylloplane. Canadian Journal of Plant Pathology-Revue Canadienne De Phytopathologie 28:52–62

Thomas CS, Marois JJ, English JT (1988) The effects of wind speed, temperature, and relative humidity on the development of aerial mycelium and conidia of *Botrytis cinerea* on grape. Phytopathology 78:260–265

Tiquia SM (2005) Microbial community dynamics in manure composts based on 16S and 18S rDNA T-RFLP profiles. Environ Technol 26:1101–1113

Tuomela M, Vikman M, Hatakka A, Itavaara M (2000) Biodegradation of lignin in a compost environment: a review. Bioresour Technol 72:169–183

Ugarte CM, Zaborski ER, Wander MM (2013) Nematode indicators as integrative measures of soil condition in organic cropping systems. Soil Biol Biochem 64:103–113

van Loon LC, Bakker P, Pieterse CMJ (1998) Systemic resistance induced by rhizosphere bacteria. Annu Rev Phytopathol 36:453–483

Walker AS, Micoud A, Remuson F, Grosman J, Gredt M, Leroux P (2013) French vineyards provide information that opens ways for effective resistance management of *Botrytis cinerea* (grey mould). Pest Manag Sci 69:667–678

Watson ME (2004) Testing compost. In: The Ohio State University Extension, Ohio. http://ohioline.osu.edu/anr-fact/pdf/0015.pdf. Accessed 26 Oct 2013.

Welke SE (2004) The effect of compost extract on the yield of strawberries and the severity of *Botrytis cinerea*. Journal of Sustainable Agriculture 25:57–68

Weltzien HC (1990) The use of composted materials for leaf disease suppression in field crops. In: Unwin R (ed) Crop Protection Inorganic and Low Input Agriculture: Options for Reducing Agrochemical Usage, vol 45. British Crop Protection Council, Farnham, UK, pp 115–120

Weltzien HC, Ketterer N (1986) Control of downy mildew, *Plasmopara viticola*, on grapevine leaves through water extracts from composted organic wastes. Journal of Phytopathology 116:186–188

Willocquet L, Colombet D, Rougier M, Fargues J, Clerjeau M (1996) Effects of radiation, especially ultraviolet B, on conidial germination and mycelial growth of grape powdery mildew. Eur J Plant Pathol 102:441–449

Zabkiewicz JA (2007) Spray formulation efficacy-holistic and futuristic perspectives. Crop Prot 26:312–319

Zhang W, Han DY, Dick WA, Davis KR, Hoitink HAJ (1998) Compost and compost water extract-induced systemic acquired resistance in cucumber and *Arabidopsis*. Phytopathology 88:450–455

Chapter 10
Microbial Biomass Improvement Following Municipal Solid Waste Compost Application in Agricultural Soil

Olfa Bouzaiane, Naceur Jedidi and Abdennaceur Hassen

Abstract Soil microbial biomass (SMB) was considered as a sensitive and as indicator of soil management especially for agricultural soil. Municipal solid waste (MSW) composting process is a promising way to reduce waste production and to obtain a stable end product such as compost available for agricultural use. However, the main requirement for the safe use or application of compost to agricultural lands is its degree of stability, which implies stable organic matter content. This practice is becoming one of the most promising ways for the reclamation and correction of organic matters loses in degraded soils. Many studies showed that MSW compost soil application could (i) improve soil physico-chemical properties, (ii) increase soil microbial biomass and activity (iii), play a biopesticide role to control soil borne diseases. Many authors investigated the different doses of MSW compost and examined their effects on soil microbial biomass available on the vertical and horizontal distribution. However MSW compost application could contaminated agricultural soil by heavy metals, toxic compounds and pathogens. Concerning the heavy metals pollution of agricultural soils is related essentially to crop quality and human health. In this review we tried to show the different investigations concerning the progress of microbial biomass following the MSW compost application in agricultural soil.

Keywords Soil microbial biomass · MSW composing process · Organic matter

10.1 Introduction

Municipal solid wastes microorganisms (fungi and bacteria) are responsible for degradation and biological transformation of organic matter. These microorganisms degradation is responsible for temperature increase in wastes (Mustin 1987). Consequently composting is known of heat treatment (sanitization of compost)

O. Bouzaiane (✉) · N. Jedidi · A. Hassen
Laboratoire Traitement et Recyclage des Eaux, Centre de Recherches et des Technologies des Eaux (CERTE), BP 24–1082, Cité Mahrajène, Tunis, Tunisie
e-mail: olfa_bz2004@yahoo.fr

D. K. Maheshwari (ed.), *Composting for Sustainable Agriculture,* Sustainable Development and Biodiversity 3, DOI 10.1007/978-3-319-08004-8_10

stabilization process of wastes with two principal steps: aerobic digestion and maturation (Hassen et al. 2001; Ben Ayed et al. 2007). This process controlled and optimized different parameters especially temperature, water content, oxygenation to produce a stable and pathogens free-compost. Pathogens were destroyed within the thermophilic phase (Epstein 1997). Heavy metals did not eliminate with heat treatment of the process (Richard 1992). Consequently various countries from European Union and the USA have providing permissible limit of heavy metals content in MSW compost for land application. However Land application of MSW compost is an excellent way of recycling both the nutrients and the organic matter contained in waste and became the promising ways to correct the degraded soil (Sanchez-Mondero et al. 2004; Bouzaiane 2007a). In agricultural systems, biotic (e.g., micro-organisms) and abiotic factors (e.g., soil acidity and water content) affect the fertility of the soil, its nutrient content and its organic matter. Soil amendment with mature MSW compost, affects both biotic and abiotic factors. In fact, the application of compost introduces new organic matter, nutrients and microbial organisms (Beffa et al. 1995) that considerably improve the texture of soils (Mays et al. 1973), and increase the soil water content. Biotic factors such as the microbial biomass activities are stimulated (Pedra et al. 2007; Roca-Perez et al. 2009). Measurement of soil microbial biomass and their extracellular enzymes after MSW compost application provide valuable information for soil quality and for a sustainable management of agricultural soils.

10.2 Soil Microbial Biomass Role and Assessment

Soil microbial biomass mainly bacteria and fungi and their extracellular enzymes activities (Tabatabai 1994) are responsible for the biological transformation that make nutrients available to plants and for sustaining soil functions. Since soil microbial communities play a critical and a key role in nutrient cycling and may be used as a sensitive and as an indicator of environmental changes or disturbance of soil management (Bouzaiane et al. 2007a). In addition soil microbial biomass (SMB) is one of the major indices applied today to study soil fertility and soil health (Sparling 1997) and conducts biochemical transformations in soil (Breland and Eltun 1999). Several studies showed that the SMB varies with soil management including the farming system (Hu et al. 1997) fertilisation (Salinas-Garcia et al. 1997), municipal solid waste application (Jedidi et al. 2004; Bouzaiane et al. 2007a) and heavy metals (Garcia-Gil et al. 2000; Fagnano et al. 2011).

Measurement of the soil microbial biomass provide valuable information for soil quality and for a sustainable management of agricultural soils. The potential influence of the SMB in a soil sample may be assessed by its amount (Anderson and Domsch 1989). Assessment of SMB can be achieved by direct methods which asses the cultivable microbes, such as the plating counts (Paul and Johnson 1977), or by several indirect methods which asses the non cultivable microbes, such as the chloroform fumigation-extraction method (CFE), the chloroform fumigation-incubation method (CFI) (Vance et al. 1987; Tate et al. 1988), and the substrate-induced respiration (SIR).

The different methods (Vance et al. 1987; Brookes 1995) have been widely used to estimate microbial biomass under different field and laboratory conditions. The microbial biomass have been estimated, in both cultivated and uncultivated soils (Vong et al. 1990), in forest soils (Gallardo and Schlesinger 1990), edaphically conditions characterized by the alternation of desiccation–rehumidification cycles (Van Gestel et al. 1991), as well as the effects of seasonally dried soils (Wu and Brookes 2005). Molecular methods such as DNA quantification method has been used for SMB estimation in different soils (Marstorp et al. 2000; Bailey et al. 2002; Leckie et al. 2004). The DNA quantification method (DNA) has been compared to the CFE method in different soils (Marstorp et al. 2000; Bailey et al. 2002; Leckie et al. 2004) and has been proposed as an alternative method of CFE to measure SMB (Marstorp et al. 2000; Anderson and Martens 2013). Bouzaiane et al. (2007b) showed that the quantification of DNA yields could be used as an alternative and a reliable method to estimate microbial biomass in wheat cultivated soil after municipal solid waste compost application.

10.3 Municipal Solid Waste Composting Process

The municipal solid waste composting process has been defined as a controlled aerobic microbial process widely used to transform organic matter contained in wastes by their microbes into a stable product consisting of a humus-like substance (Michel and Reddy 1995).

10.3.1 Composting Process

Composting is one of the most complex biotechnologies since many factors influence the optimization and the reproducibility of the process such as mechanical (wastes composition and mix), chemical (temperature of environmental conditions eg. season, pH, moisture water content, O_2 concentration, porosity, C/N ratio), and biological parameters (microorganisms composition and content).

These different factors influence the quality and the degree of stability of the end product (Mondini et al. 2002; Jedidi et al. 2004). In general the composting process occurs into three major and classic stages based on temperature parameter (Fig. 10.1) (Hassen et al. 2001; Ben Ayed et al. 2007).

Mesophilic Phase: During this phase, the temperature and the water content increased as a consequence of increase of psychrophilic and mesophilic microorganisms which improved the biodegradation of organic compounds like lipids glucose and amino acid (Mustin 1987). This phase is short in time between 20 and 30 days (Hassen et al. 2001; Ben Ayed et al. 2007), the temperature increased to reach 45 °C to improve the maximum of biodegradation.

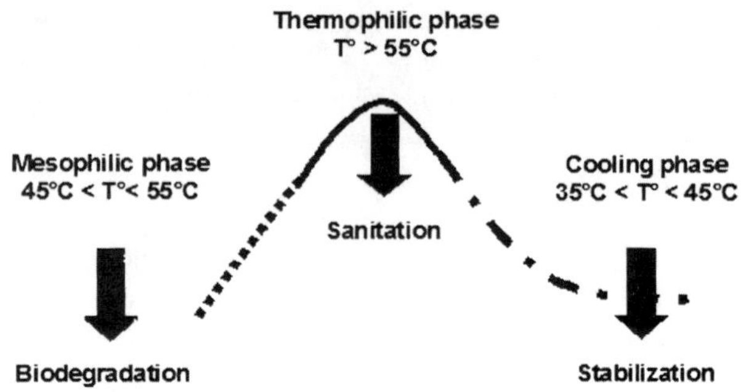

Fig. 10.1 Major classic phases of municipal solid waste composting process based on temperature parameter

Thermopilic Phase: this phase is important in time which could be occurred between 30 and 100 days of composting process (Hassen et al. 2001; Ben Ayed et al. 2007). This phase improve the development of thermophilic microorganisms according to the authors. The temperature must be maintained below 65 °C inside the window of wastes by ventilation and watering. Marrug et al. (1993) mentioned that temperature above 60 °C affect the decomposition rate of the waste organic matter as a result of microbes activities decrease. This phase is known as a phase of compost sanitation which heat sensitive pathogens like viruses, bacteria, protozoa or helminthes (Strauch 1991) could be eliminated.

Cooling or Stabilization Phase: in this phase the temperature began to decrease after 12th week according to Hassen et al. 2001 and after 111 days according to Ben Ayed et al. (2007). By the end of the process, the average temperature inside the windrow showed a reduction of values to reach approximately 30 °C. This decrease was the result of depletion of organic compounds in compost and C/N ratio tend to stabilize. The end product or compost is available for agricultural use. This practice is becoming one of the most promising ways for the remediation of degraded soils (Bouzaiane et al. 2007a). However, the main requirement for the safe use or application of compost to agricultural lands is its degree of stability, which implies stable organic matter content (Castaldi et al. 2004, 2008; Mondini et al. 2004).

Bouzaiane et al. (2011) evaluated that the microbial biomass C and N and DNA content during the municipal solid waste composting process can be used to understand the compost stability state. According to the authors biological properties could be combined to chemical properties, to indicate the compost stability.

10.3.2 MSW Compost Application Improves Agricultural Soil Properties

Application of MSW compost in agricultural soils can directly improves soil physico-chemical properties such as soil organic matter contents, soil aggregates, buffering capacity (Reeves 1997) and soil biological properties such as soil microbial biomass and activity (Pedra et al. 2007; Roca-Perez et al. 2009).

10.3.2.1 MSW Compost Application Improves Soil Physico-Chemical Properties

Compost from MSW represents an important resource of organic matter to maintain and restore soil fertility. Since it is important source of the plants nutrient and is of great value nowadays, particularly in those countries where the organic matter content of the soil is low (Castaldi et al. 2004). Bouzaiane et al. (2007a) showed that the application of MSW compost improves the organic matter of degraded soils in semiarid zone of Mediterranean countries like Tunisia. Roca-Perez et al. (2009) reported that the application of MSWC into the soil increased soil quality in two soils from Spain and increased soil organic matter, N, P and stable aggregates from both amended soils.

The soil addition with mature or stable MSW compost represents a valuable and effective tool to increase the long term of soil aggregates. Recently Spaccini and Piccolo (2013) showed that field application on three different agricultural soils of mature compost improve the distribution of water stable aggregates with the significant improvement of soil aggregate stability.

In the other hand MSW compost could reduce the adverse effects of salinity showed by Lakdhar et al. (2008) in *Hordeum maritimum* under greenhouse conditions. Plants were cultivated in pots filled with soil added with 0 and 40 t ha^{-1} of MSW compost, and irrigated twice a week with tap water at two salinities (0 and 4 g l^{-1} NaCl). According to the authors the MSW compost may be safely applied to salt-affected soils without adverse effects on plant physiology.

10.3.2.2 MSW Compost Application Improves Soil Microbial Biomass and Activity

Addition of good quality of compost may increase global microbial biomass and enhance soil enzyme activity (Albiach et al. 2000; Debosz et al. 2002; Garcia-Gill et al. 2000). The improve of this soil biological properties was study with many authors to evaluate the MSW compost effect. Perruci (1990) showed that microbial biomass, carbon, nitrogen, sulphur and phosphorus were significantly increased over a period of 12 months in a soil treated with compost of municipal solid waste.

According to Garcia-Gill et al. (2000), a long-term field experiment utilizing barley received MSW compost at 20 t ha^{-1} (C20) or at 80 t ha^{-1} (C80) were studied. The effects of these applications on soil enzyme activities and microbial biomass at crop harvest were measured after nine years in upper horizon of 0–20 cm. In comparison with the control (no amendment soil) MSW compost addition increased biomass C by 10 and 46% at application rates of 20 and 80 t ha^{-1}, respectively. The authors evaluated enzyme activities and they showed that the dehydrogenase and catalase enzymes, were higher in the MSW compost treatments by 730 (C20) and 200% (C80), respectively, indicating an increase in the microbial metabolism in the soil as a result of the mineralization of biodegradable C fractions contained in the amendments. The addition of MSW caused different responses in hydrolase enzymes. Phosphatase activity decreased with MSW (±62% at both rates) to less than that in the control treatments. Urease activity decreased by 21% (C20) and 28% (C80), possibly being affected by the heavy metals contained in the MSW. However, β-glucosidase and protease increased with MSW compost.

The use of composts in agricultural soils is a widespread practice and the positive effects on soil and plants are known from numerous studies. Ros et al. (2006) investigate a long term of crop-rotation (maize, summer-wheat and winter barley) in field experiment in Austria. The application of compost produced from urban organic wastes at rate of 175 kg N ha^{-1} yr^{-1} for 12 years. The microbial biomass C (B_C), was analyzed at different depths (0–10, 10–20 and 20–30 cm). The results showed that the continued addition of compost to soil enhances B_C compared to control soil and which this effect declined with depth.

The application of organic wastes as amendments to improve soil properties has become a very common practice, especially under Mediterranean semiarid conditions. Bouzaiane et al. (2007b) evaluated the microbial biomass C and N (B_C and B_N) by the chloroform fumigation-extraction (CFE) method and microbial biomass DNA concentration in a loam-clayey wheat cultivated soil. They obtained the highest values of microorganisms counts with MSW compost at 40 t ha^{-1}. The microbial biomasses C and N and DNA concentration increased in wheat cultivated soil amended with MSW compost at 40 t ha^{-1} in comparison with 80 t ha^{-1} and in the superficial profile (0–20 cm) than in the deep one (20–40 cm). Recently Mardomingo et al. (2013) have investigated the changes in soil microbial activity under field conditions over a one-year period after the application of a single high dose (160 Mg ha^{-1} dry mass) of municipal solid waste compost (MSWC). Measurements were made for microbial biomass carbon (MBC), basal respiration (BR) and enzymatic activities evaluated by assays of catalase (CA), dehydrogenase (DA), urease (UA), protease (PA), phosphatase (PhA) and β-glucosidase (βGA). This organic amendment produced different effects on soil microbial activity. The application of MSWC significantly increased ($p \leq 0.05$) the B_C, with the highest content observed in summer season (1369.1 ± 13.2 mg C kg^{-1}). According to the authors soil microbial activity (BR, CA, DA and hydrolase activity) remained stable throughout the one-year period in MSW compost.

Other authors were interested on the biopesticide view of organic materials that could be used to control many soil-borne plant pathogens (Boulter et al. 2000; Hoitink et al. 1993). Since the microbial biodiversity may be increased (Peacock et al. 2001) and soil borne pathogens may be reduced by stimulation of antagonistic organisms (Tilston et al. 2002) allowing less use of potentially fumigants or pesticides (Pascual et al. 2000; Ros et al. 2005). Moreover MSW compost could contained many antagonistic microorganisms such as *Bacillus sublilis*, *Trichoderma* and *Pseudomonas* (Serra-Wittling et al. 1996; Hoitink et al. 1993) that controlled wheat plants phytopathogenic like *Fusarium oxysporum*.

10.3.2.3 Effect of MSWC Heavy Metals on Agricultural Soil

Heavy metals do not degrade throughout the composting process, and frequently become more concentrated due to microbial degradation and loss of carbon and water from the compost (Richard 1992). A significant decrease of soil microbial biomass was noted after three years of MSW compost application (Bouzaiane et al. 2007b). The decrease of microbial biomass was due to heavy metals content elevation in compost at 80 t ha^{-1} treated soil. Thus according to the authors the highest rate of MSW compost induced the lowest ratio of biomass C to soil organic carbon and the lowest ratio of biomass N to soil organic nitrogen.

Carbonell et al. 2011 conducted to assess the inputs of metals to agricultural land from soil amendments. Maize seeds were exposed to a municipal solid waste (MSW) compost (50 Mg ha^{-1}) and NPK fertilizer (33 g $plant^{-1}$) amendments considering N plant requirement until the harvesting stage with the following objectives: (1) determine the accumulation of total and available metals in soil and (2) know the uptake and ability of translocation of metals from roots to different plant parts, and their effect on biomass production. According to the authors the results showed that MSW compost increased Cu, Pb and Zn in soil, while NPK fertilizer increased Cd and Ni, but decreased Hg concentration in soil. The root system acted as a barrier for Cr, Ni, Pb and Hg, so metal uptake and translocation were lower in aerial plant parts. Biomass production was significantly enhanced in both MSW and NPK fertilizer-amended soils (17%), but also provoked slight increases of metals and their bioavailability in soil. The highest metal concentrations were observed in roots, but there were no significant differences between plants growing in amended soil and the control soil. Important differences were found for aerial plant parts as regards metal accumulation, whereas metal levels in grains were negligible in all the treatments.

10.4 Conclusion

The MSW compost application to an agricultural soil can improve and maintain soil quality by decreasing the need of chemical fertilizers and pesticides, improving soil tillage, increasing soil microbial biomass and enzyme activities, increasing the organic matter of degraded soils and increasing the plants productivities. However the MSW compost could be contaminated by heavy metals, toxics compounds and pathogens that limits the use of the compost. The utilization of MSW compost in low rate should be used for sustainable agricultural soil to mitigate the cumulative effects of environmental pollution and gain public acceptance.

Acknowledgment Special thanks to all who helped in the water treatment and recycling laboratory of CERTE (Centre de Recherche et des Technologies des Eaux).This mini review expressed the views of the authors and do not necessarily reflect the opinion of CERTE.

References

Albiach R, Canet R, Pomares F, Ingelmo F (2000) Microbial biomass content and enzymatic activities after the application of organic amendments to a horticultural soil. Bioresour Technol 75:43–48

Anderson TH, Domsch KH (1989) Ratios of microbial biomass carbon to total organic carbon in arable soils. Soil Biol Biochem 21:471–479

Anderson TH, Martens R (2013) DNA determinations during growth of soil microbial biomasses. Soil Biol Biochem 57:487–495

Bailey VL, Peacock AD, Smith JL, Bolten HJ (2002) Relationships between soil microbial biomass determined by chloroform fumigation-extraction, substrate-induced respiration, and phospholipid fatty acid analysis. Soil Biol Biochem 34:1385–1389

Beffa T, Blanc M, Marilley L, Lott Fischer J, Lyon PF, Aragno M (1995) Taxonomic and metabolic microbial diversity during composting. In: de Bertoldi M, Sequi P, Lemmes B, Papi T (eds) The science of composting, vol 1. Blackies Academic and Professional, Glasgow, Scotland, pp 149–161

Ben Ayed L, Hassen A, Jedidi N, SaidiI N, Bouzaiane O, Murano F (2007) Microbial C and N dynamics during composting process of urban solid waste. Waste Manage Res 25:24–29

Boulter JI, Boland GJ, Trevors JT (2000) Compost: a study of the development process and end-product potential for suppression of turfgrass disease. World J Microbiol Biotech 16:115–134

Bouzaiane O, Cherif H, Saidi N, Jedidi N, Hassen A (2007a) Effects of municipal solid waste compost application on the microbial biomass of cultivated and non-cultivated soil in a semi-arid zone. Waste Manage Res 25:327–333

Bouzaiane O, Cherif H, Saidi N, Hassen A, Jedidi N (2007b) Municipal solid waste compost dose effects on soil microbial biomass determined by chloroform fumigation-extraction and DNA methods. Ann Microb 57(4):681–686. (ISSN 1590-4261)

Bouzaiane O, Saidi N, Ben Ayed L, Jedidi N, Hassen A (2011) Relationship between microbial C, microbial N and microbial DNA extracts during municipal solid waste composting process. Progress in biomass and bioenergy production. pp 239–252. (ISBN 978-953-307-491-7 (chapter 12))

Breland TA, Eltun R (1999) Soil microbial biomass and mineralization of carbon and nitrogen in ecological, integrated and conventional forage and arable cropping systems. Biol Fert Soils 30:193–201

Brookes PC (1995) The use of microbial parameters in monitoring soil pollution by heavy metals. Biol Fert Soils 19:269–279

Carbonell G, de Imperial RM, Torrijos M, Delgado M, Rodriguez JA (2011) Effects of municipal solid waste compost and mineral fertilizer amendments on soil properties and heavy metals distribution in maize plants (*Zea mays* L). Chemosphere 85:1614–1623

Castaldi P, Garau G, Melis P (2004) Influence of compost from sea weeds on heavy metal dynamics in the soil–plant system. Fresenius Environ Bull 13:1322–1328. (ISSN 1018–4619)

Castaldi P, Garau G, Melis P (2008) Maturity assessment of compost from municipal solid waste through the study of enzyme activities and water-soluble fractions. Waste Manage 28:534–540

Debosz K, Petersen SO, Kure L K, Ambus P (2002) Evaluating effects of sewage sludge and household compost on soil physical, chemical and microbiological properties. Appl Soil Ecol 19:237–248

Epstein E (1997) The science of composting. Technomic Publishing, Lancaster

Fagnano M, Adamo P, Zampella M, Fiorentino N (2011) Environmental and agronomic impact of fertilization with composted organic fraction from municipal solid waste: a case study in the region of Naples, Italy. Agric Ecosyst Environ 141:100–107

Gallardo A, Schlesinger WH (1990) Estimation of microbial biomass nitrogen by the fumigation-incubation and fumigation-extraction in warm temperate forest soil. Soil Biol Biochem 22:927–932

Garcia-Gil JC, Plaza C, Soler-Rovira P, Polo A (2000) Long term effects of municipal solid waste compost application on soil enzyme activities and microbial biomass. Soil Biol Biochem 32:1907–1913

Hassen A, Belguith K, Jedidi N, Cherif A, Cherif M, Boudabbous A (2001) Microbial characterization during composting of municipal solid waste. Bioresour Technol 80:185–192

Hoitink HAJ, Hadar MJY (1993) Mechanisms of suppression of soil borne plant pathogens in compost amended substrates. In: Hoitink HAJ, Keener HM (eds) Science and engineering of composting: design, environmental, microbiological and utilization aspects. The Ohio State University, Ohio, pp 601–621

Hu S, Grunwald NJ, Van Bruggen AHC, Gamble GR, Drinkwater LE, Shennan C, Demment MH (1997) Short term effects of cover crop incorporation on soil carbon pools and nitrogen availability. Soil Sci Soc Am J 61:901–911

Jedidi N, Hassen A, Van Cleemput O, M'hiri A (2004) Microbial biomass in soil amended with different types of organic wastes. Waste Manage Res 22:93–99

Lakdhar A, Hafsi C, Rabhi M, Debez A, Montemurro F, Abdelly C, Jedidi N, Ouerghi Z (2008) Application of municipal solid waste compost reduces the negative effects of saline water in Hordeum maritimum L Bioresour Technol 99:7160–7167

Leckie SE, Prescott CE, Grayston SJ, Neufeld JD, Mohn WW (2004) Comparison of chloroform fumigation-extraction, phospholipid fatty acid, and DNA methods to determine microbial biomass in forest humus. Soil Biol Biochem 36:529–532

Mardomingo IJ, Rovira P S, Casermeiro MA, de la Cruz MT, Polo A (2013) Seasonal changes in microbial activity in a semiarid soil after application of a high dose of different organic amendments. Geoderma 206:40–48

Marrug C, Grebus M, Hassen RC, Keener HM, Hoitink HA J (1993) A kinetic model of yard waste composting process. Compost Sci Util 1:38–51

Marstorp H, Guan X, Gong P (2000) Relationship between dsDNA, chloroform labile C and ergosterol in soils of different organic matter contents and pH. Soil Biol Biochem 32:879–882

Mays DA, Ternan GL, Duggan JC (1973) Municipal compost effects on crops yields and soil properties. J Environ Qual 2:89–92

Michel FC, Reddy CA & Forney LJ (1995) Microbial-degradation and humification of the lawn care pesticide 2, 4-dichlorophenoxyacetic acid during the composting of yard trimmings. Appl Environ Microbiol 61:2566–2571

Mondini C, Contin M, Leita L, De Nobili M (2002) Response of microbial biomass to air-drying and rewetting in soils and compost. Geoderma 105:111–124

Mondini C, Fornasier F, Sinicco T (2004) Enzymatic activity as a parameter for the characterization of the composting processes. Soil Biol Biochem 36:1587–1594

Mustin M (1987) Le Compost: gestion de la matière organique. Editions François Dubusc Paris, 953 p

Pascual JA, Hermander T, Garcia C, Deleij FAAM, Lynch JM (2000) Long-term suppression of *Pythium ultinum* in arid soil using fresh and composted municipal wastes. Biol Fertil Soils 30:478–484

Paul EA, Johnson RL (1977) Microscopic counting and adenosine 5′-triphosphate measurement in determining microbial growth in soils. Appl Environ Microbiol 34:263–269

Peacock AD, Mullen MD, Ringelberg DB, Tyler DD, Herdrick DB, Gale PM, White DC (2001) Soil microbial community response to dairy manure or ammonium nitrate applications. Soil Biol Biochem 33:1011–1019

Pedra F, Polo A, Ribeiro A, Domingues H (2007) Effects of municipal solid waste compost and sewage sludge on mineralization of soil organic matter. Soil Biol Biochem 39:1375–1382

Perez PA, Edel HV, Alabouvette C et Steinberg C (2006) Response of soil microbial communities to compost amendments. Soil Biol Biochem 38:460–470

Perruci P (1990) Effect on the addition of municipal solid waste compost on microbial biomass and enzyme activities. Biol Fert Soils 10:221–226

Reeves DW (1997) The list of soil organic matter in maintaining soil quality in continuous cropping systems. Soil Till Res 43:131–167

Richard TL (1992) Municipal solid waste composting: physical and biological processing. Biomass Bioenerg 3:163–180

Roca-Perez L, Martinez C, Marcilia P, Boluda R (2009) Composting rice straw with sewage sludge and compost effects on the soil-plant system. Chemosphere 75:781–787

Ros M, Hernandez MT, Garcia C, Pascual JA (2005) Biopesticide effects of green composts against Fusarium wilt on melon plants. J Appl Microbiol 98:845–854

Ros M, Hernandez MT, Garcia C (2003) Soil microbial activity after restoration of a semiarid soil by organic amendments. Soil Biol Biochem 35:463–469

Ros M, Pascual JA, Garcia C, Hernandez MT, Insam H (2006) Hydrolase activities, microbial biomass and bacterial community in a soil after long-term amendment with different composts. Soil Biol Biochem 38:3443–3452

Salinas-Garcia JR, Hons FM, Matocha JE (1997) Long term effect of tillage and fertilization on soil organic matter dynamics. Soil Sci Soc Am J 61:152–159

Sanchez-Mondero MA, Mondini C, de Nobili M, Leita L, Roig A (2004) Land application of biosolids: soil response to different stabilization degree of the treatment organic matter. Waste Manage 24:325–332

Serra-Wittling C, Houot S, Alabouvette C (1996) Increased soil suppressiveness to Fusarium wilt of flax after addition of municipal solid waste compost. Soil Biol Biochem 28:1207–1214

Spaccini R, Piccolo A (2013) Effects of field managements for soil organic matter stabilization on water-stable aggregate distribution and aggregate stability in three agricultural soils. J Geochem Explor 129:45–51

Sparling GP (1997) Soil microbial biomass, activity and nutrient cycling as indicators of soil health. In: Pankhurst C, Doube B.M, Gupta VVSR (eds) Biological indicators of soil health. CAB International, Wallingford, pp 97–119

Strauch D (1991) Survival of pathogenic micro-organism and parasites in excreta, manure and sewage sludge. Rev Sci Tech Off Int Epiz 10:813–846

Tabatabai MA (1994) Soil enzymes. In: weaver RW, Angle S, Bottomley P (eds) Methods of soil Analysis, part 2: microbiological and biochimical methods, SSSA Book series: S Soil Science. Society of America, Madison, pp 775–883

Tate KR, Ross DJ, Feltham CW (1988) A direct extraction method to estimate soil microbial C: effects of experimental variables and some different calibration procedures. Soil Biol Biochem 20:329–335

Tilston EL, Pill D, Groenhof AC (2002) Composted recycled organic water suppresses soil borne diseases of field crops. New Phytol 154:731–740

Van Gestel M, Ladd JN, Amato M (1991) Carbon and nitrogen mineralization from two soils of contrasting texture and microagregate stability: influence of sequential fumigation, drying and storage. Soil Biol Biochem 23:313–322

Vance ED, Brookes PC, Jenkinson DS (1987) Microbial biomass measurements in forest soils: determination of KC values and tests of hypotheses to explain the failure of the chloroform fumigation-incubation method in acid soils. Soil Biol Biochem 19:689–696

Vong PC, Kabibou I, Jacquin F (1990) Etude des corrélations entre biomasse microbienne et differentes fractions d'azote organique présentées dans deux sols Lorrains. Soil Biol Biochem 22:385–399

Wu J, Brookes PC (2005) The proportional mineralisation of microbial biomass and organic matter caused by air-drying and rewetting of a grassland soil. Soil Biol Biochem 37:507–515

Chapter 11
Bio-composting Oil Palm Waste for Improvement of Soil Fertility

A. W. Gandahi and M. M. Hanafi

Abstract Sources of bio-compost as agro-industrial wastes includes wide range of oil palm wastes viz. waste, biomass, palm kernels, empty fruit bunch, mill effluent, trunk and frond compost. Various composting processes are summarized in brief with distinct reference of oil palm composting covering aerated static pile, and co composting with earthworms (vermicomposting). However, in-vessel composting and windrow composting has meritorious advantages in composting. This review article refers to various significant roles played by microorganisms associated. Noteworthy study of bio-compost applications and procedures are correspondingly glosses framework of ecological, economical and agro-ecosystemic benefits.

Keywords Agro-industrial waste · Composting process · Co-composting · Inorganic fertilizer · Organic fertilizer

11.1 Introduction

Plant matter that has been decomposed and recycled for use as a fertilizer or manure is said to be bio-compost. It is considered as a key ingredient in organic farming for being full of nutrients required for the growth and development of plants. Bio-compost is applied in numerous ways mainly in gardens, landscaping, horticulture, and agriculture. Simply, the bio-composting is done by piling up wastes in the garden

M. M. Hanafi (✉)
Laboratory of Plantation Crops, Institute of Tropical Agriculture/Department of Land Management, Faculty of Agriculture, Universiti Putra Malaysia, 43400 Serdang, Selangor, Malaysia
e-mail: mmhanafi@upm.edu.my

A. W. Gandahi
Department of Soil Science, Faculty of Crop Production, Sindh Agriculture University, Tandojam 70060, Sindh, Pakistan

D. K. Maheshwari (ed.), *Composting for Sustainable Agriculture,* Sustainable Development and Biodiversity 3, DOI 10.1007/978-3-319-08004-8_11

or any outdoor place and then has to be left un-disturbed for a year or more. Bio-compost has proved to be useful for controlling soil-erosion, its efficiency in wet-land construction and as landfill cover is undoubtedly remarkable.

Currently, bio-composting process refers to monitoring of the composting. As commonly practiced, it is usually done by shredding the plant matter, addition of adequate water to maintain the required level of moisture and then to provide better aeration, the mixture is turned on regular basis. It is worth to mention that the process of decomposition is enhanced when worms and fungi added to the mixture. The complex compounds are broken into simpler ones and consequently greater amounts of heat, carbon dioxide and ammonium are produced. Later on, microbes utilize this ammonium available to the plants in the form of nitrites and nitrates. The worms and other microbes require (NO_3–), carbon (C) for energy resultantly heat is released when the microbes oxidized this C. The materials that carry higher quantity of C appear to be brown and dry. For the growth and the reproduction of more organisms nitrogen (N) is essential so that C could be oxidized. Those materials which appear to be green and wet have higher N. Availability of water in sufficient amount is required so as to maintain the moisture level and to prevent the anaerobic condition. Water also helps in reducing heat generated by the microbes during the process of oxidation of C. Bio-composting is, therefore, considered to be an efficient and easier way for the decomposition of organic wastes that yield into useful manure or fertilizer. Moreover, it is a very cost effective process.

Benefit over the new technology includes (i) Bio-compost is supplied for meeting the nutrient needs of crops, (ii) It liberates growth promoting substances and vitamins and help to maintain soil fertility, (iii) It increases crop yield by 10–20 %, (iv) It is cheaper and based on renewable energy sources, (v) It improves soil physical, chemical, biological properties, tilth and soil and soil health in general, and (vi) Supplies various macro- and micro-nutrients, provides organic matter in significant quantity and beneficial micro-organisms to the soil. Biological treatment has played prominent roles in bioremediation of wastes and contaminants. Composting is one of the biological processes that have proved to be among suitable ways of converting organic wastes into products responsible for the growth and development of plants. Transformation of various organic wastes into beneficial products through composting is viable and hazard free, more appropriately it yield into products that can successfully be used as bio-fertilizers so as to retain fertility and other condition related matters. Composting resolves a number of issues related to the utilization of agricultural wastes meant for soil amendments, malodors, human pathogens, and unnecessary physico-chemical characteristics. Mineralization took place and unavailable plant nutrients converted to available forms through the composting process, disease infestation is minimized and pathogens are destroyed, as pressed partial sterilization is done. Pollutants are detoxified and malodors are abated (Parr and Hornick 1992). The growth and activity of mixed population of bacteria is the major component on which composting, is a microbiological process depends on. Simultaneously, fungi as well as actinobacteria, which are native to the wastes, are also composted. Rising unease concerning to land degradation, menace to eco-systems from over and improper synthetic fertilizers, atmospheric contamination, soil health, soil biodiversity and hygiene have fascinated the worldwide attention to organically reuse practice like composting (Abdelhamid et al. 2004).

Now-a-days a large amount of organic waste is produced as a by-product by human activities, which has impact on the environment and has the ability to cause danger for animal and human health. Managing this huge quantity of wastes is important to protect the environment and reduce the amount of these wastes. As reported by many researchers, composting is an efficient alternative way to reduce the residue of organic wastes (Alburquerque et al. 2006; Baharuddin et al. 2009).

Composting is a biological process used to speed up the decomposition process under controlled conditions. This process converts organic wastes to a much stable product, which is free of phytotoxicity and pathogens. Compost is used safely and beneficially as organic fertilizer and soil conditioner (Arslan et al. 2008; Bustamante et al. 2008; Bernal et al. 2009). The use of compost as fertilizer improves soil structure, water-holding capacity of the soil as well as improving the aeration. Compost could also provide humus or organic matter, vitamins, hormones, and plant enzymes that are not incorporated by synthetic fertilizers. The amount of C:N ratio of soil is reduced by compost and it also perform as buffer to adjust soil pH. Composting also can kill pathogenic organisms, weeds and other unwanted seeds. Fully established compost rapidly comes into symmetry with the soil. It has the ability to be blended or mixed with different materials which increases the nutrient content of the compost fertilizers (Campitelli and Ceppi 2008; De Guardia et al. 2012).

With technology innovation and research investigations, agricultural waste is no longer an environmental concern, on the other hand, it has become a source for energy production. A remarkable potential in improving the common status of sanitation, constructive environmental measures to lessen greenhouse gas emissions, considerably enhanced the crop production, soil fertility, decreased the universal reliance on synthetic fertilizers, fossil fuel, etc. Available literature showed that Malaysia produced around 41 % of the entire global palm oil production in 2008 (Shafie et al. 2011). Being a main agricultural product, the production pattern match up to a quick augment from 2.6 Mt in 1980 to about 18.9 Mt in 2011 (Malaysian Palm Oil Board (MPOB) 2012). Malaysia palm oil industry is expecting a strong demand from the global market by a constant increase from 20.6 Mt in 2013 to 21.5 Mt in 2015 owing to the mounting worldwide demand for vegetable oil (Corley 2009). In principle, improvement in palm oil production would ultimately interpret to loads of biomass offered in the market (Umar et al. 2013).

In the palm oil manufacture practice there is generally a surplus of by-product and the utilization pace of these by-products is small particularly for palm oil mill effluent (POME), empty fruit bunch (EFB) and decanter cake (DC). The better nutrient reuse will perk up soil fertility and sustainability of palm oil production. Techniques available, such as normal composting, co-composting and vermicomposting are being practiced however, have not been exploited in its full strength as huge quantity of palm waste could be decomposed in short time and the compost made from oil palm waste could not only be applied to palm plantations but also to various crops. This will ultimately eliminate the synthetic fertilizers application (Embrandiri et al. 2013).

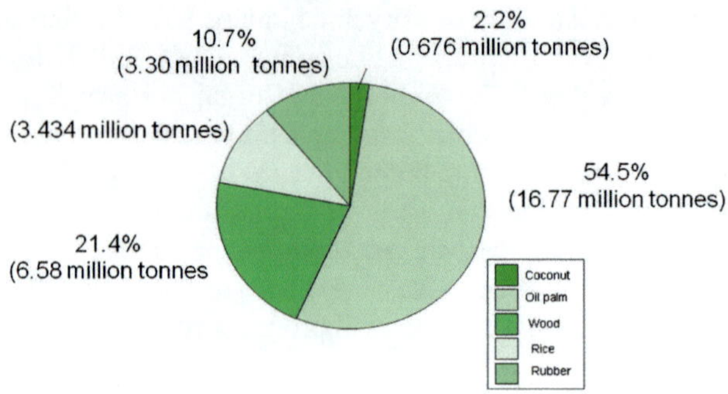

Fig. 11.1 Proportionate annual production of agricultural waste in Malaysia. (Source: ESCAP 1997)

11.2 Sources of Bio-compost

The materials inserted into compost heap possess a key effect on how sound the composting mechanism works and the quality of the final product (compost). A good composting must provide a diversity of materials and a reasonable C:N ratio. Various kinds of microbes at work in the pile could provide very high probability of achieving nutrient prosperous compost.

11.2.1 Agro-Industrial Wastes

Waste is an inevitable by-product of human actions. Better financial conditions and life style in the Asian and Pacific region had increased the quantity and density of generated waste. Agro-industrial waste disposal is a main problem in many industries around the world. The disposal of industrial wastes in the nearby areas causes environmental dangers. The recycling of wastes is a disposal mechanism and resource management.

China harvests the biggest quantities of agriculture waste and crop residues followed by India in the Asian and Pacific region. The quantities of waste that Malaysia produces from the palm oil, rubber, coconut, rice and forest products is illustrated in Fig. 11.1 (ESCAP 1997). Nutrient requirement of crops by organic manures as compost resulting from agro-industrial wastes is a major source of soil fertility and crop productivity which reduces use of chemical fertilizers (Kayikcioglu 2013).

11.2.1.1 Waste from Palm Oil Mill

Malaysia is the world's second biggest producer of palm oil in 2012, with a planted area of 5.08 million ha, produced 18.79 Mt of crude palm oil (Malaysian Palm Oil Board (MPOB) 2013) with export potential of 18 Mt in 2011. Main importers of Malaysian palm oil include China, India, USA and Pakistan.

Approximately 420 oil palm mills in Malaysia handle with the quantity of waste created in the palm oil extraction process. About 5.5 t of fresh fruit bunch (FFB) are needed by palm oil mill for extraction of 1.0 t crude palm oil. The wastes generated are 28 % EFB, 24 % fibres, 6 % shell, 3 % DC and POME. Oil palm decanter cake (OPDC) is a solid waste containing nitrogen (2.42 %), phosphates (0.51 %), potash (1.29 %), calcium (1.6 %) and magnesium (0.5 %) hence, could be used as organic fertilizer as well as soil amendment.

11.2.1.2 Biomass Wastes from Palm Oil Mills

The palm oil industry produces immense amount of wastes and its disposal is a big task (Ma 1999; Sridhar and AdeOluwa 2009; Rupani et al. 2010). In the palm oil mill the FFB are sterilized and oil fruits are removed from the branches so the EFB become residues and the fruits are pressed, the kernel is separated from the press cake. The kernels then shifted and pressed in separate mills. In palm oil plantation about 70 % of the FFB's are rotated into wastes in the form of EFB, fibers, shells and liquid effluent. These by-products may be transformed into value-added products or source of energy for extra income (Table 11.1).

11.2.1.3 Palm Kernel Shells

Palm kernel shells (PKS) are the portion left over subsequently the nut has been detached during crushing in mill. Huge and little shell sections are assorted by dust-like portion and tiny fibers. Kernel shells contain small quantity (11–13 %) moisture compared to other biomass residues from different sources.

11.2.1.4 Oil Palm Empty Fruit Bunches

The EFB's are amply available as fibrous material of solely biological source in a palm oil mill. Following the harvesting of FFBs, they are sterilized to inactivate enzymes available in pericarp and untie fruits from bunches. The sterilized bunches are fed into a gyratory drum thresher in order to remove the sterilized fruit from bunches. These bunches exclusive of fruit are called as EFB. It was estimated that 20,000–22,000 kg of EFB are produced from 100 t of FFB (Katamanee 2006). These EFB contains neither chemical nor mineral additives and have no inappropriate components. Usually EFB's are burnt which is a cause of air contamination.

11.2.1.5 Palm Oil Mill Effluent

The whole area under oil palm plantation increased in the past few years in Malaysia so the palm oil mill production also increased. As palm oil mill wastes viz. POME and POMF are the main by-products of the milling process. The fleshy mesocarp of the fruit is used to get the oil and yield is 45–56 % of FFB. The probable yield from both mesocarp and kernel is about 17 t ha^{-1} $year^{-1}$ of oil (Corley 1983).

Table 11.1 Various palm mill wastes and their common uses

Sample no	Type of waste residue	Uses
1	Fronds, trunks and leaves	Used as mulching material in the fields which helps in moisture retention
		Also used as roofing material and some are processed as furniture
2	Empty fruit bunch (EFB)	As raw material for products, such as paneling, composites, fine chemicals, pulp and paper as well as compost and bio-fertilizer (Dashiny 2009)
		Main substrate for the cultivation of *Pleurotus ostreatus* (Oyster mushroom) (Tabi et al. 2008; Rosnah et al. 2009)
		Most of it is just disposed off back into the fields as the above uses are not in large scale
3	Palm press fibre (PPF)	Fuel for the mills
		Used as a substrate for animal feed in addition to soymeal, fishmeal
		Used for making fiber boards
		Polymeric composites for building materials referred to as AGROLUMBER for products like wall panels, sub-floors, doors and furniture parts (Malaysian Palm Oil Board (MPOB) 2009; Dashiny 2009)
		Used as potting material for ornamental plants
		Currently trials are on-going utilizing fibre + Pome for vermicompositing
4	Decanter cake (DC)	Used as animal feed
		Used in combination with inorganic fertilizer to improve soil quality (Haron and Mohammed 2008)
		Currently on going work using dried, powdered form of DC as biofertilizer for vegetable gardening (Embrandiri 2013)
5	Palm kernel cake(PKC)	Suitable as feedstock because it has 48 % carbohydrate and 19 % protein (Kolade et al. 2005)
6	Shells	Used mainly for fuel
7	Palm oil mill effluent (POME)	Mainly used for Irrigation purposes but due to its acidic nature is quite toxic to flora and hence needs to be treated

In 2011, approximately 426 palm oil mills are working producing around 99.9 Mt of FFBs year^{-1} (Malaysian Palm Oil Board (MPOB) 2012).

Palm oil mill effluent is a thick brownish liquid which comprises high solids, grease and oil. The POME is source of organic waste and difficult to decomposes in natural state, but it can be converted into a value added product with the help of earthworms (Rupani et al. 2010). Palm oil mill effluent is categorized by brownish colloidal suspension containing high concentrations of organic matter, amounts of total solids (40,500 mg L^{-1}), COD (50,000 mg L^{-1}), oil and grease (4000 mg L^{-1}), BOD (25,000 mg L^{-1}) and low pH between 4 and 5 (Ma 2000). The discharge of effluents and other by-products on the lands owes to environmental pollution. Hence, there is necessity for an efficient management system for the treatment by-products.

Table 11.2 The proximate composition and mineral contents of raw POME (Habib et al. 1997)

Major constituents	Composition (%)	Macro-elements	Composition (μg/g dry weight)	Micro-elements	Composition (μg/g dry weight)
Moisture	6.99±0.14	K	8951.55±256.45	Fe	11.08±2.20
Crude protein	12.75±1.30	Na	94.57±6.45	Cu	10.76±1.04
Crude lipid	10.21±1.24	Ca	1650.09±160.45	Zn	17.58±2.10
Ash	14.88±1.35	Mg	911.95±95.50	Mn	38.81±3.65
Carbohydrate	29.55±2.44	P	14377.38±1206.88	Mo	6.45±0.40
Nitrogen free	26.39±2.33	S	13.32±1.45	Cr	4.02±0.44
Total carotene	0.019±0.001			Co	2.40±0.35
				Ni	1.31±0.30
				Se	12.32±1.35
				Si	10.50±1.80
				Sn	2.30±0.30
				Al	16.60±1.44
				B	7.60±0.60
				As	9.09±0.65
				V	0.12±0.02

Some studies showed that the vermicomposting technique for POME and POMF is a good alternative sustainable management of palm oil mill effluent absorbed with fiber effective treatment of POME and transform it into organic fertilizer (Table 11.2).

The inappropriate dumping of huge mass of agro-industrial waste creates economic as well as environmental issues. Soil health could be improved through utilization of these wastes as they got organic matter and essential elements (Khan et al. 2009). The proper reuse of organic wastes in soil can alleviate ecological risk caused by intensive farming (Ordonez et al. 2006). The microbial technology composting is frequently applied to alleviate different kinds of wastes. Through composting, volume and weight of sludge could be minimized (Abd-Rahman et al. 2003). It can decrease the mixture volume by 40–50 %, efficiently destroy the pathogens through the metabolic heat generated in the thermophilic phase, degrade hazardous organic pollutants and deliver a product that is used as a soil amendment or fertilizer (Epstein 1997). Furthermore, the compost waste is easy to handle, can be consumed as soil amendment, thus generate extra profits (Abd-Rahman et al. 2003). Composting is also practiced for recycle of waste that form chemically stable substances which can be used as soil nutrients and improving soil structure (Castaldi et al. 2005).

The composting is change of plant and animal organic substances into manure (Bharadwaj 1995). The end product is humus with good amount of plant nutrients and the by-products are CO_2, H_2O and heat (Abbasi and Ramasamy 1999). The aerobic microbes utilize organic matter as a substrate throughout the composting procedure. The substrate is decayed by these microbes, enabling it to break/convert in simpler products (Epstein 1997; Ipek et al. 2002). In composting process, compounds with C and N are changed during consecutive activities of various microorganisms to let organic matter be stable to look like humus substance (Pare et al. 1999).

11.2.1.6 Palm Oil Mill Effluent Conversion to Value Added Products

The POME full of carbohydrate, nitrogenous, protein, lipids and mineral nutrients (Habib et al. 1997) is waste as well as could be used by various industries as raw material. To convert POME in to a value added product, some technologies are developed including carotenoid for A and E vitamins production (Wahid et al. 2004), fertilizer (Basiron and Weng 2004), citric acid (Alam et al. 2008) and bio-diesel (Gutiérrez et al. 2009). These techniques solves environmental problem and also give value added products.

Through composting difficult organic deposits of animal and plant origin could be changed into manure (Singh et al. 2010) through the help of microorganisms. Such practice stabilizes various agro-industrial wastes as well as sludge (Abd-Rahman et al. 2003). The EFB produced in palm oil mill have widely been utilized for the production of organic fertilizer and this practice also lessen EFB volume up to 50 % (Chavalparit et al. 2006). In a study, Baharuddin et al. (2009) noted some physical and chemical variations while co-composting EFB with treated POME. Preliminary increase in heat was noted up to 3rd day of treatment, after that temperature remained between 50 and 62 ° during composting process. There was not too much difference in wetness from the preliminary level of 65–75 % to about 60 % at the termination of the study. The pH ranged between 7.8 and 8.0. Declined ratio of 45:12 from 45:12 of C:N was documented following 60 days of composting. The macro- and micro-nutrients content were observed in substantial quantities in the final product. Therefore, it was concluded that the consumption of EFB with POME for composting process can produce pleasing worth compost which should be incorporated to oil palm plantations as bio-fertilizer as well as soil amendment.

11.2.1.7 Oil Palm Empty Fruit Bunch

The solid wastes generated by palm oil industries can be utilized effectively for producing manure and value added products (Kavitha et al. 2013). Malaysian oil palm plantation is just on 1.97 % (5.0 million ha) of the total 253.9 million ha planted with oilseed crops globally (Malaysian Palm Oil Board (MPOB) 2012). Each year about 0.09 billion t of oil palm biomass (trunks, shells, fronds, EFB) is collected from this EFB share is around 9 % (Bari et al. 2010). These are the left over residues after the fruit bunches are pressed and oil extracted at oil mills (Ma et al. 1993; Kamarudin et al. 1997).

The EFB is the main solid waste from palm oil extraction. Together with other solid waste, such as the mesocarp fibres (from pressed fruits) and kernel shells (from fruit kernels), they are usually used as boiler fuels for the steam turbines to produce steam for sterilization of fruit and for generation of electricity. The EFB, as a lignocellulosic rich crop residue is suitable for many kinds of applications. The EFB can be converted to bio-plastic, pulp, source of enzymes, hydrogen and animal feed (Shuit et al. 2009). The EFB has been reported to be re-used in the industry to produce bio-oil, biodiesel, chemical compounds, and microorganisms (Sumathi

Table 11.3 Nutrients content in oil palm empty fruit bunch from various sources.

Sources	Nutrient content (%) on dry weight			
	N	P	K	Mg
Corley et al. (1971)	0.35	0.03	2.29	0.18
Chan et al. (1980)	0.34	0.03	2.21	0.17
Gurmit et al. (1981)	0.80	0.10	2.40	0.20
Sabrina et al. (2009)	0.70	0.20	3.16	–
Mean	0.57	0.084	2.252	0.19

Table 11.4 Physico-chemical properties of empty fruit bunch of palm oil industry. (Source: Kavitha et al. 2013)

Parameters	EFB
Organic carbon (%)	45.10
Total nitrogen (%)	0.55
C/N ratio	82.00
Total phosphorus (%)	0.02
Total potassium (%)	1.28
Total iron (mg kg^{-1})	210.00
Total zinc (mg kg^{-1})	71.00
Total copper (mg kg^{-1})	26.00
Total manganese (mg kg^{-1})	88.00
Cellulose (%)	33.00
Lignin (%)	34.00
Hemicellulose (%)	30.00

et al. 2008). However, the demand of EFB re-use by these industrial processes is still limited. Nevertheless, these processes will again create another type of waste and introduce release of greenhouse gas during the transportation of the biomass. Composting, being not the only choice, however, offers the advantage of zero waste as the EFB can be used as a nutrient source for the nearby plantation soils. The EFB is a preferred source as composting materials. The EFB has high porosity, water holding capacity and consequently high nutrient holding capacity; these characteristics are suitable features for aerobic microbial composting (Table 11.3).

The EFB is a lignocellulosic material containing 25 % lignin, 25 % hemicellulose and 50 % cellulose in cell wall. Its use as mulch for suppressing weeds in soil is best (Alam et al. 2009; Misson et al. 2009). Empty oil palm fruit bunches could be used as cheaper organic fertilizer in oil palm farms (AdeOluwa and Adeoye 2008). There is a growing need for environment friendly and cost-efficient substitute to discard the EFB in a small duration of time. Composting is the appropriate selection among the waste management plan for cost-effective and environmental beneficial plant growth (Table 11.4).

11.2.1.8 Oil Palm Trunk and Frond

The oil palm industries generate about 90 million t (Mt) of renewable biomass containing about 1.3 Mt of oil palm trunks and 8 Mt oil palm frond (OPF) annually in

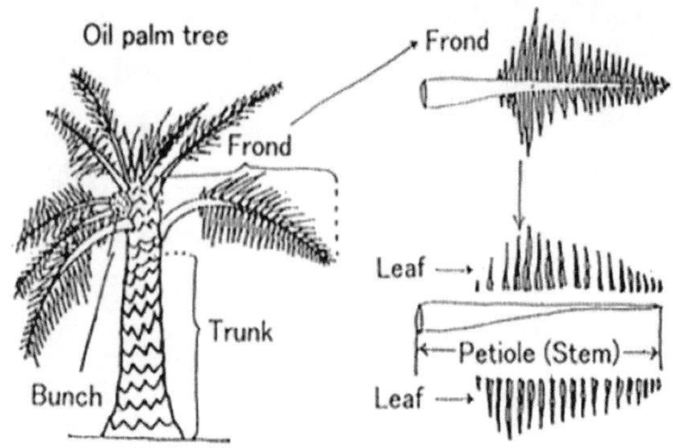

Fig. 11.2 Anatomy of an oil palm tree and oil palm found. (Source: Ishida and Abu Hasan 1997)

Malaysia (Alam et al. 2007b). The OPF are a by-product of the cultivation of oil palm trees. Since 1990s the fast growth of the palm oil sector in South-East Asia has caused rising production of fibrous wastes obtained from the oil palm fruit bunches during oil palm tree pruning and replanting practices (Dahlan 2000). About 100 kg ha^{-1} of OPF can be produced per day (Ishida and Abu Hassan 1997) with a yearly production of 11 t DM ha^{-1} (Husin et al. 1986; Lim 2000), hence it is a main waste product of oil palm plantations in addition to trunks and EFB.

Both oil palm wastes (fronds and trunks) previously destroyed by burning but this practice was prohibited because of damage to environment during 1990s. Now a day's these wastes are left on the soil to decay and fertilize the soil (Lim 2000). Oil palm fronds are low in protein but high-fiber material that is edible for herbivore livestock. The OPF are lying on soils of oil palm plantations whole the year due to regular pruning operations (Hassan et al. 1996)

The OPF are crop residues with high amounts of fiber: NDF and ADF content are in the 63–80 and 45–57 % DM range, respectively, lignin content is also high with values from 12 to 37 % DM (Aim-oeb et al. 2008; Islam et al. 2000; Khamseekhiew et al. 2001). These are poor in protein (around 7 % DM with values between 4 and 10 %) and fat (<2 %) (Islam et al. 2000). The fat is usually high in unsaturated fatty acids (Hassim et al. 2010).

The oil palm trunks are similar in composition like fronds, with less protein (<3 % DM). Empty bunches are bulky (60 % water) and with less protein (<4 % DM) and high fiber (>80 % NDF) than fronds (Hassan et al. 1996; Prasertsan and Prasertsan 1996). Oil palm wastes, mainly EFB, fronds and trunks composts have many characteristics which are superior from peat in growing media (Lin and Ratnalingam 1980).

The oil palm frond (OPF) compost has abundant potential to be used for the improvement of soilless culture system. It was effectively used to control plant diseases (Siddiqui et al. 2008; Souleymane et al. 2010). Furthermore, plant nutrients from OPF compost are slowly released to plant for longer period of time and leaching chances from the media are less. There is a rising attention toward the low and cost effective solid state bioconversion of OPF into value added product compost (Molla et al. 2004; Alam et al. 2005; Bari et al. 2009) (Fig. 11.2).

11.3 Procedure for Making Compost from Oil Palm EFB and Frond

A basic consideration of the composting system can aid fabricate a soaring class product. The microbes that are involved in composting have a small number of indispensable necessities that need to be met. Oxygen, moisture and the correct foodstuff and heat merge to produce a high-quality composting atmosphere.

11.3.1 Composting Process

Oil palm empty fruit bunches as well as frond are used for making compost. Now-a-days, usage of EFB mulching faces various limitations, the alternative use of composting EFB may therefore be useful. A preliminary study achieved a weight reduction of about 40 % but a loss of about half the nutrient value of fresh EFB. Composing of fresh EFB in 3–5 t heaps for a period of 8–10 months in the oven, indicated that the nutrient concentrations and pH generally increased but C/N ratio, temperature, weight and volume decreased with time (Lim 1989). Composting should be carried out from home gardening to advanced technology by utilizing centralized plants (Crowe et al. 2002). Complete composting process rely on biodegradable waste, economic implementation, supporting workers and required period along with desired quantity of waste product (Evans 2001). The conversion of organic residues of plant and animal into manure through the activities of several bacteria, actinobacteria and fungi are the basic principle of composting process (Bharadwaj 1995) ultimately end product is heavy in humus and providing more plant nutrients, whereas carbon dioxide, water and heat are the by-products of composting process (Abbasi and Ramasamy 1999). Initially, microorganisms decompose the substrate, breaking down it to from highly complicated form to intermediate and finally into very much useful product (Epstein 1997; Ipek et al. 2002). However, C and N are transferred through the activities of various microbes and at the end stable organic matter formed which is chemically near to humic substances (Pare et al. 1999). These transformation mostly depends on the level of availability of substrates to make it complete composting process (Marche et al. 2003). Production of chemically stabilized composted material are made up by the broken remaining of biodegradable organic compounds.

11.3.1.1 Compost from Oil Palm EFB and Frond

Composting is widely used to produce organic fertilizer from EFB and frond. The EFB, POME, frond and trunk are some of oil palm by products. Among them, average macro and secondary nutrient contents in EFB were noted as N 0.8 %, P 0.1 %, K 2.5 % and Mg 0.2 % on a dry weight basis (Gurmit et al. 1981). Furthermore, Alam et al. (2005) reported that EFB is not only a source of nutrients but also a

Fig. 11.3 Process involved in oil palm industry. (Source: http://www.uwapabriksawit.blogspot.com/2009/10/schematic-processof-palm-oil-mill.html)

potential substrate for production of cellulose and for xylose (Rahman et al. 2007) production. Meanwhile, frond of oil palm was commonly used as a ruminant food, run off water breaking in the sloping oil palm plantation area. The oil palm frond is comparatively richer in non-cellulosic polysaccharides, especially arabinoxylan; accetygroups are substitute which may probably to rabinoxyan. The lignin of oil palm fond and the wall of polysaccharides of coconut coir dust are substitute with hydroxybenzoic acids with ester and ether linkages (Fig. 11.3).

Four major types of composting are used in industry viz. aerated static piles, open static piles, turned widrows and piles and in-vessel systems (Hubbe et al. 2010). Furrow system and a windrow pile system are also used for OPEFB and POME sludge composting processes (Baharuddin et al. 2009, 2010). Conversely, for these systems, extra time for composting was needed coupled with some problems faced in controlling main traits, like water content, thermal, and oxygen level (Xiao et al. 2009). On the other hand an in-vessel composting system is comparatively better than other systems since it needed shorter area and produce enhanced agitation control, ventilation, and incorporation of the compost material (Kim et al. 2007). Whereas, Singh et al. (2009) defined that in support of in-vessel composting systems, O_2 phases are implicated; a high-rate phase in the vessel composter and a curing phase in exterior compost.

11.3.2 Aerated Static Pile

Primary composting made by biodegradation organic material system without any physical manipulation refers to as aerated static pile (ASP). Whereas, blended mixture which provide controlled aeration is usually placed on perforated piping and it may be found in windrows or in closed, open or covered containers. Although the technique may range from very small to very larger, capital intensive industrial installation but aerated system is generally used through professionally handled composting facilities.

The ASP proposed for speedy biodegradation and having good facilities for wet material and larger quantity of feed stocks. The ASP has capability to work under roof or open-air windrow composting operations and some time also refer to tunnel composting or totally enclosed in vessel composting.

11.3.2.1 Aeration

In order to accelerate decomposition process in compost heap, as well as to resolve various compost problems, it is imperative to turn and aerate the compost on regular basis. The pile requires a cyclic influx of O_2. The anaerobic state means slow-moving decomposition, lesser temperatures and potential odor problem with the pile.

11.3.2.2 Aeration System for a Closed Chamber Composting Facility

The aeration scheme utilizes fans to advance or drag air over the composting pile. Inflexible or flexile punched piping, linked with fans, bring the air. The pipes can be mounted in channels, on apex of a level, or integrated right through the mass through buildup. In large-scale systems, compulsory ventilation go together with a mechanized supervision accountable for controlling the pace and timetable of air release to the composting heap, while meters and physical monitoring practice may also be used in smaller scale process.

This composting method is beneficial for maintaining the proper moisturizing and oxygen level for microbial population and help to reduce the pathogens. Further this method also used for bio-filters to treat process air and tone down odors prior to venting (Hickman and Lanier 1999) (Fig. 11.4).

11.3.2.3 Windrow Composting

Compost production through piling wastes of animal and plant remains in lengthy rows is known as windrow composting in agriculture. These rows are usually turned to get better porosity and oxygen content, along with redistribute cooler and hotter

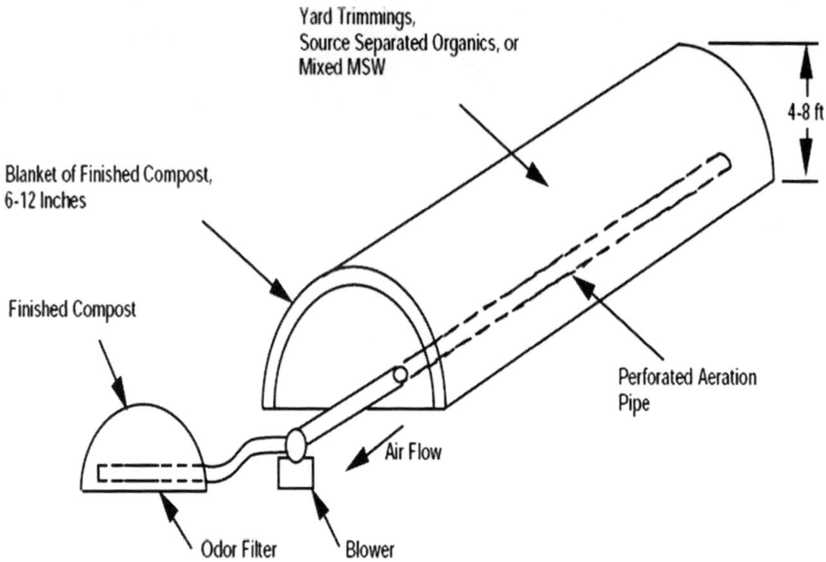

Fig. 11.4 Schematic of a static-pile forced air composting process. (Source: Hickman and Lanier 1999)

portion of the pile. Windrow composting is well known for the frame scale composting method (Natural Resource, Agriculture, and Engineering Service (NRAES) 1992). Control parameters for this composting practice include the basic ratios of C and N rich materials, the bulk agent quantity in order to guarantee air porosity, size of the pile, water level, and finally turning rate. Furthermore, the windrow temperature should be assessed regularly to observe the optimal point to turn the windrows for faster production of compost.

11.3.2.4 Compost Windrow Turners

Compost windrow turners were developed by Fletcher Sims Jr. of Canyon, Texas for large scale production of compost. They are a conventionally a big machine that straddles a windrow of 121.92 cm or more high and 365.76 cm across. Steel drum with paddles quickly turns and O_2 is introduced within compost by means of drum/paddle assembly, and waste gases formed through microbial decomposition are vented. The O_2 feeds the aerobic bacteria and hence speedup the composting mechanism.

11.3.2.5 In-vessel Composting

It is a cluster of methods which incarcerate the composting supplies inside building, container, or bucket (Natural Resource, Agriculture, and Engineering Service

Table 11.5 Comparison of composting methods. (Source: Parsons (2002))

Aerated static pile	Windrow	In-vessel
Highly affected by weather (can be lessened by covering, but at increased cost)	Highly affected by weather (can be lessened by covering, but at increased cost)	Only slightly affected by weather
Extensive operating history both small and large scale	Proven technology on small scale	Relatively short operating history compared to other methods
Large volume of bulking agent required, leading to large volume of material to handle at each stage (including final distribution)	Large volume of bulking agent required, leading to large volume of material to handle at each stage (including final distribution)	High biosolids to bulking agent ratio so less volume of material to handle at each stage
Adaptable to changes in biosolids and bulking agent characteristics	Adaptable to changes in biosolids and bulking agent characteristics	Sensitive to changes in characteristics of biosolids and bulking agents
Wide-ranging capital cost	Low capital costs	High capital costs
Moderate labor requirements	Labor intensive	Not labor intensive
Large land area required	Large land area required	Small land area adequate
Large volumes of air to be treated for odor control	High potential for odor generation during turning; difficult to capture/contain air for treatment	Small volume of process air that is more easily captured for treatment
Moderately dependent on mechanical equipment	Minimally dependent on mechanical equipment	Highly dependent on mechanical equipment
Moderate energy requirement	Low energy requirements	Moderate energy requirement

(NRAES) 1992). This system can control air flow and temperature by using metal or plastic tanks or concrete bunkers in which, with the principles of a "bioreactor" usually, the air circulation is metered via buried tubes that permit fresh air to be injected below pressure, with the exhaust being extracted through a bio-filter, with temperature and moisture conditions monitored using probes in the mass to allow upholding of most favorable aerobic decomposition situation (Table 11.5).

11.3.3 Co-composting of Oil Palm Waste

Co-composting is one of the important bio-waste treatments in the palm oil zero waste. However, improper conditions of composting may cause several problems, such as gas emission, bad odor, low quality product, production delay and high handling cost. Enhancing the efficiency of waste composting becomes a vital issue to overcome these problems (Yeoh et al. 2011). Numerous amendments have been formed for improvement in the composting mechanism, enhancing rate of degradation and compost quality to obtain about 30 C/N ratio (Costa et al. 1992), this is commonly known as co-composting.

Co-composting process has been performed using partially treated POME with EFB (Baharuddin et al. 2009). Produced compost had sufficient contents of some

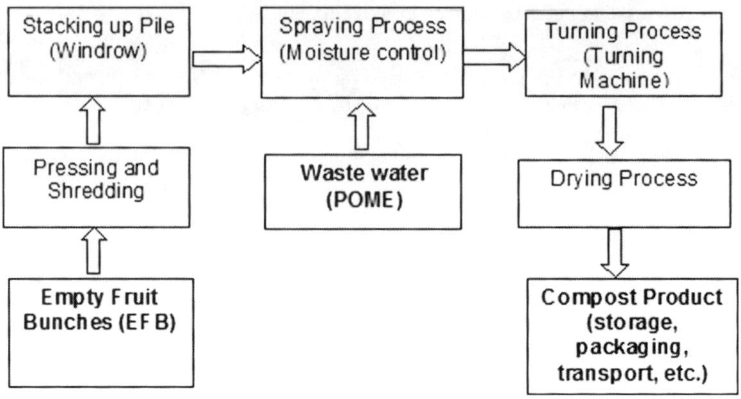

Fig. 11.5 Co-composting plant process flow. (Source: CDM-PDD 2006)

macro, secondary and micronutrients, thus it could be appropriate to be treated as fertilizer for oil palm plantation. Blending saw dusts obtained from furniture sector with POMS improve efficiency of composting and can mitigate the air pollution process (Bhamidimarri and Pandey 1996; Yaser et al. 2007). Anaerobic sludge, waste of kitchen and saw dust were collected to facilitate the composting process and finally it appeared like fertilizer value but still improvements are required to make it an ultimate substrate (Yaser et al. 2007). To overcome unexpected results of the EFB composting process, researchers have studied various methods to improve it (Thambirajah et al. 1995; Salates et al. 2004; Yaser et al. 2007; Baharuddin et al. 2009; Sabrina et al. 2009).

As a result, a pilot scale study was built to investigate the quality of co-composting of EFB of oil palm with moderately treated POME (Baharuddin et al. 2009). Vermicomposting of EFB was evaluated by Sabrina et al. (2009). Static aerated reactor was used by Yaser et al. (2007) to compost palm oil mill sludge with sawdust. In some cases, addition of ripe compost to the mass of EFB may provide a significant effect on the degradation speed (Salates et al. 2004). These researches, however, did not show that composting of EFB could be shortening to below 60 days. Further research was conducted by trying different methods or treatments to achieve a shorter composting period.

In other type of waste composting, usage of additives has been a popular method to achieve fast composting. The compost or bio-fertilizer could be produced with the inoculation of appropriate functional microbes which increase the decomposition rate, shorten the maturity period and improve the compost (or bio-fertilizer) quality (Wei et al. 2007). The degradation rate of municipal solid waste was highest at 15.671 g kg^{-1} h after seeding with a mixture inoculum of A (*Bacillus* sp.) and B (cellulolytic strains, i.e., *Trichoderma, Streptomyces* and white-rot fungi) (Xi et al. 2005). The efficiency of microbe inoculums on EFB composting therefore is a worthy subject to be explored (Fig. 11.5 and Table 11.6).

Table 11.6 Oil palm composting pattern. (Source: KYOTO Energy 2010)

Composting period	Inputs and actions
Day 1–4	EFB and POME inputs, turnings
Week 1–10	POME inputs only, turnings
Week 11–20	Curing period: no inputs, turnings

11.3.3.1 Vermicomposting

Oil palm waste disposal is a severe problem in many countries in the world. Vermicomposting could be a substitute technology for the management of oil palm waste especially EFB and frond (Sabrina et al. 2013). Excellent fertilizer made by means of earthworms by natural conversion is known as vermicomposting and in this process organic material is converted into humus-like substance called earthworm compost.

Earthworm is also identified as soil echo-system engineer (Stork and Eggleton 1992; Jones et al. 1994; Lavelle 1997) or "earth workers" (Appelhof et al. 1996), because of their capability to improve soil fertility through their active role in soil organic transformations and nutrient dynamics at various spatial and temporal scales, which ultimately cause for nutrient uptake by plants (Lavelle 1997).

Soil dwelling worms known as earth worker include *Aporrectodea calliginosa, A. trapezoids*, *A. rosea*, *A. longa*, *Microscolex dubius*, *Octolasion cyaneum*, and *Lumbricus terrestris* (Appelhof et al. 1996). Some other species of worms called as "composter".

Earthworms are insatiable feeders on organic wastes and even as make use of merely little portion for their body synthesis they send out a large part of these obsessive waste materials in a partially digested form. Those worms live in high concentrations of organic matter, such as piles and litter. The species used for vermicomposting in Australia and New Zealand (and elsewhere) includes *Eisenia fetida* (it is called "tiger worm" because of its stripes, but also known as the brandling, or manure worm); *E. andrei* ("red tiger", a close and non- striped relative of *E. fetida*); *Perionyx excavatus* (the Indian blue worm, a tropical species); and *Eudrillus eugeniae* (African night crawler, another tropical species). *Lumbricus rubellus* (red worm or dung worm) seems to be one of the few worms that can be classified as both a soil dwelling worm and a composting worm (Appelhof et al. 1996). Because earthworm intestine harbor microbes of broad array, hormones and enzymes, these partially assimilated vermicasts crumble quickly and are changed into a shape of vermicompost in a little time (Edwards and Lofty 1977; Lavelle 1988).

Vermicomposting could bring physico-chemical and biological reaction that result in alteration in the organic matter. Vermicast produced will be highly disintegrated and permeable (Edwards 1988; Edwards and Bohlen 1996). Vermicomposting could be a substitute technology to manage oil palm waste especially EFB and frond. Of the treatments studied, P-enriched vermicompost provided best results for dry matter yield of setaria grass, nutrient uptake and P availability (Sabrina et al. 2013).

Some major nutrient like NPK etc converted into available form during vermicomposting process (Ndegwa and Thompson 2001) and through this practice plant nutrient availability has been enhanced as compare to simple composting (Nagavallemma et al. 2004). Other methods are expensive whereas, this method is feasible, and cost effective for the proficient management of the organic soil waste (Hand et al. 1988; Longsdon 1994). Composting harmful pollutants in the field could minimize their negative effects on the surrounding environment, including earthworm populations (Sabrina et al. 2012).

An experiment was conducted by Sabrina et al. (2009) to investigate the effect of vermicomposting of oil palm EFB in supplying of nutrients for crop. In this study EFB and OPF were used for vermicomposting. Cow dung was incorporated as an earthworm's food. The weight, production of cocoon and earthworms mortality was documented monthly. Results showed that maximum earthworm density illustrated a superior total nitrogen, potassium and calcium in vermicompost. Hence, vermicomposting enhanced the superiority of the compost supplies in response to nutrient content.

It is well recognized that the use of vermicompost is very beneficial for plant growth and helpful for increasing yield of many crops like Black gram (urad) and soyabean (Javed and Panwar 2013), setaria grass (Sabrina et al. 2013), lilies (Mirakalaei et al. 2013), marigold (Paul and Bhattacharya 2012), french bean (Singh et al. 2011), cucumber (Azarmi et al 2009) tomato (Lazcano et al. 2009), maize (Gutièrrez-Miceli et al. 2008), sorghum (Hameeda et al. 2007), and potato (Alam et al. 2007a).

11.4 Role of Micro-organism in Bio-compost

Composting is known to be an extremely intricate process which entails microbiological degradation, mass and energy transfer phenomena and coexistence of non-steady state situation. Among the many parameters developed to measure the efficiency of composting processes are based on the survival of key degradation agents, the microorganisms. Considering that composting is an aerobic progression, the oxygen amount which is required by microorganism to stabilize the organic wastes must be kept at its optimum level. Other factors for microorganism survival are sufficient nutrient, space, water and air. These basic requirements have inspired researchers to create measurable parameters, for instance, water content, oxygen content, C and N ratio, other nutrient element contents, particle size of substrate, size of windrow, temperature, pH, electrical conductivity, humification ratio, etc. (Yeoh et al. 2011).

11.4.1 Microbial Sources

Microorganisms are everywhere in our atmosphere. Most of these microorganisms grow through the composting practice and participate in organic matter breaking down. Composting microbes are found all over the natural atmosphere including compost feedstock soil, air and water. Both feedstock as well as compost is readily decomposed by microbes during compost processing (Haug 1993). Bacteria and fungi, especially mesophilic and thermophilic, are considered as main microbial components of a compost. Most of the microorganisms involved in composting process are aerobes. Wet atmosphere is needed by compost microorganisms as they survive in the wet films neighboring organic matter particle composting.

11.4.1.1 Bacteria

Bacteria are one of the most essential types of microbes for compost (Environmental Protection Agency (EPA) 1999). A gram of compost may contain billions of bacteria in it. They utilize a broad range of enzymes to decompose organic matter chemically. At initial stage of composting, mesophilic bacteria predominate. Thermophilic bacteria control after the compost temperature surpasses 40 °. Temperature of compost has important influence on bacterial diversification. It increases when temperature lies between 50–55 °, however, it rapidly declined at greater than 60 °, when merely thermophilic bacteria continue to exist. Bacteria are usually related to the utilization of effortlessly degraded organic matter (Natural Resource, Agriculture, and Engineering Service (NRAES) 1999). They are the prevailing inhabitants right through the complete composting practice, while fungi and actinomycetes normally grow during last stages (Natural Resource, Agriculture, and Engineering Service (NRAES) 1999).

Compost has rich as well as varied microbial communities. Composition starts similarly after thermophilic phase and shifts dynamically through time. The curing phase offers a substrate and climate conducive for microbial recolonization, which can be accomplished either by inoculating post-thermophilic compost or preparing a palatable substrate that provides a competitive advantage for colonization by bacteria and fungi that offer biological control, slow-release fertility, and/or promote plant growth (Neher et al. 2013). Previous investigations have discovered that key bacterial groups in the start of the composting practice are mesophilic bacteria, such as *Lactobacillus* sp. and *Acetobacter* sp. (Golueke et al. 1954). More than 1500 full-length 16S rRNA gene sequences were analyzed and of these, above 500 were there as singletons. The majority of the sequences assessed were analogous to the bacterial species reported earlier in composts, including bacteria from the phyla Actinobacteria, Bacteroidetes, Firmicutes, Proteobacteria and *Deinococcus-thermus* (Partanen et al. 2010). In another study, variety of microbial communities and cellulolytic enzymes activities were examined through the co-composting of EFB and moderately treated POME by Baharuddin et al. (2009). They reported that composting process was dominated by uncultured bacteria species. The dominant bacterial group changed from the phylum Proteobacteria in the thermophilic stage to the phylum Chloroflexi in the maturing stage. However, latest findings in which bacterial community structure and biochemical changes during the composting of lignocellulosic oil palm EFB and POME anaerobic sludge were studied by Zainudin et al. (2013), a severe change in the bacterial community structure and diversity throughout the composting process was clearly observed which shifted between the thermophilic and maturing stages. The 16S rRNA clones belonging to the genera *Bacillus, Exiguobacterium, Desemzia,* and *Planococcus* were the dominant groups throughout composting.

11.4.1.2 Actinobacteria

These microbes are alike to fungi in fact these are bacteria which forms filaments. The colonies of the filaments are spread all over the compost pile by these bacteria, whereas, these are related to the breakdown of the interactable compounds (Yusri et al. 1995). They perform key role in the organic material degradation, like lignin, proteins, cellulose and chitin. Few species could be seen during the phase of composting by thermophillic, where as some can appear at the time of cooling phase. In the formation of organic aggregates, filaments plays vital role almost at the final stage of compost. The *Streptomyces* sp., *Nocardioides* sp. and *Thermoactinomyces* sp. are largely participating in composting process (Hubbe et al. 2010).

11.4.1.3 Fungi

Fungi plays an important role in the compost formation because of their active contribution in the decomposition and breakdown of complex organic wastes. Most of the fungi found in large number during the thermophillic and mesophillic phases of composting (Environmental Protection Agency (EPA) 1999). The hyphae of fungi maintain compost physically in to small parts, this action of hyphae improve drainage and aeration of the compost (Nancy and Elaina 1996). A gram of soil may contain 0.01–1 million of propagules of fungi. Approximately 0.7 million different species of the fungi has been described all over the world.

Oil palm waste particularly EFB is resistant to microbial degradation due to high lignin content. There were 24 fungal types isolated from the EFB compost, with 70 % of the isolates identified as *Aspergillus* sp. (Peh et al. 2011). Isolates 1c3 and 7c10 (*Diaporthe* sp.), and 5c9 (*Emericella dentata*) confirmed burly lignin-degrading activity. At mesophillic composting stage *Aspergillus fumigates* fungi are actively involved (Silva et al. 2009) whereas, at the final stage of composting process *Aspergillus fumigates*, *Emericella* sp. *Aspergillus ochraceus*, *Aspergillus terreus* and *Penicillium oxalicum* fungi are active (Dias et al. 2009).

11.5 Usage of Bio-compost

Previously before 1980s few countries had maintained the soil fertility through the incorporation of various types of manures viz. farmyard manure, green manure, compost etc (Yang and Hansen 1997). Soil quality has been affected due to imbalanced use of the synthetic fertilizer and intensive cropping system. This may cause low crop production and may affect of the physical properties of the soil and ultimately pose environmental threat (Reynolds et al. 2002). As an organic source compost can be used instead of peat in horticulture and agriculture which can maintain the moisture level and prevent the crop from disease and provide nutrients to crop plant (Liu 2000).

Recently researchers are taking interest for oil palm EFB. Many companies working in cultivation of oil palm are developing the methods to return EFB to field due to its positive effects of mulching EFB on the oil palm production (Loong et al. 1987; Gurmit et al. 1989; Lim and Chan 1989). There are different uses of compost in agriculture, such as a source of fertilizer, soil ameliorant, top dressing for hay crops and pastures and poting media (Oviasogie et al. 2010). Soil water and nutrient retention capacity could also be improved by composting as it helps to make soil porous to provide favorable soil environment for root penetration also extend the organic matter and lower the bulk density of the soil (Anon 2002).

11.5.1 Main Uses of Bio-compost

Composting has an essential role to play in closing the organic loop, returning organic matter and precious essential nutrients to soil. There is a growing interest to use composts within the agricultural and horticultural sector at present. Compost in together the short- and long-term offer vast profit to soil as it is a contributor of plant nutrients as well as creating a superior plant growing atmosphere within an integrated soil fertility system.

11.5.1.1 Microorganisms

Upon addition of compost in to the soil, the value of micro organisms could be seen in soil. During the composting process the harmful pathogens of human and plants are destroyed. Different functions of the remaining beneficial microorganisms are observed in plant and soil health (Barker 1997).

11.5.1.2 Nutrient Conversion

The conversion of the organic nitrogen in to inorganic form is done by microbes which are mostly found in soil and compost, and this process is known as mineralization. Mineralization process ensures the availability of the nutrients to plants. Population of the beneficial microorganisms can be affected by the use of toxic pesticides like methyl bromide to the soil/plants. Nutrient cycling microorganisms can re-inoculate these soils with the help of compost. However, poorly treated compost may reverse the mineralization process in soil (Barker 1997).

11.5.1.3 Pollutant Degradation

Organic pollutants in contaminated water and soil could be reduced by fully established compost. Many types of contaminant, such as chlorinated and non-chlorinated hydrocarbons, solvents, pesticides and petroleum products could also be degraded by the compost. Microorganisms present in the compost helps to break the contaminants in to substances which pose less harmful environmental effects.

11.5.1.4 Compost as a Source of Organic Matter

There are various sources of soil organic matter, such as plant/crop residues, cover crops and compost. Huge availability of organic matter present in the compost performs valuable functions in the soil (Francis and Xiying 2007).

11.5.1.5 Compost as Microbial Food

Fungi and bacteria use organic matter as source of food and release nutrients form soil. Hence, compost supplies a source of together microorganisms and their energy as well as home for microorganisms.

11.5.1.6 Nutrients and Water Retention

Compost is not only a nutrient source but also, its organic material can grasp onto various essential elements via its cation exchange capacity. There is negative charge on molecules of compost, they magnetize and hold onto positively charged ions, like NH_4^+, K^+, Ca^{2+} and Mg^{2+}.

11.5.1.7 Improvement in the Physical Properties of Soils

Structure of soil improved by addition of compost as this practice reduces bulk density of the soil. Root penetration could be improved by the application of the compost in compressed soils. Bonding between the organic matter and water molecules increase the water holding capacity of the soil and ultimately soil structure as well as water movement and hence water requirement will be reduced. Compost application could safeguard the soil surface from wind and water erosion. It increases the rate of infiltration, lower the water runoff and increase the surface moisture. It supplies sufficient oxygen to roots, improves soil aeration and removes excess CO_2 from the root zone. Compost boosts the soil temperature directly through its dark color, which augment heat assimilation by the soil. It also helps modest soil temperature and avoids quick variations of soil temperature, thus, extend a good atmosphere for root expansion.

11.5.2 Benefits of Bio-compost

1. Rich and quality manure from crop and animal residues present at the farm. Beneficial microorganisms and nutrients exist in the manure.
2. Due to the continuous application of the bio-compost physical, chemical and biological properties of the soil can be improved.
3. Due to the improvement in soil fertility, quality products can be obtained from the crop.
4. Bio diversity of soil can be improved by enhancing soil organic matter content by the application of compost as mulching or as in potting media is also beneficial in many ways.

Compost contains almost all essential plant nutrients. Nutrient level could be tested in the compost and soil to realize what kind of other nutrients may be required for plants.

- Macro- and micro-nutrients are present in the compost while they may be absent in chemical fertilizers.
- Compost release nutrients gradually for months or years unlike chemical fertilizers.
- Soil applied with compost retains fertilizer. Reduce fertilizer run off and ultimately reduce waterways pollution.
- Compost neutralizes both the acid and alkaline soils, make nutrients availability to plants by maintaining the neutral pH.

Compost perform role as binding agent of soil particles by which soil structure is improved. These soils posses good pore space percentage which holds air, moisture and nutrients.

- Compost makes physical condition of sandy soil good enough to maintain water and nutrients.
- Tightly bound clay particles could be loosened by the application of compost which helps roots to penetrate and spread easily.
- Compost prevents soil from being eroded by improving soil structure and also prevents soil spattering on plants.
- Compost bound nutrients tightly to prevent form leaching making them more available to plants.
- Compost makes soil favorable for the growth of plants.

Compost is used as food for the soil organisms like bacteria, fungi, insects, worms and many others which support plant growth.

- Bacteria in compost convert organic material into inorganic form thus increase the concentration of available nutrients in soil. Few species of bacteria are able to convert atmospheric nitrogen to plant available form.
- Compost rich soils contain many beneficial insects, worms and other organisms which keep soil aerated.
- Compost may reduce the diseases and insect pest attack.
- Root system could be improved by compost application that decline runoff.

Table 11.7 Summary of the potential benefits of compost use-on-land in the short-,mid-, and long-term retrieved from the literature review (adapted from Martínez-Blanco et al. 2013)

Benefit	Indicator (unit)	Short-term (<1 year)		Mid-term (<10 years)		Long-term (<100 years)	
		Min.	Max.	Min.	Max.	Min.	Max.
Nutrient supply	N mineralized (% of N applied)	5	22	40	50	20	60
	P mineralized (% of P applied)	35	38	90	100	90	100
	K mineralized (% of K applied)	75	80	100	100		
Carbon sequestration	C sequestered in soil (% of C applied)	40	53		30	2	16
Weed, pest, and disease suppression	Weed suppression (–)	ns	ns	–	–	–	–
	Pest and disease suppression (–)	nad	nad	–	–	–	–
Crop yield	Crop yield gain[a] (% from mineral fertilizers)[b]	−138	0	−71	52	–	–
Soil erosion	Soil loss[a] (%)[b]	–	–	−5	−36	–	–
	Soil structural or aggregate stability[a] (%)	29	41	0	63	–	–
Soil moisture content	WHC[a] (%)	0	50	–	–	–	–
	PAW[a] (%)	0	34	–	–	–	–
Soil workability	Soil bulk density[a] (%)[b]	−2.5	−21	−0.7	−23		−20
Soil biological properties and biodiversity[c]	Microbial diversity[a] (%)[b]	–	–	–	–	−2	4
	Microbial biomass[a] (%)	22	116	10	242	3.2	100
	Microbial activity[a] (%)	0	344	–	264	0	43
Crop nutritional quality	Crop nutritional quality (–)	nad	nad	–	–	–	–

WHC water holding capacity, *PAW* plant available water, *ns* no significant differences, *nad* no average data because of complexity of available dataset, – no reported benefits

[a] Change in the indicator

[b] Negative value indicates a decrease in the indicator

[c] The ranges of benefit for three of the more used indicators are presented

- Use of chemical fertilizer could be reduced by compost application.
- Compost suppresses pesticides application as it got helpful microbes which may safeguard plants form pests and diseases (Table 11.7).

11.6 Compost Application

Soil health could be maintained by organic manure. Organic manures could not be equated with synthetic fertilizers. It provides all nutrients to soil but in small quantity. Five tons of enriched bio-compost is recommended per hectare land usually applied as basic dose before sowing of crop.

11.6.1 Combined Application of Bio-compost with Inorganic Fertilizers

The easiest way to differentiate between compost and fertilizer is that compost provides food to soil while fertilizer feeds the plants. Fertilizer applied to soil for increased nutrient supply, however, instead of feeding the soil food web, the constituents in the fertilizers are proposed to fill the requirement of rapid growing plants. Whereas, suggested dose of compost can be relatively common, fertilizer application rates depends on the plant need. Organic or synthetic fertilizers perform well for vegetables, but organic manures have proved to be friendlier to the soil food web. Chemical fertilizers may also feed composting, but their continuous application may turn soil chemical properties unbalanced and depress microorganisms.

The compound fertilizer prepared through compost is welcome extraordinarily at present since peasants know that excessive use of chemical fertilizers is unsuitable for the soil and the environment (Wei et al. 2007). Different agro-ecosystems and diversity of organic material could be used in the system, including shrubs, trees, and cover crops. Compost presents a challenge for extension and research activities in soil fertility management (Palm et al. 2001).

Organic fertilizers and compost can go collectively. The organic matter in compost sponges up the fertilizer elements until they are required by plants. Compost also supplies micronutrients that are needed in small quantity, such as boron. Fertilizer could be used exclusive of compost; however, this practice will not increase the fertility of soil. Continuous incorporation of compost makes soils generally crumbly and darker in color and often needed a smaller quantity of fertilizer as compared to soil that has not yet gained from normal helpings of compost.

Recently, the use of organic material as fertilizer has been receiving increasing attention in the oil palm industry. Current practice of the application of EFB either through direct application or in the form of compost provides additional plantation benefits. Effort to use POME as a direct fertilizer is also gaining attention as an organic fertilizer. This is also a practice that leading companies have been incorporating into plantation practices for many years. The POME can be applied as a liquid or when mixed with EFB for composting. Organic material functions thus not only as a source of nutrients, but also as material to improve soil quality, especially in regards to physical nature (texture) and biological fertility.

The EFBs are an organic substance containing nutrients of great potential to be used as fertilizer for a plantation. Each ton of EFB contains nutrients equivalent to 3 kg urea, 0.6 kg PR, 12 kg MOP and 2 kg Kieserite (Loong et al. 1987). However, the application of EFB or EPB compost is not recommended to be a complete substitute for total inorganic fertilizer needs. In general, EFB application of up to as much as 40 t EPB per hectare still needs to be combined with the application of inorganic fertilizers by as much as 60 % of the recommended normal dose (Rahutomo et al. 2007). In many cases, the application of EFB compost is preferred to save transportation costs. Hence, 40 t of EFB is equivalent to 10 t of EFB compost. The benefits of applying organic fertilizers in oil palm are well understood and the materials are largely available in the plantations, as trunks, fronds, EFB and POME. Smallholder farmers, however, have little or no access to EFB and POME as these are the waste from palm oil mill extraction processes. In some cases the farmers may be able to purchase the EFB or EFB compost which is mixed with POME, but in general they are competing against the demand of the plantations owning the mills for organic fertilizers.

A study was conducted to evaluate combined dose of synthetic fertilizer, bio-fertilizer in combination with compost for the yellow sarson (*Brassica campestries* cv. B9) (Datta et al. 2009). The treatment used T_1 for 40 % less N fertilizer, 25 % less P fertilizer, and K fertilizer constant + 12 kg ha^{-1} bio-fertilizer (Azophos) and organic manure (compost) at 5 Mt ha^{-1}, showed the maximum chlorophyll accumulation fresh weight, highest seed, test weight of seeds and highest seed yield. A comparison between all the morphological, anatomical, physiological and biochemical parameters due to application of chemical fertilizer, bio-fertilizer and compost alone and in combination and their impact on soil microorganism, flora and fauna will throw sound environmental information. On the other hand AdeOluwa and Adeoye (2008) assessed the effect of various forms of EFB compost on oil palm nursery. Compost was applied at 4.8 g N $plant^{-1}$ in the soil. Results revealed that oil palm EFB composting application blending with cow dung increased the oil palm seedlings performance. Therefore, oil palm seedlings could be treated with EFB compost as an environmental friendly substitute to mineral fertilizer.

Application of mineral fertilizers is the most common means of improving soil fertility among farmers. However, the positive effects of mineral fertilizers on soil for crop production last only for a short time. In the long run, mineral forms of N fertilizers (urea and ammonium sulphate) can lead to decreasing base saturation, acidification, and a drop in soil pH (Phicot et al. 1981). A cationic imbalance in the soil of the Okomu oil palm plantation, Benin, Nigeria has been linked this to the problem of intensive application of mineral fertilizer (Ogendengbe 1991). This situation has triggered the problem of under fertilization on many farms. As a result, crop performance has been reduced. Organic fertilizers have the potential to correct almost all negative impacts of mineral fertilizers on soil. Efforts targeting increases in agricultural production should be backed up with environmentally friendly fertilizer application practices that should guarantee safety and sustainability of the soil natural resources.

Oil palm is a crop of national economic importance. Mineral fertilizers (urea and N-P-K-Mg 12-12-17-2) are the conventional fertilizers in raising oil palm. However, oil palm EFB and cow dung (from oil palm/livestock integration), usually available year-round, seem to be underutilized. These materials, if composted and used as organic fertilizer in oil palm production, could increase yield and also eliminate problems associated with intensive mineral fertilizer application. Thus, this investigation focused on determining effective combinations of oil palm EFB and cow dung in composting for raising oil palm in the nursery.

11.6.2 Bio-compost Handling and Storage

The compost has various physico-chemical properties that permit it to be applied in diverse ways. Final product must be tested before they are handled for utilization to make sure (i) worker protection, (ii) prevention of ecological deprivation, (iii) maintenance of the composting practice, and (iv) verification of produce attributes. Product attributes are associated to safety obligation and to the marketing and utilization of the compost. Key utilizations of compost are to apply it to the soil or to use it as mulch.

The physico-chemical properties of compost rely on the kind/type of material originally utilized (Bary et al. 2004) generally termed as feedstock. Any negligence in proper handling of compost, its constituents can be injurious to the soil environment and human health. As a result, because properties of compost can differ greatly, tests have been developed to quantify different essential parameters of the compost. Thus, the key intention of testing compost is to find out the concentrations of components and properties of the compost to assess its quality.

Being a living thing, compost is crammed with organisms that need aeration, wetness and food. Knowledge about compost storage is imperative and its nutritive value could be enhanced by storing it on the ground. It may possibly be pile up in a compost bin. It is crucial to regulate the moisture levels throughout storage period of compost, since it may turn into moldy upon soggy; however, it should not dehydrated entirely. A single easiest technique of compost storage is on the floor covered with a tarp or plastic sheeting. This will avoid excess wetness due to rainfall or snow runoff and also allow a bit of dampness to seep in and keep the pile moist. In this way the worms can get into the pile and put down their rich castings behind.

If space is an issue in compost storage; compost bin could be used which will keep the compost evenly humid and turned, but many of user have a constant batch of compost going and the bin is needed for the next generation of rich soil amendment. Under such condition, compost could be stored in plastic bags. To get most excellent outcome, optimum moisture of the compost must be ensured.

Compost should preferably be applied at the earliest possible time because there are chances of loss of nutrients from compost if it is stored for longer period. However, it may be enriched if stored for longer period or may be mixed with an almost finished batch of compost. Such practice will increase number of organisms and could maintain the compost viable. Composting is an uninterrupted practice, thus sustaining a given maturity or quality involves biological activity and storage period to be reduced. Moisture level of the compost directly affect the biological activity, accordingly, permitting compost to dry out will allow storage amid nominal change to the quality. During compost storage, continuous whirling the compost exclusive of water addition is necessary until moisture levels come close to 30%, it then be pushed into big heaps and be left with no further turning (Paulin and O'Malley 2008).

11.7 Conclusion

Disposal of agro-industrial waste is a key concern in numerous industries around the world. The inappropriate discard of large magnitude of such waste poses environmental troubles. Because these wastes encompass an elevated content of organic matter and mineral elements, they can potentially be utilized to reinstate soil fertility. Oil palm is one of the world's fast mounting crops. Indonesia and Malaysia are the two biggest oil palm producing countries. Oil palm industries provide lignocellulosic biomass, such as oil palm trunks, fronds, EFB's, palm shells and POME. Composting is a controlled aerobic conversion of mixed organic materials into a form so as to be appropriate for incorporating to soil. At present, the exploitation of oil palm EFB, fronds, and trunks as compost is getting consideration by investigators. Use of composting as well as vermicomposting technology is an efficient waste management option. Compost encourage beneficial microorganisms availability in soil, increases nutrient conversion in soil, act a source of organic matter, helps in nutrients and water retention, improves physical properties of soils and is an efficient pollutant degrader. Through reducing the load on inorganic fertilizer, it will also heighten the economy. Combined application of bio-compost, bio-fertilizer and synthetic fertilizers are need of the day for obtaining excellent value produce. Knowledge about appropriate composting technology, handling and its storage is imperative for increasing its nutritive value.

References

Abbasi SA, Ramasamy EV (1999) Biotechnological methods of pollution control. Universities Press, Orient Longman, Hyderabad

Abdelhamid MT, Horiuchi T, Oba S (2004) Composting of rice straw with oilseed rape cake and poultry manure and its effects on faba bean *(Vicia faba* L.) growth and soil properties. Bioresour Technol 93:183–189

Abd-Rahman R, Kalil MS, Abu ZY (2003) Composting palm oil mill sludge-sawdust: effect of different amount of sawdust. Proceedings of national workshop in conjunction with ARRPET workshop on wastewater treatment and recycling 2: removal chloroorganics and heavy metals. ISBN:983-2982-04–9

AdeOluwa OO, Adeoye GO (2008) Potential of oil palm empty fruit bunch (EFB) as fertilizer in oil palm (*Elaeis guineensis* L. Jacq.) nurseries. Cultivating the future based on science: 2nd conference of the International Society of Organic Agriculture Research ISOFAR. Modena, Italy. http://orgprints.org/12808/

Aim-oeb J, Harnbunchong A, Insung O, Engkagul A (2008) Chemical composition and ruminal dry matter and organic matter degradability of oil palm fronds. In: Proceedings of the 46th Kasetsart University annual conference. Kasetsart, Kasetsart University

Alam MZ, Mahamat ME, Muhammad N (2005) Production of cellulase from oil palm biomass as substrate by solid state biodegradation. Am J Appl Sci 2:569–572

Alam MN, Jahan MS, Ali MK, Ashraf MA, Islam MK (2007a) Effect of vermicompost and chemical fertilizers on growth, yield and yield components of potato in barind soils of Bangladesh. J Appl Sci Res 3(12):1879–1888

Alam MZ, Kabbashi NA, Razak AA (2007b) Liquid state bioconversion of domestic waste water sludge for bioethanol production. In: Ibrahim F, Abu Osman NA, Usman J, Kadri NA (eds) Biomed 06 IFMBE Proc. vol 15, pp 479–482

Alam MZ, Jamal P, Nadzir MM (2008) Bioconversion of palm oil mill effluent for citric acid production: statistical optimization of fermentation media and time by central composite design. World J Microbiol Biotechnol 24:1177–1185

Alam MZ, Mamun AA, Qudsieh IY, Muyibi SA, Salleh HM, Omar NM (2009) Solid state biodegradation of oil palm empty fruit bunches for cellulase enzyme production using a rotary drum bioreactor. Biochem Eng J 46:61–64

Alburquerque JA, Gonzalvez J, Garcia D, Cegarra J (2006) Composting of a solid olive-mill by-product ("alperujo") and the potential of the resulting compost for cultivating pepper under commercial conditions. Waste Manage 26(6):620–626

Anon (2002) Malaysian palm oil board: review of the Malaysian oil palm industry 1999. Ministry of Primary Industries, Malaysia

Appelhof M, Webster K, Buckerfield J (1996) Vermicomposting in Australia and New Zealand. Biocycle 37(6):63–66

Arslan EI, Obek E, Kirbag S, Ipek U, Topal M (2008) Determination of the effect of compost on soil microorganisms. Int J Sci Technol 3(2):151–159

Azarmi R, Giglou MT, Taleshmikail RD (2009) Influence of vermicompost on soil chemical and physical properties in tomato (*Lycopersicum esculentum*) field. Afr J Biotechnol 7(14):2397–2401

Baharuddin AS, Wakisaka M, Shirai Y, Abd AS, Abdul Rahman NA, Hassan MA (2009) Co composting of empty fruit bunches and partially treated palm oil mill effluents in pilot scale. Int J Agric Res 4:69–78

Baharuddin AS, Lim SH, Md Yousof MZ, Rahman NA, Umikalsom MS, Hassan MA, Wakisaka M, Sakai K, Shirai Y (2010) Effects of palm oil mill effluent (POME) anaerobic sludge from 500 m^3 of closed anaerobic methane digested tank on pressed-shredded empty fruit bunch (EFB) composting process. Afr J Biotechnol 9(16):2427–2436

Bari NM, Alam MZ, Muyibi SA, Jamal P, Mamun AA (2009) Improvement of production of citric acid from oil palm empty fruit bunches: optimization of media by statistical experimental designs. Bioresour Technol 100:3113–3120

Bari MN, Alam MZ, Muyibi SA, Jamal P, Mamun AA (2010) Effect of particle size on production of citric acid from oil palm empty fruit bunches as new substrate by wild *Aspergillus niger*. J Appl Polym Sci 10(21):2648–2652

Barker AV (1997) Composition and uses of compost. In: ACS Symposium Series. vol 668. American Chemical Society, Washington, pp 140–162

Bary AI, Cogger CG, Myhre EA (2004) Yard trimmings as a source of nitrogen for crop production. Compost Sci Util 12(1):11–17

Basiron Y, Weng CK (2004) The oil palm and its sustainability. J Oil Palm Res 16(1):1–10

Bernal MP, Alburquerque JA, Moral R (2009) Composting of animal manures and chemical criteria for compost maturity assessment: a review. Bioresour Technol 100(22):5444–5453

Bhamidimarri SMR, Pandey SP (1996) Aerobic thermophilic composting of piggery solid wastes. Water Sci Technol 33:89–94

Bharadwaj KKR (1995) Improvements in microbial compost technology: a special reference to microbiology of composting. In: Khanna S, Mohan K (eds) Wealth from waste. Tata Energy Research Institute, New Delhi, pp 115–135

Bustamante MA, Paredes C, Marhuenda-Egea FC, Perez-Espinosa A, Bernal MP, Moral R (2008) Co-composting of distillery wastes with animal manures: carbon and nitrogen transformations in the evaluation of compost stability. Chemosphere 72:551–557

Campitelli P, Ceppi S (2008) Chemical, physical and biological compost and vermicompost characterization: a chemometric study. Chemom Intell Lab Syst 90(1):64–71

Castaldi P, Alberti G, Merella R, Melis P (2005) Study of the organic matter evolution during municipal waste solid composting aimed at identifying suitable parameters for the evaluation of compost maturity. Waste Manage 25:209–213

CDM-PDD (2006) Clean development mechanism. Project design document form (CDM-PDD). Version 03. http://www.google.co.in/search?q=CDM-PDD+2006&oq=CDM-PDD+2006&aqs=chrome.69i57.2228j0j7&sourceid=chrome&espv=210&es_sm=93&ie=UTF-8#

Chan KW, Watson I, Lim KC (1980) Use of the oil palm waste material for increased production. In: Pushparajah E, Chin SL (eds) Proceedings of conference of soil science and agriculture development in Malaysia. Kuala Lumpur, Malaysia, pp 213–241

Chavalparit O, Rulkens WH, Mol APJ, Khaodhair S (2006) Option for environmental sustainability of the crude palm oil industry in Thailand through enhancement of industrial ecosystem. Environ Dev Sustain 8:271–287
Corley RHV (1983) Potential productivity of tropical perennial crops. Exp Agric 19(3):217–237
Corley RHV (2009) How much palm oil do we need? Environ Sci Policy 12(2):134–139
Corley RHV, Gray BS, Ng SK (1971) Productivity of the oil palm (*Eleais guineensis* Jacq.) in Malaysia. Exp Agric 7(2):129–136
Costa F, Garcia C, Hernandez T (1992) Residuos organicos urbanos: manejoy utilizacion. CSIC Centro de Edafología y Biología Aplicada del Segura, pp 181
Crowe M, Nolan K, Collins C, Carty G, Donlon B, Kristoffersen M, Brogger M, Carlsbaek M, Hummelshoj RM, Thomsen CL (2002) Part 3: technology and market issues. European Environment Agency (EEA), Copenhagen
Dahlan I (2000) Oil palm frond, a feed for herbivores. Asian-Aust J Anim Sci 13:300–303
Damanhuri H (1998a) Ikan Chaetodontid sebagai ikan penunjuk kepelbagaian terumbu karang perairan Pulau Bidong Terengganu. PhD thesis, Fakulti Sains dan Teknologi, Malaysia
Danmanhuri MA (1998b) Hands-on experience in the production of empty fruit bunches (EFB) compost. In: CETDEM Malaysian organic farming seminar, Petaling Jaya, Selangor, pp 50–61
Dashiny G (2009) Oil recovery from palm oil solid wastes. PhD thesis. University Malaysia, Pahang
Datta JK, Banerjee A, Saha MS, Gupta S, Mondal NK (2009) Impact of combined exposure of chemical fertilizer, bio-fertilizer and compost on growth, physiology and productivity of *Brassica campestris* in old alluvial soil. J Environ Biol 30(5):797–800
De Guardia A, Petiot C, Benoist JC, Druilhe C (2012) Characterization and modelling of the heat transfers in a pilot-scale reactor during composting under forced aeration. Waste Manage 32(6):1091–1105
Dias ES, Guimaraes SE, Siqueira FG, Silva R, Batista LR (2009) Allergenic and toxigenic fungi in the compost of cultivation of *Agaricus brasiliensis*. Sci Agrar 10:507–511
Edwards CA (1988) Breakdown of animal, vegetable and industrial organic wastes by earthworms. In: Edwards CA (ed) Earthworms in waste and environmental management. SPB Acad Pub, The Hague, pp 21–31
Edwards CA, Bohlen PJ (1996) Biology and ecology of earthworms. Chapman & Hall, London
Edwards CA, Lofty JR (1977) Biology of earthworms. 2nd edn. Chapman & Hall, London
Embrandiri A (2013) Application of palm oil mill wastes (decanter cake) as soil amendments: appraisal of potential effects on selected vegetables and soil (unpublished dissertation)
Embrandiri A, Mahamad HI, Rajeev PS (2013) Palm oil mill wastes utilization; sustainability in the Malaysian context. Int J Sci Res Pub 3(3):1–7
Environmental Protection Agency (EPA) (1999) Alternative technologies/uses for manure. Office of Wastewater Management, USA
Epstein E (1997) The science of composting. Technomic publishing, Lancaster
ESCAP (1997) Guidelines and manual on land-use planning and practices in watershed management and disaster reduction. Economic and social commission for Asia and the Pacific, United Nation, Bangkok
Evans G (2001) Biowaste and biological waste treatment. James & James, London
Francis JL, Xiying H (2007) A review of composting as a management alternative for beef cattle feedlot manure in southern Alberta, Canada. Bioresour Technol 98(17):3221–3227
Golueke CG, Card BJ, McGauhey PH (1954) A critical evaluation of inoculums in composting. Appl Microbiol 2(1):45
Gurmit S, Manoharan S, Kanapathy K (1981) Commercial scale bunch mulching of oil palm. In: Proceedings of International conference in the oil palm in agriculture in the eighties, Kuala Lumpur, pp 367–377
Gurmit S, Manoharan S, Toh TS (1989) United plantations approach to palm oil mill by-product management and utilization. In: PORIM international palm oil development conference, PORIM, Kuala Lumpur, pp 225–234

Gutiérrez LF, Sánchez OJ, Cardona CA (2009) Process integration possibilities for biodiesel production from palm oil using ethanol obtained from lignocellulosic residues of oil palm industry. Bioresour Technol 100(3):1227–1237

Gutièrrez-Miceli FA, Moguel-Zamudio B, Abud-Archila M, Gutiérrez-Oliva VF, Dendooven L (2008) Sheep manure vermicompost supplemented with a native diazotrophic bacteria and mycorrhizas for maize cultivation. Bioresour Technol 99:7020–7026

Habib MAB, Yusoff FM, Phang SM, Ang KJ, Mohamed S (1997) Nutritional values of chironomid larvae grown in palm oil mill effluent and algal culture. Aquaculture 158:95–105

Hameeda B, Harini G, Rupela OP, Reddy G (2007) Effect of composts or vermicomposts on sorghum growth and mycorrhizal colonization. Afr J Biotechnol 6(1):009–012

Hand P, Hayes WA, Frankland JC, Satchell JE (1988) The vermicomposting of cow slurry. In: Earthworms in waste and environmental management. SPB Academic Publishing, The Hague, pp 49–63

Haron K, Mohammed AT (2008) Efficient use of inorganic and organic fertilizers for oil palm and potential utilization of decanter cake and boiler ash for biofertilizer production. In: Proceedings of the 2008 national seminar on biofertilizer, biogas and effluent treatment in the oil palm industry: waste to wealth, Malaysia, pp 21–32

Hassan OA, Ishida M, Shukri IM, Tajuddin ZA (1996) Oil–palm fronds as roughage feed source for ruminants in Malaysia. In: Extension bulletin. Food and Fertilizer Technology Center for the Asian and Pacific Region, pp 1–8

Hassim HA, Lourenco M, Goel G, Vlaeminck B, Goh YM, Fievez V (2010) Effect of different inclusion levels of oil palm fronds on in vitro rumen fermentation pattern, fatty acid metabolism and apparent biohydrogenation of linoleic and linolenic acid. Anim Feed Sci Technol 162:155–158

Haug RT (1993) The practical handbook of compost engineering. Lewis Publishers, London

Hickman J, Lanier H (1999) The principles of integrated solid waste management. American Academy of Environmental Engineers, Annapolis, MD

Hubbe MA, Mousa MN, Carmen S (2010) Composting as a way to convert cellulosic biomass and organic waste into high-value soil amendments: a review. BioResource 5(4):2808–2854

Husin M, Hassan AH, Mohammed AT (1986) Availability and potential utilization of oil palm trunks and fronds up to the year 2000. PORIM occasional paper (20) p 17

Ipek U, Obek E, Akca L, Arslan EI, Hasar H, Dogru M, Baykara O (2002) Determination of degradation on radioactivity and its kinetics in aerobic composting. Bioresour Technol 84:283–286

Ishida M, Abu Hasan O (1997) Utilization of oil palm frond as cattle feed. Jpn Agric Res Q 31:41–47

Islam M, Dahlan I, Rajion MA, Jelan ZA (2000) Productivity and nutritive values of different fractions of oil palm (*Elaeis guineensis*) frond. Asian-Aust J Anim Sci 13(8):1113–1120

Javed S, Panwar A (2013) Effect of biofertilizer, vermicompost and chemical fertilizer on different biochemical parameters of *Glycine max* and *Vigna mungo*. Rec Res Sci Technol 5(1):40–44

Jones CG, Lawton JH, Sachak M (1994) Organisms as ecosystem engineering. Oikos 69:373–386

Kamarudin H, Mohamad H, Arifin D, Johari S (1997) An estimated availability of oil palm biomass in Malaysia, In: Occasional paper (37). PORIM, Malaysia

Katamanee A (2006) Appropriate technology evaluation for oil palm product utilization in Krabi province. PhD thesis, Mahidol University. www.li.mahidol.ac.th/thesis/2549/cd388/4637145.pdf

Kavitha B, Rajannan G, Jothimani P (2013) Compost maturity test for empty fruit bunch of palm oil industry solid waste. Sch Acad J Biosci 1(3):98–101

Kayikcioglu, HH (2013) Effects of composts from agro industrial wastes on microbial activity of a typic xerofluvent soil under Mediterranean conditions, SE Turkey. Geomicrobiol J 30(9):228–236

Khamseekhiew B, Liang JB, Wong CC, Jalan ZA (2001) Ruminal and intestinal digestibility of some tropical legume forages. Asian-Aust J Anim Sci 14(3):321–325

Khan MAI, Ueno K, Horimoto S, Komai F, Tanaka K, Yoshitaka O (2009) Physicochemical, including spectroscopic and biological analyses during composting of green tea waste and rice bran. Biol Fertil Soil 45:305–313
Kim JD, Park JS, In BH, Kim D, Namkoong W (2007) Evaluation of pilot-scale in-vessel composting for food waste treatment. J Hazard Mater 154:272–277
Kolade O, Coker AO, Sridhar MKC, Adeoye GO (2005) Palm kernel waste management through composting and crop production. J Environ Health Res 5(2):81–85
Kolade OO, Coker AO, Sridhar MKC, Adeoye GO (2006) Palm kernel waste management through composting and crop production. J Environ Health Res 5(2):81
KYOTO Energy (2010) CDM project 2527: co-composting of EFB and POME project. Project monitoring, July 2009 to July 2010
Lavelle P (1988) Earthworms and the soil system. Biol Fertil Soil 6:237–251
Lavelle P (1997) Faunal activities and soil processes: adaptive strategies that determine ecosystem function. Adv Ecol Res 24:93–132
Lazcano C, Arnold J, Tato AJ, Zaller JG, Domínguez J (2009) Compost and vermicompost as nursery pot components: effects on tomato plant growth and morphology. Span J Agric Res 7(4):944–951
Lim KH (1989) Trials on composting EFB of oil palm with and without prior shredding and liquid extraction. PORIM oil palm conference, Malaysia
Lim BA (2000) Feature on palm oil fuel ash. In: The New Straits Times. Thursday, 28 Dec 2000. Malaysia, pp 3–4
Lim KC, Chan KW (1989) Towards optimizing empty fruit bunch application in oil palm. PORIM Int Palm Oil Cong-Agric PORIM Bangi, pp 235–242
Lin KO, Ratnalingam R (1980) Fuel value of some Malaysia vegetation. Planter (Kuala Lumpur) 56:40–48
Liu J (2000) Composting and use of compost as a soil amendment. PhD thesis, Natural Resource and Environment Management Department, University of Hawaii, p 138
Longsdon G (1994) Worldwide progress in vermicomposting. Biocycle 35:63–65
Loong S, Mohd GN, Letchumanan A (1987) Optimising the use of EFB milch on oil palm on two different soils, Proc 1987 int oil palm/palm oil conf, Kuala Lumpur, Malaysia
Ma AN (1999) Treatment of palm oil mill effluent. In: Singh G, Lim KH, Leng T, Kow DL (eds) Oil palm and the environment: a Malaysian perspective. Malaysian Oil Palm Growers Council, Kuala Lumpur, pp 113–126
Ma AN (2000) Environmental management for the palm oil industry. Palm Oil Dev 30:1–10
Ma AN, Cheah SA, Chow MC (1993) Current status of palm oil processing waste management. In: Yeoh BG, Chee KS, Phang SM, Isa Z, Idris A, Mohamed M (eds) Waste management in Malaysia: current status and prospects for bioremediation. pp 111–136
Malaysian Palm Oil Board (MPOB) (2009) Over view of the Malaysian palm oil industry 2009. http://econ.mpob.gov.my/economy/Overview_2009.pdf
Malaysian Palm Oil Board (MPOB) (2012) Overview of the Malaysian oil palm industry (2011). http://bepi.mpob.gov.my/images/overview/Overview_of_Industry_2011.pdf
Malaysian Palm Oil Board (MPOB) (2013) Overview of the Malaysian oil palm industry 2012. Economics and industry development division. Malaysian Palm Oil Board. http://bepi.mpob.gov.my/index.php/publications.html/overview_of_Industry_2012.pdf
Marche T, Schnitzer M, Dinel H, Pare T, Champagne P, Schulten HR, Facey G (2003) Chemical changes during composting of a paper mill sludge-hardwood sawdust mixture. Geoderma 116(3–4):345–356
Martínez-Blanco J, Lazcano C, Boldrin A, Muñoz P, Rieradevall J, Møller J, Anton A, Christensen TH (2013) Assessing the environmental benefits of compost use-on-land through an LCA perspective. In: Lichtfouse E (ed) Sustainable agriculture reviews. Springer, Netherlands, pp 255–318
Mirakalaei SMM, Ardebill ZO, Mostafavi M (2013) The effects of different organic fertilizers on the growth of lilies (*Lillium longiflorum*). Int Res J App Basic Sci 4(1):181–186

Misson M, Haron R, Kamaroddin MFA, Amin NAS (2009) Pre-treatment of empty palm fruit bunch for production of chemicals via catalytic pyrolysis. Bioresour Technol 100:2867–2873
Molla AH, Fakhrul-Razi A, Alam MZ (2004) Evaluation of solid-state bioconversion of domestic waste water sludge as promising environmental friendly technique. Water Res 38:4143–4152
Nagavallemma KP, Wani SP, Lacroix S, Padmaja VV, Vineela C, Rao MB, Sahrawat KL (2004) Vermicomposting: recycling wastes into valuable organic fertilizer. Global theme on agroecosystems report no. 8. http://oar.icrisat.org/3677/1/172-2004.pdf
Nancy T, Elaina O (1996) Compost microorganisms. Cornell University Composting Page. 1996. http://www.cfe.cornell.edu/compost/microorg.html
Natural Resource, Agriculture, Engineering Service (NRAES) (1992) On-farm composting. In: Rynk R (ed) NRAES Cooperative Extension, Ithaca
Natural Resource, Agriculture, Engineering Service (NRAES) (1999) Introduction: chapters 1–4 and 6. In: A field guide to on-farm compost. Natural Resource, Agriculture, and Engineering Service (NRAES), Publication no. NRAES–114
Ndegwa PM, Thompson SA (2001) Integrating composting and vermicomposting in the treatment and bioconversion of biosoilds. Bioresour Technol 76(2):107–112
Neher DA, Weicht TR, Bates ST, Leff JW, Fierer N (2013) Changes in bacterial and fungal communities across compost recipes, preparation methods, and composting times. PLoS ONE 8(11): e79512. doi:10.1371/journal.pone.0079512
Ogendengbe HD (1991) The nutrient status of Okomu oil palm plantation. MSc thesis, University of Ibadan, Nigeria
Ordonez C, Tejada M, Benitez C, Gonzalez JL (2006) Characterization of a phosphorus–potassium solution obtained during a protein concentrate process from sunflower flour application on ryegrass. Bioresour Technol 97:522–528
Oviasogie PO, Aisueni NO, Brown GE (2010) Oil palm composted biomass: a review of the preparation, utilization, handling and storage. Afr J Agric Res 5(13):1553–1571
Palm CA, Catherine NG, Robert JD, Georg C, Ken NG (2001) Organic inputs for soil fertility management in tropical agroecosystems: application of an organic resource database. Agric Ecosys Environ 83:27–42
Pare T, Dinel H, Schnitzer M (1999) Extractability of trace metals during co-composting of biosolids and municipal solid wastes. Biol Fertil Soils 29:31–37
Parr JF, Hornick SB (1992) Utilization of municipal wastes. In: Metting FB (ed) Soil microbial ecology: applications in agricultural and environmental management. Marcel Dekker, New York, pp 545–559
Parsons W (2002) Organic fruit and vegetable production: is it for you? VISTA on the agri-food industry and the farm community. Agriculture Division of Statistics Canada, Ottawa
Partanen P, Hultman J, Paulin L, Auvinen P, Romantschuk M (2010) Bacterial diversity at different stages of the composting process. BMC Microbiol 10(1):94
Paul S, Bhattacharya SS (2012) Vermicomposted water hyacinth enhances growth and yield of marigold by improving nutrient availability in soils of north bank plain of Assam. Res Rev: J Agric Sci Technol 2:36–46
Paulin B, O'Malley P (2008) Compost production and use in horticulture. Bulletin 4746, pp 16
Peh KL, Kadir J, Othman, R. Ting ASY (2011) Fungal species from oil palm empty fruit bunches (EFB) and their lignin-degrading potential. Proc of the International Congress of the Malaysian Society for Microbiology. 8-11 December 2011, Penang. Malaysia, pp 61–64
Phicot JMP, Sedego MP, Arrivers HF (1981) Evaluation de la fertilite dun sol ferrugineux tropical sous l influence de femures milnerales et oganiques. L' Agron Trop 36(2):122–133
Prasertsan S, Prasertsan P (1996) Biomass residues from palm oil mills in Thailand: an overview on quantity and potential usage. Biomass Bioenergy 11(5):387–395
Project design document form CDM-PDD (2006) CDM project Organic Waste Composting at Takon Palm Oil Mill, Malaysia, as described in the PDD version 02 dated 30th June 2008—version 03.1

Rahman SHA, Choudhory JP, Ahmad AL, Kamaruddin AH (2007) Optimization studies on acid hydrolysis of oil palm empty fruit bunch fiber for production of xylose. Bioresour Technol 98:554–559

Rahutomo S, Sutarta ES, Siregar HH, Darlan NH, Winarna (2007) Kemajuan budidaya kelapa sawit di Indonesia. In: Latif S, Purba H (eds) 90 Tahun Penelitian Kelapa Sawit Indonesia. Pusat Penelitian Kelapa Sawit, Medan

Ratnalingam R (1980) Fuel value of some Malaysia vegetation. Planter (Kuala Lumpur) 56:40–48

Reynolds WD, Bowman BT, Drury CF, Tan CS, Lu X (2002) Indicators of good soil physical quality: density and storage parameters. Geoderma 110:131–146

Rosnah MS, Astimar AA, Hasamudin W, Hassan W, Gapor MT (2009) Solid-state characteristics of microcrystalline cellulose from oil palm empty fruit bunch fibre. J Oil Palm Res 21:613–620

Rupani PF, Singh RP, Ibrahim H, Esa N (2010) Review of current palm oil mill effluent (POME) treatment methods: vermicomposting as a sustainable practice. World Appl Sci J 11(1):70–81

Sabrina DT, Gandahi AW, Hanafi MM, Mahmud TMM, Nor Azwady AA (2012) Oil palm empty-fruit bunch application effects on the earthworm population and phenol contents under field conditions. Afr J Biotech 11(19):4396–4406

Sabrina DT, Hanafi MM, Mahmud TMM, Abdul Aziz NA (2009) Vermicomposting of oil palm empty fruit bunch and its potential in supplying of nutrients for crop growth. Compost Sci Util 17:61–67

Sabrina DT, Hanafi MM, Gandahi AW, Mahmud TMM, Abdul Aziz NA (2013) Effects of earthworms, *Arbuscular mycorrhizae*, and phosphate rock on setaria grass (*Setaria splendida*) and phosphorus availability in soil. Aus J Crop Sci 7(13):2136–2144

Salates S, Siregar FA, Caliman JP, Liwang T (2004) Ligno-cellulose composting: case study on monitoring oil palm residuals. Compost Sci Util 12:372–382

Shafie SM, Mahlia TMI, Masjuki HH, Andriyana A (2011) Current energy usage and sustainable energy in Malaysia: a review. Renew Sustain Energy Rev 15(9):4370–4377

Shuit SH, Tan KT, Lee KT, Kamaruddin AH (2009) Oil palm biomass as a sustainable energy source: a Malaysian case study. Energy 34(9):1225–1235

Siddiqui Y, Meon S, Ismail R, Rahmani M, Asghar A (2008) Bio-efficiency of compost extracts on the wet rot incidence, morphological and physiological growth of okra (*Abelmoschus esculentus* (L.) Moench). Sci Horticult 117:9–14

Silva CF, Azevedo RS, Braga C, Silva R, Dias ES, Schwan RF (2009) Microbial diversity in a bagasse-bassed compost prepared for the production of *Agaricus brasiliensis*. Braz J Microbiol 40(3):590–600

Singh YK, Ajay S, Kalamdhad MA, Kazmi AA (2009) Maturation of primary stabilized compost from rotary drum composter. Resour Conserv Recycl 53(7):386–392

Singh RP, Hakimi Ibrahim M, Esa N (2010) Composting of waste from palm oil mill: a sustainable waste management practice. Rev Environ Sci Biotechnol 9:331–344

Singh RP, Embrandiri A, Ibrahim MH, Esa N (2011) Management of biomass residues generated from oil palm mill: vermicomposting a sustainable option. Resour Conserv Recycl 55:423–434

Souleymane BK, Antoine D, Russell JT, Hani A, Tyler JA (2010) Suppressive effect of non-aerated compost teas on foliar fungal pathogens of tomato. Biol Control 52:167–173

Sridhar MKC, AdeOluwa OO (2009) Palm oil industry residue. In: Nigam PS, Pandey A (eds) Biotechnology for agro-industrial residues utilization. Springer Science, Berlin, pp 341–355

Stork NE, Eggleton P (1992) Invertebrates as determinants and indicators of soil quality. Am J Alternat Agric 7:38–47

Sumathi S, Chai SP, Mohamed AR (2008) Utilization of oil palm as a source of renewable energy in Malaysia. Renew Sustain Energy Rev 12:2404–2421

Tabi MAN, Zakil AF, Fauzai MF, Ali N, Hassan O (2008) The usage of empty fruit bunch (EFB) and palm press fibre (PPF) as substrates for cultivation of *Pleurotus ostreatus*. J Teknol 49:189–196

Thambirajah JJ, Zulkifli MD, Hashim MA (1995) Microbiological and biochemical changes during the composting of oil palm empty fruit bunches: effect of nitrogen supplementation on the substrate. Bioresour Technol 52:133–144

Umar MS, Philip J, Tania U (2013) Strengthening the palm oil biomass renewable energy industry in Malaysia. Renew Energy 60:107–115

Umikalsom MS, Ariff AB, Zulkifli HS, Tong CC, Hassan MA, Karim MIA (1997) The treatment of oil palm empty fruit bunch fibre for subsequent use as substrate for cellulase production by *Chaetomium globosum kunze*. Bioresour Technol 62:1–9

Wahid MD, Abdullah SNA, Henson IE (2004) Oil palm achievements and potential: new directions for a diverse planet. In: Proc of the 4th international crop science congress. Brisbane, Australia

Wei Z, B Xi, Y Zhao, S Wang, H Liu, Y Jiang (2007) Effect of inoculating microbes in municipal solid waste composting on characteristics of humic acid. Chemosphere 68:368–374

Xi BD, Zhang GJ, Liu HL (2005) Process kinetics of inoculation composting of municipal solid waste. J Hazard Mater 124:165–172

Xiao Y, Zeng GM, Yang ZH, Shi WJ, Huang C, Fan CZ, Xu ZY (2009) Continuous thermophilic composting (CTC) for rapid biodegradation and maturation of organic municipal solid waste. Bioresour Technol 100:4807–4813

Yang HS, Hansen BH (1997) Analysis of impact of farming practices on dynamics of soil organic matter in northern China. Eur J Agron 7:211–219

Yaser AZ, Abd Rahman R, Kalil MS (2007) Co-composting of palm oil mill sludge-sawdust. Pak J Biol Sci 15:4473–4478

Yeoh CY, Chin NL, Tan CS, Ooi HS (2011) Acceleration effects of microbial inoculum on palm oil mill organic waste composting. Compost Sci Util 19:135–142

Yusri A, Rasol AM, Mohammed O, Azizah H, Kume T, Hashimoto S (1995) Biodegradation of oil palm empty fruit bunch into compost by composite micro-organisms. Proceedings of the EU-ASEAN conference combustion of solids and treated product, Hua-Hin, Thailand

Zainudin MHM, Hassan MA, Tokura M, Shirai Y (2013) Indigenous cellulolytic and hemicellulolytic bacteria enhanced rapid co-composting of lignocellulose oil palm empty fruit bunch with palm oil mill effluent anaerobic sludge. Bioresour Technol 147:632–635

Chapter 12
Decomposition of Organic Materials into High Value Compost for Sustainable Crop Productivity

Dinesh Kumar Maheshwari, Shrivardhan Dheeman and Mohit Agarwal

Abstract In developing countries, augmenting sustainable agriculture and up keeping of soil fertility with intensified agricultural practices which include use of organic and chemical fertilizer with their negative penalties, affect biodiversity of soil and contribute to agricultural soil quality decline. The application of compost aids agricultural sustainability by means of various processes like green manuring, leaf manuring, mulching and composting on attention of scientific community. Composting as microbial decomposition process bears community dynamics of various microbial genera hence, it is influenced by microbial population and balanced through their physiological activity. The understanding of the factors like temperature, pH, oxygen level etc. influencing decomposition of organic wastes and performance. These factors are the key issue over sustainable waste management practices on which the high value compost prepared by improved techniques and applied in soil recycling which was found productive for agricultural production.

Keywords Mulching · Recycling · Soil fertility · Organic manure

12.1 Introduction

Millions of farmers in developing countries need adequate infrastructural resources for enhancement of the agricultural crop productivity. The adequate agro-technology and proper management ensure to increase crop productivity and soil management practices in order to get sustainability and adaptable crop production for long duration. The practice of composting as one of the old established traditional methods for improving the fertility of soil/land owned by small farmers of the country. Although, crop rotation and mass introduction of microbial inoculants introduced into soil also yield better productivity (Bernard et al. 2012). The practices of

D. K. Maheshwari (✉) · S. Dheeman · M. Agarwal
Department of Botany and Microbiology, Gurukul Kangri University, Haridwar, India
e-mail: maheshwaridk@gmail.com

D. K. Maheshwari (ed.), *Composting for Sustainable Agriculture,* Sustainable Development and Biodiversity 3, DOI 10.1007/978-3-319-08004-8_12

composting converts all kind of organic waste into humus, provide an opportunity to the farmers to make better use of organic wastes and refuse which are already present either at the farm or neighborhood. The organic materials that are usually considered as waste are either be thrown away or burnt but now-a-days proved to be suitable material for the preparation of composting material for agricultural crops.

According to "agriculture education 1964" it is estimated that 15 million ton of compost can be obtained from forest litter annually, without adversely affecting the natural regeneration of the forests 15 million ton of forest litter manure present throughout the country contain 0.75, 0.73 and 0.75 million ton of N, P_2O_5 (phosphates) and K_2O (oxides of potassium) respectively. Forest litter can thus be collected and is used for composting to quite a reasonable extent. On the other hand, variety of aquatic plants is known for their utilization pattern as compost has not been considered yet. Among two aquatic weeds, *Ipomoea aquatica* and *Eichhornia crassipes* are the most common floating weeds growing luxuriantly in several parts of the world. The chemical compositions of these weeds reveal a significant amount of holocellulose in them which could be enzymatically converted into various important products including biofuels, single cell protein etc. by microbial enzymes (Kour et al. 1993). Hubbe et al. (2010) reviewed plant-derived cellulosic materials as organic wastes to produce beneficial amendment compost. With major consideration on composting, emphasizing sources of energy seems to be essential to drive the biological transformations. Lignin, humus and the recalcitrant organic matter of good water-holding, ion exchange, and bulking capabilities contribute soil health and productivity. Recycling and composting of organic wastes with objective to understand the dynamics of microbial biomass-organic carbon mineralization into CO_2 studied by Zhang et al. (2012) contributed a unique set of estimated parameters of different organic fractions, with the exception of some compost because of a poor simulation of the cellulosic and soluble pools.

The significance of organic manures as a source of plant nutrients to increase the fertility level of soil has been well established (Stockdale et al. 2002; Bationo et al. 2007). The organic matter content is low in the cultivated soils of tropics and subtropics regions. The low organic matter is mainly due to high temperature and intense microbial activity. Therefore, the soil humus has to be replenished through periodic addition of organic manures. Under intensive cultivation practices, soils are losing their organic matter much faster than that replenished by farm yard manures (FYM), composts or green manures. It is, therefore, application of agrochemicals to be the only way to increase the fertility of large acreage of Indian soils substantially and rapidly (Raychaudhary 1972). But, due to excessive application of agrochemicals that create microbial resistance, disturb soil ecology, responses to ground water population and impart chemical residue in plant parts besides high cost, their availability to the farmers at reasonable cost is fast declining. There is thus, an urgent need for utilizing organic manures or waste for supplementing or substituting chemical fertilizers in the present context. The adoption of agrochemical adopted inoculant strain has an enormous potential for the betterment of crop production and fertility restoration of soil (Kumar et al. 2009; Maheshwari 2013).

Organic materials are available as byproducts of farming and allied industries. Derived from the plant and animal origin they are considered as the source of immense practical value for crop productivity. Composting is the best method for converting these organic residues into the inputs for growing crops (Gaur 1984).Organic manures are bulky in nature that supply the plant nutrients in small quantities hence, termed as bulky manures, e.g., farmyard manures, rural and town composts, night soil, green manures etc. Whereas those containing higher percentage of major plant nutrients like nitrogen, phosphorus and potassium are termed as concentrated organic manures, e.g., oil cakes, blood and meat meals, fish meals, guano and poultry manures etc. It is recognized that the bulky organic manures plays a significant role in increasing the crop yield via supply of the both macro as well as micro-nutrients to the plants and in improving the physical, chemical and biological properties of the soil which greatly constitutes to its fertility and water holding capacity. It ameliorates the soil structure i.e., sandy soils become compact while clayey soil becomes more open. The crumb structure of soil is due to humus.

The process and application of organic manures or composts in the agricultural fields is a traditional approach of composting. Generally the mature crop residues like wheat straw, rice straw, sugarcane bagasse, saw-dust etc. are used for which the production of compost, have high C/N ratio and contain highly resistant lignified tissues that are difficult to degrade. Thus, breakdown or microbial conversion of these complex organic materials require long time and the final product in form of compost is bulky with low in major plant nutrients.

12.2 Recycling of Organic Materials for the Maintenance of Soil Fertility

The process of recycling for maintenance of the soil fertility is as old as agriculture itself. When early nomad shepherd settled down as farmers, they realized the significance of compost. Organic residues of plant and animal origin have been considered as a source of immense practical value for enhancement of crop productivity both directly and indirectly. While indirect enhancement of plant growth enhancement of plants may occur due to immobilizing nutrient availability and plant growth promoting activity of diverse microorganisms associated with rhizosphere. Which facilitate direct growth promotion with production of growth promoting hormone and solubilizing essential nutrient for plant availability (Maheshwari et al. 2013). Such microorganisms promote plant growth provisioning bio-control activity against several fungal pathogens (Kumar et al. 2012).These microorganisms called as PGPR's utilize several organic compounds as their sole source of carbon supply and have influence in composting. Organic residues including green manures, animal and municipal waste serves as effective source for mobilization of nutrient for improving plant growth and enhancement of soil fertility. In fact, tropical and subtropical soils in India are generally deficient in organic matter and plant nutrients because of their rapid loss of these components in the biodegradation processes. To make up these losses, extensive utilization of organic residues in agriculture through various processes such as green manuring, green leaf manuring, mulching and composting had attracted the attention of many workers (Mandal et al. 2003; Elfstrand et al. 2007; Möller and Stinner 2009).

12.2.1 Mulching

The effects of crop residues on soil productivity and as surface mulches are quite promising. To define it more precisely this is a simple method for recycling of crop residues if incorporation of organic matter in the soil or composting be applied. Several scientist perform experiments and have reported that straw mulches are helpful in conserving soil moisture, thereby saving irrigation, control of weeds and eventually increases the crop yield (Gaur 1986). Mukharjee and Gaur (1984) reported that various mulched treatments increased the ammonical and nitrate nitrogen, available phosphorus, humus content and microbial population over unmulched control. Paddy-straw as soil mulched significantly increased the grain yields of different crops such as wheat, pea, green gram and maize over unmulched treatment (Gaur 1978; Gaur and Mukharjee 1980). Mukharjee and Guar (1984) further reported that application of FYM and as amendments in soil under mulched condition increased the grain yield because of increase in ammonical and nitrate nitrogen, available phosphorus and humus content of soil and crop yield. Several recent scientific contributions standardize the effectiveness and importance of mulching in soil recycling (Fan et al. 2005; Li et al. 2006; Adekalu et al. 2007; Huang et al. 2008; Bezborodov et al. 2010; Li et al. 2012).

Decomposition of nutrients released from mulch with five different leguminous species confirmed that low quality and high amount of organic C as mulch application act as limiting factor for the quantity of energy available for microorganisms and increases the nutrient immobilization for biomass decomposition, which results in competition for nutrients to the crop plants (Cattanio et al. 2008). Recent research by Silverman et al. (2006) have shown that mulching significantly increased available water capacity by 18–35 %, total porosity by 35–46 % and soil moisture retention at low suctions from 29 to 70 %. At high suctions, no differences in soil moisture content were observed between mulch levels. Soil bulk density was not affected by mulch rate. High correlations obtained between mulch rate and soil mean weight diameter ($R^2=0.87$) and percent stable aggregates ($R^2=0.84$). Mulching the soil surface with a layer of plant residue is an effective method of conserving water and soil because it reduces surface runoff, increases infiltration of water into the soil and retard soil erosion. The effectiveness of using elephant grass (*Pennisetum purpureum*) as mulching material facilitates runoff and soil loss decreased with the amount of mulch used and increased with slope. Mulching the soils with elephant grass residue may benefit late cropping (second cropping) by increasing stored soil water for use during dry weather and help to reduce erosion on sloping land (Xia et al. 2013).

12.2.2 Composting

Composting is one of the oldest solid waste treatment methods known to man since Biblical or Vedic period. The scientific study of composting was initiated by

Hutchinson and Richards (1921) at Rothamsted Experiment Station in U.K. Later Howard and Wad (1931) did considerable work and developed a systematized concept of compost process occurred during the period of 1920–1930 in India. They observed that best compost was prepared when crop residues including different types of plant materials are layered one over the other and are turned twice during composting creating aerobic conditions. The method known as "Indore method of composting" named after the work place of Howard, after some modification in the process it was widely accepted in India and South Africa and found suitable for the decomposition of sisal wastes, wastes of tea, coffee, rubber coconut and oil palms. Frequent turning helped to maintain aerobic conditions for rapid decomposition and thus shortening the composting period (Vimal and Talashilkar 1986).

The history and development in composting has been reviewed by Gaur (1982). The "Bangalore method" of composting for the treatment of night soil was developed by Acharya (1939). Earlier, Waksman et al. (1939a, b, c) carried out basic research on aerobic decomposition of vegetable residues and animal wastes. They investigated the influence of temperature on the rate of decomposition and the role of individual group of microorganisms and the effect of mixed cultures as compared with that of pure culture in the decomposition of substrates. Earlier, Rodale (1948) propagated the concept of organic methods of farming in United States. Subsequent development such as sheet composting, shredding of materials for rapid decomposition, digester composting, addition of the rock phosphate plus sulfur for enhancing the composting process, and role of different species of earthworm in the humification of straw has also been well documented (Gaur 1982).

A systematic research on the composting and recycling is still in progress in India and lot of work has been carried out in the past on shortening the length of time or reducing the period of composting of agro residues such as wheat straw, rice straw and other residues or dry leaves etc., the inoculation of nitrogen fixing bacteria in the process of composting and after maturation have been proved significant in the area of compost science and utilization (Gangwar and Pathak 1959; Seal and Eggins 1976; Dalzell 1979; Yadav et al. 1982; Gaur et al. 1982; Crawford 1983; Gaur et al. 1984; Mahmood et al. 1985; Negi 1985; Vimal and Talashilkar 1986; Gaur 1987, 1990; Joeggensen and Meyer 1990). Recently, Pepe et al. (2013) observed microbial diversity concerning the composition and dynamics of the microflora during the composting of municipal solid wastes, agro-industrial residue/waste or organic material. There are several diverse group of N_2 fixing: ammonia oxidizing bacteria (AOB) and nitrite-oxidizing bacteria (NOB) bearing pectinolytic, amylolytic and aerobic cellulolytic microorganisms. Some uncommon N_2 fixing genera i.e., *Stenotrophomonas, Xanthomonas, Pseudomonas, Klebsiella, Alcaligenes, Achromobacter, Caulobacter* etc. employed as improver of quality of compost (Partanen et al. 2010).

Recent studies on the progress of composting include detail investigations on the factors affecting the process of biodegradation and applications of municipal waste for composting (Itävaara et al. 2002; Alvarez et al. 2009; Kumar 2011; Gautam et al. 2012; Muktadirul et al. 2013). Earlier, Thambirajah (1991) studied the effect of various factors affecting the process of composting i.e., substrate, microorganisms, temperature, pH, cellulose, lignin and ash and chemicals like phosphorus, potassium, calcium magnesium, zinc and iron. This was seen that most important factors that affect the composting process are concerned with the interrelationship between the microorganisms and the organic environment. Thus, all possible factors that affect the microbial activity should be taken in to consideration for the process of optimization. This is found significant for the production of good quality of compost in order to get in shortest possible time. Ferrari et al. (1991) demonstrated that for successful production of marketable compost from sludge, the raw materials from the sewage waste must be collected separately. So, as to avoid the contaminations with inert materials and toxic elements that impede or drastically limit its use. Similarly, effective use of solid waste considered quite suitable (Chakarbarty et al. 1991) and the process of yard waste composting was demonstrated by Reinhart (1991). On the other hand, Das and Ghatnekar (1979) developed a simple and efficient technique for the degradation of green leafy materials of some terrestrial and aquatic weeds in polythene and PVC bags. They observed that different leafy materials individually as well as in different combinations decomposed uniformly into black color manure called bio-dung due to its dung like appearance within a short span of 15–20 days. The similar method was modified and optimized later for the degradation of the green leafy materials of some wild growing plants like *Clerodendrum inerme, Parthenium hysterophorus, Lantana camara* and *Leucaena leucocephala* (Joshi et al. 1989a, b). Effects of temperature fluctuations corresponding to ambient temperature in sun, shade and room observed on the degradation of leafy materials (Joshi and Thakre 1991). Cleveland et al. (2006) showed that organic matter decomposition for nutrient regulation in tropical rain forests did not effects mass loss during decomposition in nutrient-poor, wet ecosystems, but, ultimately regulate CO_2 losses and hence (storage) by limiting microbial mineralization of dissolve organic matter (DOM) leached from the litter layer of soil.

Management of municipal solid waste compost is utmost importance in ensuring environmental health and composting is an important aspect of managing solid waste. As stated earlier, the application of good quality high value compost showed positive influence on soil productivity due to replenishing nutrients and maintainenance of optimum moisture content. On the contrary, bad quality compost may lead an negative impact on soil health and integrity by accumulation of heavy toxic pollutants (metal) and immobilization of nitrogen in soil. It is, therefore, an urgent need to establish quality parameters of compost for application purposes particularly for crops. The guidelines related to compost related aspects have been given by several workers but, based on the recommendation of Bureau of Indian Standard (BIS) (2013) specification have been published on the compost prepared from municipal solid waste. Such standard prescribed the requirement, method of sampling etc. enlist for municipal solid waste compost of manure grade. Compost under this standard is defined as "substance made of one or more unprocessed waste material of biological nature (plant and animal) and that may include unprocessed mineral material altered through microbiological decomposition". Requirement to standards include description of physical properties including particle size. Standard packaging and marking for characterization as maximum bulk density of compost (gm cm^{-3}) should be 1.00, moisture by mass 25 %, conductivity (dsm^{-1}) 4.0, C:N ratio 20:1, Arsenic (as

Table 12.1 Microbial population during mesophilic composting of agro-industrial waste. (Adopted and modified from: Pepe et al. 2013)

Microbial functional group	Microbial count (cfu g^{-1}) at different time interval		
	0 days	23 days	123 days
Total aerobic bacteria	6.8–6.9	7.0–7.3	8.1–8.2
Total anaerobic bacteria	7.5–6.8	4.0–5.7	7.1–7.1
Mold and yeast	5.3–5.0	4.9–6.4	5.2–5.0
Actinobacteria	7.2–7.4	3.7–5.5	3.6–3.7
Proteolytics	4.0–4.6	3.9–4.8	3.5–3.3
Ammonifiers	4.3–5.6	3.8–3.9	5.0–5.0
N_2 fixing aerobic	6.2–6.4	3.2–5.5	3.4–3.7
N_2 fixing anaerobic	4.5–3.8	1.9–4.7	2.2–3.0
Ammonia oxidizing	2.9–<0.5	3.3–<0.5	<0.5–<0.5
Nitrite oxidizing	2.3–<0.5	3.7–5.3	4.7–6.4
Nitrate reducing	2.8–2.9	4.6–4.4	4.2–4.3
Amylolytics	5.8–5.6	5.6–7.8	7.9–8.1
Pectinolytics	6.9–7.7	5.3–7.9	5.5–8.0
Cellulolytics	5.0–4.9	5.5–6.5	7.3–8.0

As_2O_3) mg kg^{-1} on dry mass basis 10.00, cadmium (as Cd) 5.00, Chromium (as Cr) 50.00, Zinc (as Zn) 100.00, and mercury (as Hg) should be 0.15. Similarly, minimum total organic carbon by total dry mass should contain 40 %, total nitrogen 0.8 %, total phosphorus (P_2O_5) 0.4 %, total potassium (as K_2O) 0.4 % and sum of all i.e., Total Organic carbon, nitrogen, phosphate and potassium may present at 1.5 %.

12.3 Ecology of Composting

Since composting is mainly involves different genera of microorganisms, community dynamics is essential for the process of decomposition. Several groups of cellulytic, pectinolytic, amylolytic and ligninolytic bacteria, fungi and actinobacteria producing different enzyme are actively involved in process of compost formation. The bacterial community dynamics of compost, from mesophilic to thermophilic depends on several factors that directly or indirectly influence the ecology of compost (Boulter et al. 2000; Taiwo and Oso 2004).

Hargerty et al. (1999) reported that there was maximum increase in microbial population in the early stages of composting which was dependent on intial substrate used and environmental conditions of composting. Pepe et al. (2013) reported population dynamics of some functional microorganisms involved in composting at different time interval (Table 12.1). The four major classes of bacteria associated in compost exclusively available for agricultural benefits are firmicutes, β-proteobacteria, γ-proteobacteria and actinobacteria as stated by Ntougias et al. (2004). The total majority of microorganisms involved in composting is summarized in Table 12.2.

Table 12.2 Diversity of microorganisms of different physiology in composting process. (de Bartoldi et al. 1983; Ryckeboer et al. 2003b; Pepe et al. 2013)

Thermophillic	*Absidia Ramosa, Allescheria terrestris, Mucor pusiles, Chaetomium thermophilum, Talaromyces thermophillus, Aspergillus fumigatus, Humicola insolens, H. languginosa, Lenzites* sp., *Penicillium-dupontii, Sctytalidium thermophilum, Sporotricum thermophilic, Thermoascus aurantiacus and Micelia sterlia*	*Achromobacter* sp., *Acinetobacter* sp., *Actinomyces, Alcaligenes faecalis, Bacillus amyloliquifaciens, Bacillus coagulase, B. circulans, B. mycoides, B. pallidus, B. pumilus, B. schlegelie, B. smihii, B. subtilis, Brevibacillus brevis, B. thermoruber, Citrobacter freundii, Desulfotomaculum thermosapovorans, Entrobacter cloacae, Geobacillus stereothermophilus, G. Thermodenitrificans, G. thermoglucosidasivs, Hydrogenobacter* sp., *Methanothermobacter thermpautotrophicum, Micromonospora* sp., *M. vulgaris, Nocardia brasiliensis, Norcardia* sp., *Paenibacillus lentimorbus, P. macerans, Plectridia* sp., *P. strutzeri, Pseudonocardia* sp., *Saccromonospora* sp., *Staphylococcus intermedius, Streptomyces fradiae, Streptomyces griseoflavus, S. rectus, Streptomyces* sp., *S. thermoviolaceus, S. thermovulgaris, Symbiobacterium* sp., *S. thermophilum, Thermoactinomyces* sp., *T. vulgaris, Thermobifida fusca, Thermocrispum agreste, Thermocrispum* sp., *T. municipale, Thermomonospora curvata, Thermomonospora* sp., *T. viridis, Thermopolyspora polyspora, T. gluaca, Thermopolyspora* sp., *Thermus* sp. and *T. thermophiles*
Mesophillic	*Allescheria terrestris, Chatomium* sp., *Dasyscipha* sp., *Emericella nidulans, Mollisia* sp., *Thermoascus aurantiacus, Armillaria mellea, Clitopilus insitus, Coprinus cinereus, Fomes* sp., *Lentinus lepideus, Lenzites trabea, Pleurotus ostreatus, Polyporus versicolor, Alternaria tenuis, Artrobotrys oligospora, Aspergillus amstelodami, Aspergillus* sp., *Botryotrichum piluliforum, Cephaliophora tropica, Cephaliosporiosis alpine, Cephalosporium* sp., *Cladosporium herbanum, Dorathomyces* sp., *Geotricum candidum, Gliocladium* sp., *Gliomastox murorum, Graphium* sp., *Harpographium* sp., *Leptographium lundbergii, Paccilomyces* sp., *Penicillium* sp., *Phlyctaena* sp., *Rhinocladiella artrovirens, Scopulariopsis brevicaulis, Sporotrichum thermophile, Stachybotrys* sp. and *Trichoderma viridae*	*Achromobacter xyloxidans, Acidovorax facilis, Acidovorax* sp., *Alcaligenes* sp., *Amphibacillus xylanus, Arthrobacter ilicis, Arthrobacter* sp., *Azotobacter, Bacillus badius, B. cereus, B. megaterium, Bacillus* sp., *B. sphaericus, B. thuringiensis, Bacterioides* sp., *Bradyrhizobium* sp., *B. agri, B. Latersporus, Brevundimonas diminuta, Brevundimonas* sp., *Caryophanon latum, Caulobacter* sp., *Cellulomonas* sp., *Chromatobacterium* sp., *Chyseobacterium balustinum, Citrobacter* sp., *Clostridium* sp., *Clostridium thermocellum, Comamonas testosteroni, Corynebacterium jeikeuim, C. Striatum, Curtobacterium flaccumfaciens, Cytophaga* sp., *desulfotomaculum thermosapovorans, Entrobacter cloacae, Enterobacter* sp., *Enterococcus gallinarum, Enterococcus* sp., *Escherichia coli, Flavimonas oxyzihabitans, Flavobacterium johnsoniae, Flavobacterium mizutaii, Flavobacterium* sp., *Janthinobacterium lividum, Klebsiella* sp., *Kocuria varians, Methylobacterium extorquens, Methylobacterium organophilum, Methylobacterium* sp., *Microbacterium flavcescens, Micrococcus luteus, Moraxella bovis, Nitrobacter* sp., *Nitrosomonas* sp., *Nitrobacter* sp., *Nitrosomonas* sp., *Nocardia otitidiscaviarium, Paenibacillus pabuli, P. polymyxa, Pantoa agglomerans, Paracoccus denitrificans, P. versutus, Paucimonas lemoignei, Phyllobacterium rubiacearum, Propionibacterium* sp., *Proteus hauseri, P. mirabilis, Pseudoaltermonas haloplenktis, Pseudomonas aeruginosa, P. alcaligenes, P. balearica, P. flourescens, P. mendocina, P. pseudoalcaligenes, P. putida, Pseudonocardia asaccharolytica, P. sulfidoxydans, P. thermophila, Psychrobacter immobilis, Rathayibacter rhodochrous, Rhodococcus* sp., *Rhodovulum adriatinum, Serratia entomophila, S. marcescens, Serratia* sp., *Sphingobacterium thalpophilum, Staphylococcus* sp., *Stenotrophomonas maltophila, Streptomyces violaceoruber, Terrabacter* sp., *Variovorax paradoxus* and *Xanthobacter* sp.

During composting acidic pH < 5.0 environment is facilitated by fungal microorganisms optimally at temperature ranges between 22.5 and 45 °C (Finstein and Morris 1975). At this range of temperature fungal activity limits in term of fungal cell destruction without affecting spores (Boulter et al. 2000). As compost reached at 60 °C and above thermophilic microorganisms dominate the composting process.

Composting of organic materials or waste influenced by microbial population having multifarious chemical and biological properties. Several enzymatic activity involves cellulase, xylanase and protease have provision for decomposition. Further, generally, C:N ratio index confirms the maturity of compost. As a result of decomposition, heat output and temperature decreased. At this duration, mesophilic microorganisms that grows at < 40 °C recolonize in compost. Therefore, suppression of pathogens and/or disease is largely induced during curing, because most bio-control agents recolonize compost.

12.4 Principles of Composting

Composting and recycling is becoming an efficient treatment for organic wastes where in the available literature revealed the limited attention has been paid on microbiological aspects of compost science (Ryckeboer et al. 2003a; Benito et al. 2003; Tiquia 2005; Liu et al. 2011). Newer techniques and complementary methods to characterize microbial diversity and evolution during composting may undoubtedly develop the future prospectus with recent concern in this applied aspect of composting of agricultural apprehension. An outline of the composting process is given in Fig. 12.1. Composting of organic matter includes food waste, agro-wastes, plant biomass residues, animal manures etc. The process involved various groups of bacteria, fungi, protozoan, rotifers etc. which cater the process on variable temperature, pH, oxygen concentration, carbon nitrogen ratio etc. these factors influence the dynamics of decomposer community as well as process of compost formation of high value. During compost or recycling of agricultural wastes process is affected by environmental stresses (Peigné and Girardin 2004). Some environmental burden of composting can be resolved out by using such assessment i.e., gas emission (CH_4, CO_2, N_2O, NH_3 etc.) and volatile organic compounds (VOCs). Colón et al. (2010) studied and confirmed that stable and final compost have not any remaining pathogen and phytotoxic elements, though it has higher amount of C:N ratio. The good quality of compost can be ascertained by particle size of compost, moisture and amount of impurities mixed. The process involve long duration of accomplishment ranges between 6 to 12 months (Kakezawa et al. 1990). The microorganisms obtain moisture from waste, oxygen from the air and nutrients from the organic wastes. Give off carbon dioxide, moisture and energy (Groenhof 1998): these reproduce themselves and eventually die with the reduction in oxygen index as a principal factor for microbial survival (Scaglia et al. 2000). Some of the energy is used for growth and movement; the rest is given off as heat which one tries to conserve in a compost heap. As a result, the heap passes through warming-up, increases temperature,

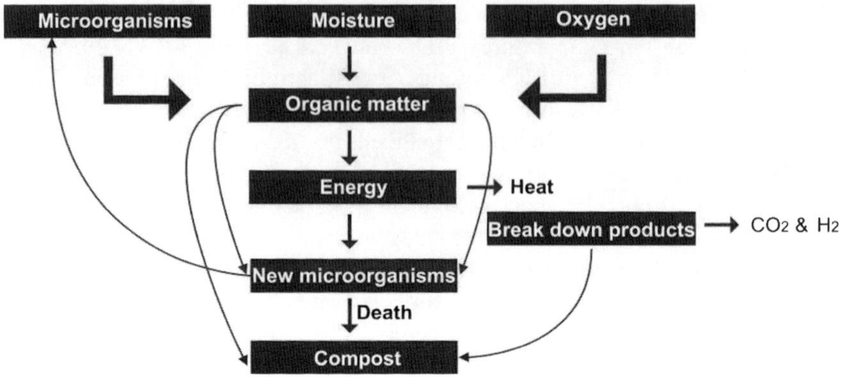

Fig. 12.1 Schematic diagram representation of compost process

that cool down resulting to obtain maturing stages. Thermal techniques been successfully developed for assessment of compost maturity based on comparative method via evaluating organic material (OM) transformation of lignocellulosic materials in-process (Dell'Abate et al. 1998, 2000). The final compost product, humus, consists of the resistant parts of the organic wastes, some breakdown products, plus dead and living micro-organisms. For the process to give a satisfactory compost product, the organisms must be given optimum conditions of food, air, moisture and temperature (Pivato et al. 2013).

12.5 Enrichment Process

Non harvested plant residues particular non-lignified tissues of such as leaves, stems, branches and roots constitute the residues from the production of food and fiber. These plant residues are recycled in soil through composting, mulching or direct incorporation into soil. Since, plant residues contain all the elements, their addition increase the nutrients status of the soil and crop productivity (Vimal and Talashilkar 1986). Compost is the store house of major plant nutrients (NPK) as well as micronutrients. It safeguards the rural and public health and maintains the quality of environment at extreamly low cost (Gaur 1984).

Nitrogen, phosphorus and potassium are the major plant nutrients derive from the soil for plant growth and development. These nutrients are supplied to the soil either through mineral fertilizers or by supplementing the soil with compost. Of these three nutrients, nitrogen is most important for growth as it is the key block of protein molecule upon which life is based, and thus it is the most indispensable component of the protoplasm of plant, animals and microorganisms. Despite of its critical role in the plant nutrition, nitrogen is assimilated almost entirely in the inorganic state as nitrate or ammonium. On the other hand, the bulk of nitrogenous materials found in the soil added in the form of plants. Microorganisms degrade these complex organic materials and release inorganic and available forms of nitrogen in the soil. While degrading organic materials, microorganisms also assimilate some of the major plant nutrients for the production of their cell and protoplasm (Alexander 1983).

Similar to nitrogen, phosphorus and potassium are also important inorganic nutrients and plays an important physiological role in the accumulation and release of energy during cellular metabolism. All the element are also added to the soil either as chemical fertilizers or organic manures and transformed in the soil through microorganisms. Thus in order to overcome microbial immobilization to increase the nutrient status of composts and to reduce the time of composting, it was suggested by many workers to enrich the compost with inorganic fertilizers like phosphates, rock phosphates, urea, ammonium sulphate or potassium salts to be added to the crop residues with wide C/N ratio prior to degradation (Gaur 1984). Crawford (1983) reported that composted farm and garden waste contain 0.4 to 3.5% nitrogen, 0.3 to 3.5% phosphorus (P_2O_5) and 0.5 to 2.0% potassium (K_2O). Vimal and Talashilkar (1986) reported that the normal nutrient content of the farm compost (N—0.5%, P_2O_5—0.2%, K_2O—0.5%) is low and imbalanced which makes the organic manures unsuitable for majority of the crops (Allision 1973). On the otherhand, Gaur (1984) expressed that the rate and extent of liberation of nitrogen from applied organic manures is adequate for normal plant growth when their C/N ratio is less than 10:1 or nitrogen content is more than 2.5%. Thus the low quality of manures like dry dung, paddy and groundnut husks, raw dust, tobacco wastes, wheat straw and any other materials can utilized effectively by the addition of urea or ammonium sulfate to add 2% more nitrogen and rise the total nitrogen content to 2.5%. In 1952, Kaila studied on humification used chopped or ground rye straw supplemented with urea and potassium phosphate. It is interesting in this study that enrichment results into short term decomposition at higher temperature (50–60 °C) and obtain better yield of valuable product in short span of time. Das and Khan (1967) observed that when urea was supplied to soil it was hydrolyzed quickly and the considerable part of ammonia tended to volatilize it into atmosphere but application of urea below the surface reduced such loss to a considerable level as evidenced by Chattarjee and Maiti (1977). The loss of nitrogen was due to leaching denitrification and immobilization. However, organic manures alone cannot meet the demand of growing crops in the fields. For the maximum production from the high yielding varieties, the nutrient contents in the leaves must be maintained at desired level throughout the growth of crops since their yield showed relationship with nitrogen content in the leaves and the leaf area duration of the crop from flowering to maturity. Thus they concluded that the best method of using organic manures was in conjunction with fertilizers so that organic manures make the nutrients slowly available to crop and minimize the losses of the organic manures, the efficiency of fertilizer use can also appreciably be improved (Lu et al. 2011). Chakraborty et al. (2011) studied the effect of NPK fertilizers and green manures on soil characteristics. The organic manures are allowed to decompose in the presence of fertilizers either applied mineral nitrogen got transformed into amino acids or more green manure protein could get hydrolyzed providing more amino acid in the presence of fertilizers (Boland et al. 2013).

Recently, Hu et al. (2011) studied that increased rate of decomposition of amorphous cellulose and hemicellulose fractions occur quickly by microbial conversion processes at ambient conditions of temperature and pH. The number of cellulolytic microorganisms increased with the addition of rock phosphate that too conserved nitrogen by decreasing the number of denitrifying bacteria (Hassimi et al. 2013; Mehta et al. 2013). Although, with addition of calcium phosphates, nitrogen fixation may increase (Tariq et al. 2013). There is generally sufficient phosphate in the organic wastes for meeting out the requirement of microorganisms, which need about 5 to 20 % only as much P as N. However, experiments at IARI, New Delhi in India have shown that application of rock phosphate is important for nitrogen conservation and enrichment of organic manures (Gaur et al. 1982). They also reported that the addition of rock phosphate at 1.0 % (w/w) increase the decomposition of paddy straw by involving cellulolytic fungi. Increased amount of rock phosphate (2–3 %) showed slight reduction in the intensity of decomposition. Khan and Bhatnagar (1977) observed the solubilization of insoluble phosphate by microorganisms and reported that *Aspergillus niger, Pencillium* spp. and other phosphate solubilizing microorganisms remains were most active in the soils. The solubilization of rock phosphate was found to be stationary corresponding increase in the concentration of rock phosphate up to a level of 4.0 %. After increasing the concentration the rock phosphate inhibited by its own solubilization which was attributed to the exploration of fluoride. In fact, sodium fluoride inhibited the microbial growth. The effect of organic matter in reducing the intensity of phosphate fixation by the soil sequioxides and maintenance of soil fertility by the use of organic manures along with super phosphate was advocated by Datta and Srivastava (1963). Organic matter applied to soil at the rate of 1 % proved more effective in increasing the available P from the native as well as added P sources (Vyas and Motiramani 1971). It was observed that both organic acid and humus and organic fractions of decomposing organic matter was more efficient releasing phosphorus from rock phosphate and tri-calcium phosphate and reducing phosphorus fixation in soil (Gaur 1969; Pareek and Gaur 1973). The quality of compost prepared from mixture of paddy straw grass and water-hycinth was improved when rock phosphate was supplemented to it with or without pyrite. They concluded that enrichment of compost occurred with the increase in the concentration of calcium and other micronutrients particularly of iron, manganese and zinc. The amendment of the finished compost with 1 % rock phosphate and inoculation with nitrogen fixing of finished urban compost and fresh moist sludge in a 2:1 proportions and amendments with 1 % rock phosphate resulted in compost enrichment in nitrogen with a C/N ratio less than 10 (Kolay 2007).

Investigators have reported that calcium phosphate increases the rate of decomposition and nitrogen conservation (Arzamasova and Kuzmenkova 1962) but more than 2 % of calcium inhibited the rate of decomposition (Gaur et al. 1980). Addition of optimum amount of calcium salts of both acids and alkaline soils includes beneficial effect, thus bring about a condition suitable for plant growth and development (Hausenbuiller 1963). Ali (1970) while studying the effect of few acid soils, addition of calcium salts increase the availability of nitrogen and phosphorus. Similarly, increase in nitrogen was reported by Srivastava et al. (1971). They observed that maximum availability of nitrogen and phosphorus obtained by the addition of calcium carbonate and calcium acetate respectively.

The compost prepared by improved application of agrochemicals enhance the productivity of agricultural crop (Tillman et al. 2002). From the history, Raychaudhary (1972) stated that it is possible to increase the soil fertility of large acreages of Indian soils substantially and rapidly through the use of commercial fertilizers. But, addition of plant nutrients to the soil thorough the application of commercial fertilizers not only make for better yield from the crops receiving them but when carried out in conjunction with fertility maintaining practices leads to more or less increase in the fertility status. Bid et al. (1974) studied the effect of different forms of nitrogenous and phosphoric fertilizers on the yield of tomato and concluded that ammonium sulfate gave highest yield when averaged over the phosphoric sources, while treatment with ammonium bicarbonate in combination with single superphosphate gave significantly the lowest yield than those of others and modified the technique for increase the yield of crop. However, Kulkarni and Nagaraj (1988) observed increase in the nitrogen contents of wheat straw occurred due to the addition of bacterial suspension and due to the partial role of nitrogen fixation by bacteria. Similarly, in the second treatment when fungi and bacteria were inoculated with straw there was an absolute increase in the nitrogen content of straw, i.e., 26 % over control and 16 % over previous treatment (bacterial suspension). Gaur et al. (1980) described the enrichment of compost through addition of chemical fertilizers (Nitrogen and Phosphorus) as well as through action and inoculation of fungi. Earlier, Banagar et al. (1989) studied the preparation of nitrogen and phosphorus enriched paddy straw compost and its effect on the yield and nutrient uptake of wheat.

Enriched organic manure is possible to obtain in short time. Losses occurring in composting are avoided and the total quantities of organic matter and nitrogen taken initially are converted and utilized. The decomposition in the presence of adequate nitrogen incorporated did not cause any adverse effect. The sufficient nitrogen in excess of microbial requirement is liberated for optimal growth of crops (Campbell et al. 1995; Liu et al. 2010; Jacobsen et al. 2012). Variations in the preparations of this enriched manures include incorporation of more nitrogen to obtain concentrated manures containing 5 to 7 % nitrogen and it supplementing this with superphosphate and potassium sulfate to get mixed with organic fertilizers of any desired grade to suit specific requirement of soil or crop (Burnham 2011).

12.6 Factors Affecting Decomposition of Waste and Compost Process

The composting is based on the decomposition and is a part of natural life cycle that is caused by microorganisms. The factors influence the quality of compost have been described.

12.6.1 pH

The pH greatly influence composting and affects the microbial community level during composting (Lei and Vandergheynst 2000). Different agro-composting residue has different range of pH. Sewage sludge as raw material for compost have pH value 8.1, however, organic fraction of municipal solid waste has pH 7.6 (Manios 2004). In the early phase of composting pH remain low but during the process, it increase upto 8.5. It has been reported that the production of lactic acid and acetic acid in the biomass degradation process leads to lower down the pH in the range of 4.2–5.5 (Hultman 2009). Kim et al. (2004) found that the pH 7.5 is the ideal pH for composting process of organic matter decomposition. Sundberg et al. (2004) studied the effect of the temperature and pH (4.6–9.2). On respiration rate ensuring the early phase of composting in two different composts. The respiration rate was strongly reduced at 40 °C at pH below 6.0 compared to compost with a higher pH at lower temperature. The high temperature and low pH is possibly adverse factors in composting. Low pH confirmed as inhibiting factors in transition from mesophilic to thermophilic phase in composting.

12.6.2 Temperature

The understanding of the factors influencing decomposition of organic wastes and performance are the major issues over sustainable waste management practices. Among all influencing factors temperature encompass prime important characteristic of thermophilic microorganisms active in compost. They are generally variably observed in the temperature range from 54 to 70 °C (Miyatake and Iwabuchi 2005). It was seemingly interested factors that temperature influence rate of mineralization and rate of leaching of dissolved organic carbon (DOC) and nitrogen (DON) (Chodak et al. 2001). Fierer et al. (2005) showed that the temperature sensitivity of litter decomposition vary depending on litter type and extent of decomposition and stated that it is the fundamental principle of enzyme kinetics. It is included that the temperature sensitivity of microbial decomposition is inversely proportional to that of litter carbon quality. Temperature response of decomposition in both fresh litter and soil organic matter (SOM) directly related to the chemical composition of the constituent organic matter, explaining 90 and 70 % of the variance in litter and SOM respectively (Erhagen et al. 2013). Miyatake and Iwabuchi (2006) investigated the relationship between temperature and microbial activity in composting using oxygen uptake rate, specific growth rate and enzymatic activity during composting. It is resulted that the enhancement of specific growth rate and enzymatic activity and volatile solids reduction induced at 54 °C in composting.

12.6.3 Carbon Nitrogen Ratio

Change in organic C, total N, C:N ratio, activities of cellulose, xylanase and protease and microbial populations showed a definite role in composting of different organic waste including sugarcane trash, cattle dung, press mud etc. C:N ratio and CO_2 evolved from finished compost can be taken as the most reliable indices of

Table 12.3 Important regulations required for C:N ratio maintenance in composting process

Factor	Regulation and importance
Carbon	Microbial cells use carbon source as sole source of energy for building cellular block(s)/unit(s)
Nitrogen	Necessary for microbial growth and function as these are crucial component of cellular metabolites
Ideal C:N ratio	30:1
C:N ratio <30:1	Nitrogen content supplied excess and lost ammonia as resultant odorous compost
C:N ratio >30:1	Insufficient nitrogen. Slow growth of microorganisms. Compost remain degradative and slow processed

Table 12.4 C:N ratio of selected raw compost materials including organic materials

Organic materials	C:N ratio
Hairy vetch/alfalfa	10:1–15:1
Rye (seedling)	12:1–15:1
Sweet clove	14:1–16:1
Food waste	14:1–16:1
Grass clippings	18:1–20:1
Rye (flowering)	20:1–21:1
Fruit waste	38:1–36:1
Dry leaves	50:1–56:1
Corn stalks	60:1–72:1
Straw	250:1–500:1
Sawdust	2:1–3:1
Liquid manure	10:1
Chicken dung	12:1
Grass cutting	13:1–23:1
Cow/pig/horse dung	20:1–25:1
Feather, hair	30:1
Paper and cardboard	250:1–500:1

compost maturity (Goyal et al. 2005). The initial carbon to nitrogen (C/N) ratio is prime factor influences compost quality (Michel et al. 1996). Important regulations are important to maintain the best integrity and product quality of compost (Table 12.3). In general, C/N ratio within the range of 25–30 is considered ideal for composting (Kumar et al. 2010). Composting at lower initial C/N ratio can thus increase the amount of manure treated, but can also increase the loss of nitrogen as ammonia gas. C:N ratio can be affected with aeration rate and moisture content during composting (Table 12.4). It is interesting to note that gaseous emission from composting can be effects the C:N ratio. Lower C:N ratio caused higher NH_3 and CH_4 emissions. The intial moisture contents influence gaseous emission significantly. Hence, high moisture content lead the low C:N ratio (Jiang et al. 2011).

In the early stages of composting total N decrease due to loss of N in the form of ammonia which in turn depends upon the type of material and its C:N ratio (Goyal et al. 2005). The composting material with low C:N ratio results in more N losses than high C:N ratio varies within the range of 13.9:15.1–15.9:16.2.

12.7 Role of Composting in Agricultural Production

Organic waste compost have several benefits to application in agricultural crops as well as on soil structural fertility and plant growth and development (Murwira 1995; Esse et al. 2001). The mature compost differs in its quality and stability, it depends upon the composition of raw materials used for the compost production (Poincelot 1974, Gaur and Singh 1995; Ranalli et al. 2001). Microbial succession plays a key role in composting process and appearance of some microorganisms reflects the quality of maturing compost (Ishii et al. 2000; Ryckeboer et al. 2003b). Compost stability is one of the significant approaches of compost quality. It relates to the degree to which the organic matter has been stabilizing during the composting process (Weppen 2002).

Lack of adequate nutrient supply and poor soil structure are the principle constraints to crop productivity under low inputs. Agricultural use of organic residues offers an attractive method for their safe disposal and a valuable source of organic amendments and nutrients. Application of organic residue material or composted organic residue influence the soil structure and fertility with special addition of nutritional content perticularly major amount of trace elements (Courtney and Mullen 2008). On the other hand, intensified agricultural practices include use of organic and chemical fertilizer with their negative consequences that affects the biodiversity of soil microorganisms and adversaly affecting soil quality. To ensure sustainable agriculture, and for evaluating the effects of management practices on soil, D'Hose et al. (2014) developed multiparameter index that includes a wide range of soil properties linked to important soil functions so as to categorise crop sustainability and productivity. Accordingly, long term farm compost (FC) amendments resulted in strengthening the soil quality. Dukare et al. (2011) investigated the potential of antagonistic bacteria amended in compost preparations. These facilitated suppression of diseases caused by plant pathogenic fungi *Fusarium oxysporum, Pythium debaryanum, P. aphanidermatum* and *Rhizoctonia solani*. This study revealed out importance and efficacy of microbe-fortified compost for use in control of phytopathogenic fungi. On the other hand, effectiveness of plant harmones blended with recycled organic waste observed for improving growth and yield of wheat (Zahir et al. 2007). In their studies huge amount of organic waste as compost was converted into a value-added product for improving growth, yield and nutrient uptake in wheat. On the contrary, implementation of composting practices may impact reverse (damage) due to overzealous use of other fertilizers. Moss et al. (2002) examined that the non-recyclable petroleum based fertilizers associated with the "green revolution" displaced the use of manures and human waste as compost have not been find much potential in crop improvement. Martínez-Blanco et al. (2014) reviewed that compost through life cycle assessment (LCA) ascertained nutrient supply, carbon sequestration, pest and fungal disease suppression, increase crop yield, decrease soil erosion, retention of soil moisture, increase soil work ability, enhance soil biological properties and biodiversity and enhance crop nutritional quality. Compost application could contribute to increase food availability and therefore, efforts should be made to alleviate the socio-economic constraints to the adoption of compost technology

(Ouédraogo et al. 2001). The compost application increases soil organic carbon content, hot-water extractable carbon, earthworm's numbers, microbial biomass and reduced soil bulk density and crop yield (D'Hose et al. 2012).

12.8 Conclusion

Augmenting suitable practices of agricultural management and up keeping of soil fertility using compost is befitting in providing the agricultural sustainability in crop production. The role of microbial decomposition balanced through factors influencing organic matter decomposition and compost preparation in improved way applied in soil recycling found to be beneficial for agricultural production. In this context decomposition of organic material or waste by suitable microorganisms with concern to produce high value compost and incorporation in alternation of chemical farming practices. Composting of organic waste material and utilization in farming practices is an alarming issue for future prespectives of research. which could mitigate the pollutants indirectly and benefit the agro-industry directly by adopting environment friendly approaches for agricultural sustainability.

References

Acharya CN (1939) Comparision of different method of composting waste materials. Indian J Agric Soil 9:565–572

Adekalu KO, Olorunfemi IA, Osunbitan JA (2007) Grass mulching effect on infiltration, surface runoff and soil loss of three agricultural soils in Nigeria. Bioresour Technol 98:912–917

Alexender M (1983) Introduction to soil microbiology, 2nd edn. Willey Eastern Limited, New York, p 467

Ali S (1970) Ph.D thesis. Allahabad University, Allahabad, India

Allision FE (1973) Soil organic matter and its role in the crop production. Elsevier Scientific Publication Co. Inc. New York, p 455

Alvarez JVL, Larrucea MA, Bermúdez PA, Chicote BL (2009) Biodegradation of paper waste under controlled composting conditions. Waste Manage 29:1514–1519

Arzamasova ZA, Kuzmenkova AM (1962) Effect of mineral additives on the decomposition of (city) refuse and its microflora. Sb Nauch Rab Akad Kemmun Khez 14:103

Banagar KC, Shankar S, Kapoor KK, Kukreja K, Mishra MM (1989) Preparation of nitrogen and phosphorus enriched paddy straw and its effect on the yield and nutrient uptake by wheat (*Triticum aestivum* L.). Biol Fertil Soils 8:339–342

Bationo A, Kihara J, Vanlauwe B, Waswa B, Kimetu J (2007) Soil organic carbon dynamics, functions and management in West African agro-ecosystems. Agric Syst 94:13–25

Benito M, Masaguer A, Moliner A, Arrigo N, Palma RM (2003) Chemical and microbiological parameters for the characterisation of the stability and maturity of pruning waste compost. Biol Fertil Soils 37:184–189

Bernard E, Larkin RP, Tavantzis S, Erich MS, Alyokhin A, Sewell G, Lannan A, Gross SD (2012) Compost, rapeseed rotation, and biocontrol agents significantly impact soil microbial communities in organic and conventional potato production systems. Appl Soil Ecol 52:29–41

Bezborodov GA, Shadmanov DK, Mirhashimov RT, Yuldashev T, Qureshi AS, Noble AD, Qadir M (2010) Mulching and water quality effects on soil salinity and sodicity dynamics and cotton productivity in Central Asia. Agric Ecosyst Environ 138:95–102

Bid NN, Bandopadhyay DPK, Dhua SP (1974) Efficacy of different forms of nitrogenous and phosphatic fertilizers on tamato (*Lycopersicon esculentum* Mill). Technology 11(2–3):300–302
BIS (2013) Draft Indian Standard; municipal solid waste compost, manure grade—specification ICS No. 13.030. 10, 65.080, Doc No.: FAD 7 (2189) C, Soil Quality and Fertilizers Sectional Committee, FAD 7, 15 Aug 2013
Boland MJ, Rae AN, Vereijken JM, Meuwissen MP, Fischer AR, van Boekel MAJS, Rutherfurd SM, Gruppen H, Moughan JP, Hendriks WH (2013) The future supply of animal-derived protein for human consumption. Trends Food Sci Technol 29(1):62–73
Boulter JI, Boland GJ, Trevors JT (2000) Compost: a study of the development process and end-product potential for suppression of turfgrass disease. World J Microbiol Biotechnol 16(2):115–134
Burnham JC (2011) U.S. Patent No. 7,947,105. U.S. Patent and Trademark Office, Washington
Campbell CA, Myers RJK, Curtin D (1995) Managing nitrogen for sustainable crop production. Fertil Res 42:277–296
Cattanio JH, Kuehne R, Vlek PLG (2008) Organic material decomposition and nutrient dynamics in a mulch system enriched with leguminous trees in the Amazon. Rev Bras Ciência do Solo 32:1073–1086
Chakarbarty R, Halder RK (1991) The management and effective use of solid waste of Calcutta municipal corporation. Proceeding of the seventh international conference on solid waste management and secondary materials. 4D. Philadelphia
Chakraborty A, Chakrabarti K, Chakraborty A, Ghosh S (2011) Effect of long-term fertilizers and manure application on microbial biomass and microbial activity of a tropical agricultural soil. Biol Fertil Soil 47(2):227–233
Chattarjee BN, Maiti S (1977) Timing and application techniques for efficient use of fertilizers. Fertil Technol 14(1–2):55–59
Chodak M, Borken W, Ludwig B, Beese F (2001) Effect of temperature on the mineralization of C and N of fresh and mature compost in sandy material. J Plant Nutr Soil Sci 164:289–294
Cleveland CC, Reed SC, Townsend AR (2006) Nutrient regulation of organic matter decomposition in a tropical rain forest. Ecology 87:492–503
Colón J, Martínez-Blanco J, Gabarrell X, Artola A, Sánchez A, Rieradevall J, Font X (2010) Environmental assessment of home composting. Resour Conserv Recycl 54:893–904
Courtney RG, Mullen GJ (2008) Soil quality and barley growth as influenced by the land application of two compost types. Bioresour Technol 99:2913–2918
Crawford JH (1983) Composting of agricultural wastes—a review. Process Biochem 18:14–18
D'Hose T, Cougnon M, De Vliegher A, Van Bockstaele E, Reheul D (2012) Influence of farm compost on soil quality and crop yields. Arch Agron Soil Sci 58:S71–S75
D'Hose T, Cougnon M, De Vliegher A, Vandecasteele B, Viaene N, Cornelis W, Bockstaele EV, Reheul D (2014) The positive relationship between soil quality and crop production: a case study on the effect of farm compost application. Appl Soil Ecol 75:189–198
Dalzell HW (1979) Composting in tropical countries, principles of composting. The International Institute of Biological Husbandry ltd.(IIBH), England, pp 1–38
Das CR, Ghatnekar SD (1979) Replacement of cow-dung by fermentation of aqauatic and terrestrial plants for use as fuel, fertilizer and biogas plant feed. Research Planning Workshop on Energy for Rural Development, Indian Institute of Management of Ahemabad, pp 1–19
Das AC, Khan JA (1967) Ureolytic bacteria in soils, part III—volatilization loss of ammonia from urea on soil. Appl Technol 4:16–18
Datta NP, Srivastava SC (1963) Influence of organic matter on the intensity of phosphate bonding in some acid soils. J Indian Soc Soil Sci 11:189
De Bartoldi M, Vallini G, Pera A (1983) The biology of composting: a review. Waste Manage Res 1:157–176
Dell'Abate MT, Canali S, Trinchera A, Benedetti A, Sequi P (1998) Thermal analysis in the evaluation of compost stability: a comparision with huification parameters. Nutr Cycl Agroecosyt 51:217–224
Dell'Abate MT, Bendetti A, Sequi P (2000) Thermal method of organic matter maturation monitoring during a composting process. J Therm Anal 61:389–386

Dukare AS, Prasanna R, Chandra DS, Nain L, Chaudhary V, Singh R, Saxena AK (2011) Evaluating novel microbe amended composts as biocontrol agents in tomato. Crop Prot 30(4):436–442
Elfstrand S, Hedlund K, Mårtensson A (2007) Soil enzyme activities, microbial community composition and function after 47 years of continuous green manuring. Appl Soil Ecol 35:610–621
Erhagen B, Oquist M, Sparrman T, Haei M, Ilstedt U, Hedenström M, Schleucher J, Nilsson MB (2013) Temperature response of litter and soil organic matter decomposition is determined by chemical composition of organic material. Global Chang Biol 19(12):3858–3871
Esse PC, Buerkert A, Hiernaux P, Assa A (2001) Decomposition of and nutrient release from ruminant manure on acid sandy soils in the Sahelian zone of Niger, West Africa. Agric Ecosyst Environ 83(1):55–63
Fan M, Jiang R, Liu X, Zhang F, Shihua L, Zeng X, Christie P (2005) Interactions between non-flooded mulching cultivation and varying nitrogen inputs in rice-wheat rotations. Field Crop Res 91:307–318
Ferrari G, Vitailoi M, Stefanati A, Ravsa G (1991) Production of organic fertilizers from waste treatment. Proceeding of the seventh international conference on solid waste management and secondry materials. 4c Philadelphia
Fierer N, Craine JM, McLauchlan K, Schimel JP (2005) Litter quality and the temperature sensitivity of decomposition. Ecology 86:320–326
Finstein MS, Morris ML (1975) Microbiology of municipal solid waste composting. Adv Appl Microbiol 19:113–151
Gangwar BM, Pathak AN (1959) Decomposition of organic matter: mineralization of organic and phosphorus in soil. Indian J Agric Sci 30(2):58–90
Gaur AC (1969) Studies on availability of phosphate in soil as influenced by humic acid. Agrochemica 14:62
Gaur AC (1978) Recycling and utilization of organic wastes as fertilizers. Proc India/FAO/Norway seminar on the development of complementary use of mineral and organic materials in india. Ministry of Agriculture, New Delhi, p 109
Gaur AC (1982) A practical manual of rural composting, project field document no. 15 FAO, p 102
Gaur AC (1984) Organic manures. Publication and Information Division, ICAR, New Delhi, p 159
Gaur AC (1986) Recycling of crop residue—recent trends in India. Izv Akad Nauk SSSR Ser Biol 2:210–217
Gaur AC (1987) Recycling of organic waste by improved technique of composting and other methods. Resour Conserv 13:157–174. Elsevier Science Publishers, B. V. Amsterdam
Gaur AC (1990) Phosphate solubilizing micro-organisms as biofertilizer. Omega Scientific Publishers, Lajpat Nagar, Delhi, India p 176
Gaur AC, Mukahrjee D (1980) Recycling of organic matter through mulch in relation to chemical and microbiological properties of soil and crop yields. Plant Soil 56:273
Gaur AC, Singh G (1995) Recycling of rural and urban wastes through conventional and vermicomposting. In: Tandon HLS (ed) Recycling of crop, animal, human and industrial wastes in agriculture. pp 31–49
Gaur AC, Sadasivam KV, Magu SP, Mathur SP (1980) Progress report of IARI center in annual workshop of all India coord. Project on Microbial Decomposition and Recycling of Farm and City Wastes, Nagpur, India
Gaur AC, Sadashivam KV, Mathur RS, Magu SP (1982) Role of mesophilic fungi in composting. Agric Wastes 4(6):453–460
Gaur AC, Neelakantan S, Dargan KS (1984) Organic manure. Publication and Information Division, ICAR, New Delhi, p 159
Gautam SP, Bundela PS, Pandey AK, Jamaluddin, Awasthi MK, Sarsaiya S (2012) Diversity of cellulolytic microbes and the biodegradation of municipal solid waste by a potential strain. Int J Microbiol 2012:1–12
Goyal S, Dhull SK, Kapoor KK (2005) Chemical and biological changes during composting of different organic wastes and assessment of compost maturity. Bioresour Technol 96:1584–1591
Groenhof A (1998) Composting: reunissance of an old-age technology. Biologist 451:164–167
Hargerty DJ, Pavoni JL, Heer JE (1999) Solid Waste Manangement Van Nostrant Reinhold, New York, pp 2–13

Hassimi MS, Hamdali H, Ouhdouch Y, Pinelli E, Merlina G, Claude RJ, Hafidi M (2013) Moroccan rock phosphate solubilization during a thermo-anaerobic grassland waste biodegradation process. Afr J Biotechnol 12(49):6859–6865
Hausenbuiller RL (1963) Principle of soil science. Orient Longsman Ltd., Calcutta
Howard A, Wad YD (1931) The waste product of agriculture; their utilization as humus. Oxford University Press, London
Hu Z, Liu Y, Chen G, Gui X, Chen T, Zhan X (2011) Characterization of organic matter degradation during composting of manure–straw mixtures spiked with tetracyclines. Bioresour Technol 102(15):7329–7334
Huang Z, Xu Z, Blumfield TJ, Bubb K (2008) Effects of mulching on growth, foliar photosynthetic nitrogen and water use efficiency of hardwood plantations in subtropical Australia. For Ecol Manage 255:3447–3454
Hubbe M, Nazhad M, Sánchez C (2010) Composting as a way to convert cellulosic biomass and organic waste into high-value soil amendments: a review. BioResources 5:2808–2854
Hultman J (2009) Microbial diversity in the municipal composting process and development of detection methods. Ph. D. Thesis Vikki Graduate School in Bioscience. University of Helsinki, Finland
Hutchinson HB, Richards EH (1921) Artificial farmyard manure. J Min Agric 28:398
Ishii K, Fukui M, Takii S (2000) Microbial succession during a composting process as evaluated by denaturing gradient gel electrophoresis analysis. J Appl Microbiol 89(5):768–777
Itävaara M, Karjomaa S, Selin JF (2002) Biodegradation of polylactide in aerobic and anaerobic thermophilic conditions. Chemosphere 46:879–885
Jacobsen SE, Jensen CR, Liu F (2012) Improving crop production in the arid Mediterranean climate. Field Crop Res 128:34–47
Jiang T, Schuchardt F, Li G, Guo R, Zhao Y (2011) Effect of C/N ratio, aeration rate and moisture content on ammonia and greenhouse gas emission during the composting. J Environ Sci 23(10):1754–1760
Joeggensen RG, Meyer B (1990) Chemical change in organic matter decomposing in and on forest Rendzina under beech (*Fagus sylyatica* L.) J Soil Sci 41:17–27
Joshi P, Thakre RP (1991) Composting of leafy biomass into farm manure. Proceedings of the seventh international conference on solid waste management and secondary materials. 3B:Philadelphia PA, USA
Joshi P, Thakre RP, Kate T (1989a) Biodgradation of leaves of *Clerodendrum inerme* L. by mycoflora. Proceedings of the seventy sixth session of the Indian Science Congress, Madurai (Abstract)
Joshi P, Thakre RP, Kate T (1989b) A note on the preparation of green leaf manures. All India UGC sponsored seminar on recent advances in plant science, Nagpur, 25 Sept, 5 (Abstract)
Kaila A (1952) Humification of straw at various temperatures. Suom Maataloust Seur Julk Acta Agr Fenn 78(3):1–32
Kakezawa M, Mimura A, Takahara Y (1990) A two-step composting process for woody resources. J Ferment Bioeng 70:173–176
Khan AR, Bhatnagar RM (1977) Studies on solubilization of insoluble phosphate by microorganisms, Part I. Solubilization of Indian phosphate rocks by *Aspergillus niger* and *Penicillium* sp. Fert Technol 14:329–333
Kim TI, Jeong KH, Ham JS, Yang CB, Chung IB, Kim MK, Kim KN (2004) Isolation and characterization of cellulose secreting bacterium from cattle manure: application to compost. Compost Sci Util 12(3):242–248
Kolay AK (2007) Manures and fertilizers. Atlantic Publishers & Distributors, Daryaganj, New Delhi, India p 173
Kour J, Maheshwari DK, Mehta P (1993) Microbial degradation of aquatic biomass by *Trichoderma viride* 992 and *Aspergillus wentii* 669 with reference to the physical structure. J Basic Microbiol 33(1):19–25
Kulkarni SM, Nagaraj R (1988) Microbial increase of nitrogen in wheat straw. Biol Wastes 25:145–147
Kumar S (2011) Composting of municipal solid waste. Crit Rev Biotechnol 31:112–136

Kumar S, Pandey P, Maheshwari DK (2009) Reduction in dose of chemical fertilizers and growth enhancement of sesame (*Sesamum indicum* L.) with application of rhizospheric competent *Pseudomonas aeruginosa* LES4. Eur J Soil Biol 45(4):334–340

Kumar M, Ou Yan L, Lin JG (2010) Co-composting of green waste and food waste at low C/N ratio. Waste Manage 30:602–609

Kumar P, Dubey RC, Maheshwari DK (2012) *Bacillus* strains isolated from rhizosphere showed plant growth promoting and antagonistic activity against phytopathogens. Microbiol Res 167(8):493–499

Lei F, Vandergheynst JS (2000) The effect of microbial inoculation and pH on microbial community structure changes during composting. Process Biochem 35:923–929

Li Y-S, Wu L-H, Lu X-H, Zhao L-M, Fan Q-L, Zhang F-S (2006) Soil microbial biomass as affected by non-flooded plastic mulching cultivation in rice. Biol Fertil Soils 43:107–111

Li FL, Yuan J, Sheng GD (2012) Altered transfer of heavy metals from soil to Chinese cabbage with film mulching. Ecotoxicol Environ Saf 77:1–6

Liu J, You L, Amini M, obersteiner M, Herrero M, Zehnder AJB, Yang H (2010) A high-resolution assessment on global nitrogen flows in cropland. Proc Natl Acad Sci U S A 107:8035–8040

Liu J, Xu X hong, Li H tao, Xu Y (2011) Effect of microbiological inocula on chemical and physical properties and microbial community of cow manure compost. Biomass Bioenergy 35:3433–3439

Lu HJ, Ye ZQ, Zhang XL, Lin XY, Ni WZ (2011) Growth and yield responses of crops and macronutrient balance influenced by commercial organic manure used as a partial substitute for chemical fertilizers in an intensive vegetable cropping system. Phys Chem Earth, Parts A/B/C 36(9):387–394

Maheshwari DK (2013) Bacteria in agrobiology: crop productivity. Springer, Heidelberg, p 507

Maheshwari DK, Agarwal M, Dheeman S, Saraf M (2013) Potential of rhizobia in productivity enhancement of *Macrotyloma uniflorum* L. and *Phaseolus vulgaris* L. cultivated in the western Himalaya. In: Maheshwari DK, Saraf M, Aeron A (eds) Bacteria in agrobiology: crop productivity. Springer, Heidelberg, pp 127–165

Mahmood T, Azam F, Malik KA (1985) Decompostiton and humification of plant residues by some soil fungi. Biotechnol Lett 7(3):207–212

Mandal UK, Singh G, Victor U, Sharma K (2003) Green manuring: its effect on soil properties and crop growth under rice–wheat cropping system. Eur J Agron 19:225–237

Manios T (2004) The composting potential of different organic solid wastes: experience from the island of Crete. Environ Int 29:1079–1089

Martnez-Blanco J, Lehmann A, Muñoz P, Antón A, Traverso M, Rieradevall J, Finkbeiner M (2014) Application challenges for the social LCA of fertilizers within life cycle sustainability assessment. J Cleaner Prod 69:34–48

Mehta CM, Palni U, Franke-Whittle IH, Sharma AK (2013) Compost: its role, mechanism and impact on reducing soil-borne plant diseases. Waste Manage 34(3):607–622

Michel FC, Forney LJ, Huang AJF, Drew S, Czuprendski M, Lindeberg JD, Reddy CA (1996) Effects of turning frequency, leaves to grass mix ratio and windrow vs. pile configuration on the composting of yard trimmings. Compost Sci Util 4:126–143

Miyatake F, Iwabuchi K (2005) Effect of high compost temperature on enzymatic activity and species diversity of culturable bacteria in cattle manure compost. Bioresour Technol 96:1821–1825

Miyatake F, Iwabuchi K (2006) Effect of compost temperature on oxygen uptake rate, specific growth rate and enzymatic activity of microorganisms in dairy cattle manure. Bioresour Technol 97(7):961–965

Möller K, Stinner W (2009) Effects of different manuring systems with and without biogas digestion on soil mineral nitrogen content and on gaseous nitrogen losses (ammonia, nitrous oxides). Eur J Agron 30:1–16

Moss LH, Epstein E, Logan T (2002) Comparing the characteristics, risks and benefits of soil amendments and fertilizers used in agriculture. Proc Water Environ Fed 2002(3):602–623

Mukherjee D, Gaur AC (1984) Effect of incorporation of organic materials on some chemical properties and yield of paddy. Indian Agriculturist 28:215–219

Muktadirul B, Chowdhury AKM, Akratos CS, Vayenas D V, Pavlou S (2013) Olive mill waste composting: a review. Int Biodeterior Biodegradation 85:108–119
Murwira HK (1995) Ammonia losses from Zimbabwean cattle manure before and after incorporation into soil. Trop Agric 72(4):269–273
Negi SS (1985) A review on the utilization of agricultural wastes as feeds for livestock and experimental safeguards against erroneous conclusions. Agric Wastes 13:93–103
Ntougias S, Zervakis GI, Kavroulakis N, Ehaliotis C, Papadopoulou KK (2004) Bacterial diversity in spent mushroom compost assessed by amplified rDNA restriction analysis and sequencing of cultivated isolates. Syst Appl Microbiol 27(6):746–754
Ouédraogo E, Mando A, Zombré NP (2001) Use of compost to improve soil properties and crop productivity under low input agricultural system in West Africa. Agric Ecosyst Environ 84:259–266
Pareek RP, Gaur AC (1973) Release of phosphate from tricalcium phosphate and rock phosphate by organic acids. Curr Sci 42:278
Partanen P, Hultman J, Paulin L, Auvinen P, Romantschuk M (2010) Bacterial diversity at different stages of the composting process. BMC Microbiol 10:94
Peigné J, Girardin P (2004) Environmental impacts of farm-scale composting practices. Water Air Soil Pollut 153:45–68
Pepe O, Ventorino V, Blaiotta G (2013) Dynamic of functional microbial groups during mesophilic composting of agro-industrial wastes and free-living (N_2)-fixing bacteria application. Waste Manage 33:1616–1625
Pivato A, Raga R, Vanin S, Rossi M (2013) Assessment of compost quality for its environmentally safe use by means of an ecotoxicological test on a soil organism. J Mater Cycl Waste Manage doi:10.1007/s10163-013-0216-8
Poincelot RP (1974) Scientific examination of principles and practice of composting. Compost Sci
Ranalli G, Bottura G, Taddei P, Garavani M, Marchetti R, Sorlini C (2001) Composting of solid and sludge residues from agricultural and food industries. Bioindicators of monitoring and compost maturity. J Environ Sci Health, Part A 36(4):415–436
Raychaudhary SP (1972) Soil fertility and fertilizer Applications. Technology 9(2–3):240–244
Reinhart DR (1991) Yard waste composting demonstration programme in central Florida. Proceedings of the seventh international conference on solid waste management and secondary materials. 6B: Philadelphia, PA
Rodale S (1948) A complete book of composting. Academic Press Inc. New York, p 500
Ryckeboer J, Mergaert J, Coosemans J, Deprins K, Swings J (2003a) Microbiological aspects of biowaste during composting in a monitored compost bin. J Appl Microbiol 94:127–137
Ryckeboer J, Mergaert J, Vaes K, Klammer S, De Clercq D, Coosemans J, Insam H, Swings J (2003b) A survey of bacteria and fungi occurring during composting and self-heating processes. Ann Microbiol 53:349–410
Scaglia B, Tambone F, Generini PL, Adani F (2000) respiration index determination:dynamics and static approaches. Compost Sci Util 8(2):90–98
Seal KJ, Eggins HOW (1976) The upgrading of agricultural wastes by thermophilic fungi. Food from wastes. Applied Science Publishers, London, pp 58–78
Silverman J, Sorenson CE, Waldvogel MG (2006) Trap-mulching Argentine ants. J Econ Entomol 99(5):1757–1760
Srivastava KK, Ali S, De SK (1971) Change in the measure of available nitrogen and phosphorus in acid soil in presence of calcium salts. Technology 8(2):149–153
Stockdale E, Shepherd M, Fortune S, Cuttle S (2002) Soil fertility in organic farming systems—fundamentally different? Soil Use Manage 18:301–308
Sundberg C, Smårs S, Jönsson H (2004) Low pH as an inhibiting factor in the transition from mesophilic to thermophilic phase in composting. Bioresour Technol 95:145–150
Taiwo LB, Oso BA (2004) Influence of composting techniques on microbial succession, temperature and pH in a composting municipal solid waste. Afr J Biotechnol 3(4):239–243
Tariq N, Arshad M, Jamshed HR, Ahmed N (2013) Process of producing bio-organo-phosphate (bop) fertilizer through continuous solubilization of rock phosphate by a composting bioprocess

and bioaugmentation with phosphorus solubilizing microorganisms U.S. Patent Application 13/789:407
Thambirajah JJ (1991) Composting of agricultural wastes factors that deternmine the sucesss or failure of the process. Proceedings of the seventh international conference on solid waste management and secondery material 3B: Philadelphia, PA
Tillman D, Cassman KG, Matson PA, Naylor R, Polasky S (2002) Agricultural sustainability and intensive production practices. Nature 418(6898):671–677
Tiquia SM (2005) Microbiological parameters as indicators of compost maturity. J Appl Microbiol 99:816–828
Vimal OP, Talashilkar SC (1986) Crop residue and plant growth. J Sci Indian Res 45:78–89
Vyas MK, Motiramani DP (1971) Effect of organic matter, silicates and moisture levels on availability of phosphate. J Indian Soc Soil Sci 19(1):39–43
Waksman SA, Cordon TC, Hulpoi N (1939a) Influence of temperature upon the microbial populations and decomposition processes in composts of stable manures. Soil Sci 47:83
Waksman SA, Umbreit WW, Cordon TC (1939b) Thermophilic actinomycetes and fungi in soils and in composts. Soil Sci 47:37–61
Waksman SA, Cordon TC (1939c) Thermophilic decomposition of plant residue in composts by pure and mixed cultures of microorganisms. Soil Sci 47:217
Weppen P (2002) Determining compost maturity: evaluation of analytical properties. Compost Sci Util 10(1):6–15
Xia L, Hoermann G, Ma L, Yang L (2013) Reducing nitrogen and phosphorus losses from arable slope land with contour hedgerows and perennial alfalfa mulching in Three Gorges Area, China. Catena 110:86–94
Yadav KS, Mishra MM, Kapoor KK (1982) The effecet of fungal inoculation of composting. Agric Wastes 4:329–333
Zahir ZA, Iqbal M, Arshad M, Naveed M, Khalid M (2007) Effectiveness of IAA, GA3 and kinetin blended with recycled organic waste for improving growth and yield of wheat (*Triticum aestivum* L.). Pak J Bot 39(3):761–768
Zhang Y, Lashermes G, Houot S et al (2012) Modelling of organic matter dynamics during the composting process. Waste Manage 32:19–30

Chapter 13
Compost: A Tool to Sustainable Urban and Peri-Urban Agriculture in Sub-Saharan Africa?

Blaise Pascal Bougnom, Onana Boyomo, Dieudonné Nwaga, Jean Justin Essia Ngang and François Xavier Etoa

Abstract Africa, with around one billion inhabitants is the world's second largest and most-populous continent after Asia, the majority of this population live in cities and generate urge amount of municipal solid waste (MSW) which constitute a serious threat to the health of the residents and the environment. Urban and Peri-urban agriculture (UPA), an activity that was developed in African cities in response to the structural adjustments programmes in 1990s is becoming more and more intensive with the continuous raising population. Composting MSW represents an economically, environmentally and socially advantage that could inspire all stakeholders involved in urban planning, within an integrated municipal waste management system, since the MSW that is daily generated has a significant percentage of organic waste that could be easily recycle within the cities in order to close the carbon cycle.

Keywords Urban agriculture · Sub-Saharan Africa · Sustainable development

13.1 Introduction

According to the United Nations (2008), it is estimated that 76.9 % of population in low and middle income countries will live in the cities by 2020. The world's urban population is projected to grow by more than a billion people between 2010 and 2025, while the rural population will hardly grow at all. In 1900, worldwide, the ratio of rural dwellers to urban dwellers was 1:6.7, and this ratio is projected to be 3:2 by 2025. Africa, with more than one billion inhabitants located in 61 countries is the world's second largest and most-populous continent after Asia, with approximately 60 % rural and 40 % urban dwellers; the urban population growth rate is 6.6 % per annum while the rural growth rate is static (0 %) (Earthtrends 2008). The world's urban population exceeded its rural population for the first time in 2008; the rapid

B. P. Bougnom (✉) · O. Boyomo · D. Nwaga · J. J. Essia Ngang, · F. X. Etoa
Faculty of Science, Department of Microbiology,
University of Yaounde 1, P. O. Box 812, Yaounde, Cameroon
e-mail: bpbougnom@uy1.uninet.cm; bpbougnom@yahoo.fr

D. K. Maheshwari (ed.), *Composting for Sustainable Agriculture,* Sustainable Development and Biodiversity 3, DOI 10.1007/978-3-319-08004-8_13

increase of urban population is source of worries since it leads to a rapid increase in urban poverty, environmental pollution and food insecurity (Satterthwaite et al. 2010). The low income of this very large urban population worldwide put their health and nutritional status at risk from any staple food price rise (Cohen and Garrett 2009). To meet the food needs, agricultural activities has been developed by urban dwellers in the cities, 200 million peoples are reported to be employed in urban farming and related enterprises, contributing to the food supply of 800 million urban dwellers (UNDP 1996).

Urban and Peri-urban agriculture (UPA) is the keeping of livestock or the production of crops for sale or consumption within the city limits; it is well established that UPA addresses the three global goals on food security (FAO 2008), i.e., (i) sustainable increases in food production and availability, (ii) economic and social progress and, (iii) sustainable management and use of natural resources.

UPA require large amount of inputs, including plant nutrients and should cope with proper management of urban towns, on the other hand, the management of municipal solid waste (MSW) is one of the most serious environmental problems that is facing the cities in sub-Saharan Africa since MSW generated, when inefficiently managed and responsible of bad odours, water pollution, proliferation of pests that constitute a serious threat to the health of the residents. Landfilling of organic waste is reported to be the major contributor to the increase of Greenhouse gas (GHG) emission in sub-Saharan Africa; developing countries were responsible for about 29 % of these emissions in the year 2000, and this share is predicted to increase to 64 % in 2030 and 76 % in 2050 (Monni et al. 2006).

Achieving food security in the cities requires increasing the productivity while taking into account the sustainability of the farming system. Chemical fertilizers are generally recommended to subsistence farmers, but their high cost prevents its use, and moreover, they are found to be responsible of water and groundwater pollution from nitrates sources and the decline of soil quality; therefore, a policy of substituting inorganic to organic fertilizers should be seriously envisaged since the amount of organic waste that is daily produced in the cities is not a limiting factor.

Composting of MSW appears to be a possibility to cope with the management of organic wastes in the cities as well as to provide UPA with organic fertilizers.

The intensification practices of agriculture in an urban environment, waste management and sanitary issues related to agriculture, as well as institutional interactions between urban farmers and municipalities, together create important challenges to urban governance (Parrot et al. 2009). The transformation of UPA into a legitimate and viable economic activity in addressing food insecurity, unemployment and poverty while ensuring sustainable environmental management and minimal health risks required adequate policies and programmes (Page 2002; Zezza and Tasciotti 2010).

The present chapter evaluates the potential that represents composting MSW for sustainable UPA.

13.2 Urbanization and Urban and Peri-Urban Agriculture

During the first decade of the twenty-first century, sub-Saharan African cities have faced increasing levels of poverty and deteriorating of urban environments as a result of rapid urban growth following massive rural exodus and high birth rate. Many factors are pointed out, including the on-going impact of externally imposed structural adjustment programs, the massive rise in both staple food prices and oil prices (Ravallion et al. 2007; Potts 2009). According to the UN-Habitat (2006), sub-Saharan Africa was the least urbanized part of the world by 2005, with 36 % of its 750 million people living in cities and towns. However, between 1990 and 2005, the urban population growth was by far the highest: 4.58 % per annum; with an estimated 200 million people that were living in slums. The rapid urbanization faces many challenges, namely problems of basic infrastructures, water and food supply. Rapid and uncontrolled urbanization in sub-Saharan Africa, results in serious decline of environmental quality and agricultural resources which lead to food insecurity.

UPA can be defined as the production of crops and rearing of livestock in urban and peri-urban areas, it is a source of food and employment in urban areas, involving almost all social groups (Yeung 1988; Obudho and Foeken 1999; Page 2002). UPA is an old phenomenon, but recently it has expanded in many cities, especially those in developing countries. In sub-Saharan Africa, UPA supports a significant number of households in terms of income and food supply (Mougeot 2000). In 2006, the number of urban farmers in the world was estimated to be over 200 million, supporting a total population of over 700 million people (Mougeot 2006). UPA in sub Saharan Africa was portrayed as an ingenious, heroic response to the rigours of structural adjustments which provided a key coping strategy for the urban poor in response to inadequate, unreliable and irregular access to food supplies (Sawio 1993; Smit et al. 1996); nowadays, it is fully integrated in the cities living and has to be integrated into agricultural policies and urban planning. Agricultural activities usually take place along roadsides, rivers and river valleys, in wetlands, in the middle of roundabouts, in open spaces and parks, under power lines and within backyards of residential plots.

The real impact of urban agriculture in term poverty alleviation and socio-economic development in urban areas are still to be debated since they are those who argue that the practice is beneficial to the urban poor (Memon and Lee-Smith 1993), while others views urban agriculture as of little or no consequence as a strategy for poverty alleviation and rather as a harmful activity, no suitable in urban life (Kulaba 1989).

13.2.1 Benefits of Urban and Peri-Urban Agriculture

In 1996, at the united international conference on human habitats, UPA was recognized for the first time for its contribution to the health and welfare of the fast growing population worldwide. It is generally admitted that UPA contributes to the health and well-being of a community by reducing hunger, strengthening access to food, improving nutrition, and improving environmental conditions that affect health; these benefits are both quantitative and qualitative, since increasing food quantities reduces hunger, while improving food quality fosters better health and nutrition (Yeung 1988; Freeman 1991; Lachance 1993; Zezza and Tasciotti 2010). UPA has been recognized as a key activity in combating poverty by playing a significant role in the provision of food and generation of income for the urban poor. UPA production systems cover 5% (urban) to 36% (peri-urban) of a city's total food supply and up to 90% of its demand for perishable vegetables (Drechsel and Dongus 2010); it is not just a source of income since, it can provide direct access to a larger number of nutritionally rich foods (vegetables, fruit, meat) and a more varied diet, increase the stability of household food consumption against seasonality or other temporary shortages, increase the time mothers spend caring for their children, as opposed to non-agricultural activities that are more likely to be located further away from home (Maxwell et al. 1998; Armar-Klemesu 2001; Egal et al. 2001; Maxwell 2003).

UPA builds safe, healthy and green environment in the neighbourhoods, schools and abandoned areas, it has the potential to recycle urban waste, and it benefits the economy, environment, and well-being of residents who enjoy its products. UPA plays a role in programs and projects that target health and nutrition, the environment, enterprise development, income generation, water and sanitation, youth and women, and food production and supply (Zezza and Tasciotti 2010; Dossaa et al. 2011).

13.2.2 Problems of Urban and Peri-Urban Agriculture

Despite the positive aspects of UPA, it presents some constraints. UPA is commonly perceived by some as an activity that is marginal, temporary, archaic, and harmful to farmers, consumers, the environment, the urban land economy, and the appearance of a city (Kulaba 1989). The main worries range from various health and environmental risks to land tenure insecurity, because of its close proximity to densely populated areas sharing the same air, water, and soil. The threat to the health of residents is directed toward possible contamination of producers (related to their activity) and consumers (via the food chain); the growing crops can be contaminated with (1) pathogenic organisms (mainly microorganisms) through irrigation with wastewater; (2) by uptake of heavy metals from contaminated soils, air or water; and (3) by residues of agro-fertilizers and pesticides (Lock and van Veenhuizen 2001; Van Veenhuizen 2006). The use of agro-fertilizers and pesticides in densely populated areas poses a serious threat to the environment, which degradation directly affects the health of the cities dwellers. Chemical fertilizers and pesticides are well known to persist in soil and water; they will therefore cause pollution through: accumulation in runoff, horticultural crops, and soils; seepage into aquifers; accumulation of heavy metals and organic compounds in aquatic life; direct contact; and airborne chemicals. The accumulation of heavy metals and organic compounds in aquatic life leads to eutrophication. Agriculture in the city can have a negative impact on green space and biodiversity if it replaces forested land, wetlands, lakes, or other biologically rich natural environments. The direct consequence of this loss of biodiversity is the increase of local warming and its consequences.

The close proximity of UPA to larger population concentrations increases the risk of spreading both communicable and non-communicable diseases.

13.3 Soil Constraints in Sub-Saharan Africa

Sub-Saharan Africa is the only region in the world where food production per capita did not increase since 40 years, with 180 million inhabitants lacking access to enough food for safe and productive life (Sanchez 2002). Soil fertility degradation has been described as the single most important constraint to food security in sub-Saharan Africa. Loss of soil fertility is associated to soil depletion, wind and water surface soil erosion, poor rainfall distribution, and restricted fallow periods necessary to restore soil fertility and low rates of fertilizer application (Floret and Pontanier 2001; Zougmoré 2003; Bationo et al. 2004; Camara and Heinemann 2006). Soil fertility decline has a negative impact on other natural resources like the degradation of water quality in rivers and lakes, owing to the increased sediment loads that affects fisheries and water quality, which directly impacts human and animal health.

The sustainability of urban agriculture ought to be improved and integrated into formal city planning; therefore soil fertility management in cities should be an important issue that has to be integrated to any important goal to underpin sustainable crop production with respect to the environment and health (Satterthwaite et al. 2010); that can be done with the participation of all stakeholders implicated in UPA by effectively integrating the complexity of interactions among bio-physical, social, economical and political factors, with respect to the paradigm of Integrated Soil Fertility Management (ISFM) (2012), which is a holistic approach to soil fertility research in Africa, integrating soil science with other disciplines including agronomy, ecology, economics and participatory social sciences. ISFM integrates organic and inorganic nutrient sources, soil water management, soil conservation and biological interventions. However, because of the sensitiveness of UPA that is carried out in highly populated areas, organic nutrient sources should be more promoted because of their impacts in terms of ecological benefits.

13.4 Urban Wastes Management Policy in Sub-Saharan Africa

Urban waste management is one of the most relevant issues that should address decision-makers in sub-Saharan Africa because of the social, political, and economical implications; which is not an easy task, since cities are mainly confronted with rapid urbanization caused by rural to urban migration overstretching resources and problems of dysfunctional solid waste management facilities and services (Yhdego 1995). MSW constitute one of the most crucial public health and environmental problems in African cities (Achankeng 2003; Rotich et al. 2006), putting municipalities and local governments under heavy pressure to find sustainable and cost-effective waste management policies to the urban waste problem without compromising environmental goals (Yhdego 1995; Diaz et al. 2007).

Municipalities generate huge quantities of wastes, with the average per-capita residential waste generation (in kg/day) from about 0.51 to 1.45; waste is generated everywhere from small houses to large industries. The main sources of wastes in sub-Saharan Africa are households, markets, institutions, streets, public areas, commercial areas and manufacturing industries; the rate of waste generation is high since the population and use of resources are higher in urban areas; urban residents generate two to three times more solid waste than their fellow rural citizens (Couth and Trois 2010). MSW undergoes a life cycle with generation, collection, transformation, and disposal as the main steps. The efficient execution of each of these stages requires taking many decisions at the strategic, tactical, and operational levels (Henry et al. 2006). The most common methods used for MSW management include landfilling, incineration, composting, recycling, mechanical–biological treatment, and waste-to-energy. The disposal of MSW in sanitary landfills is still the main waste management method applied by municipalities in sub-Saharan Africa. Landfilling has many negative impacts since it causes a profound strain on the environment, e.g., contamination of ground water resources, organic and inorganic pollution of nearby lakes and rivers, the carbon dioxide and methane (with high global warming potential) release from incineration plants, contributing to global warming (Couth and Trois 2012). Disposal of waste in landfills is reported to contribute to global GHG emissions approximately by 4 % (Bogner et al. 2004, 2008). GHG emissions per person from urban waste management activities are greater in sub-Saharan African countries than in other developing countries, and are constantly increasing since the populations are more and more urbanized (Couth and Trois 2011).

Methane (CH_4) gas generated through landfilling is 21 times more potent as a GHG than the natural carbon dioxide (CO_2) produced through composting the biogenic wastes (Seng et al. 2013). Since, global warming is more and more becoming important environmental issue, alternatives methods that are ecological and environmental sound have to be promoted. A much more efficient and cost effective way to control GHG emissions from waste is to stabilize the waste via composting and to use the composted material as a soil improver/organic fertilizer or as a component of growing media; compost that could be supplement with other ecological material (e.g., wood ash) when dealing with acidic soils (Bougnom et al. 2009, 2010, 2011).

Intensification of UPA in the tropics aiming at satisfying the increasing food demand requires large amounts of inputs. Urban farmers in the tropics are generally subsistence farmers and cannot cope with the higher cost of agrochemicals

Table 13.1 Summary of waste production and waste composition from some cities in sub-Saharan Africa

Country/city	Waste production (kg/hd/year)	Waste composition (% organic)	References
Addis Ababa/Ethiopia	91.98	60	Couth and Trois (2011)
Arusha/Tanzania	531	65	Couth and Trois (2011)
Louga/Senegal	110–250	50	Collivignarelli et al. (2007)
Maputo/Mozambique	182	–	Hunger and Stretz (2006)
Nairobi/Kenya	260	–	Muniafa and Otiato (2008)
Uganda	200	71	Okot-okumu and Nyenje (2011)
Windhoek/Namibia	242	47	Couth and Trois (2011)
Yaounde/Cameroon	288	75	Ndoumbé et al. (1995); Parrot et al. (2009)

(fertilizers, pesticides, and insecticides), which, when massively used are reported to pose a serious threat to the environment, human and animal health.

Among the beneficial environmental impacts of UPA, we have the potential to recycle organic waste within the cities, so the question is how that can be properly done and how the cities can benefit from a suitable MSW management policy. Combining organic/inorganic soil fertility management is specifically the recommended approach in African Western or Eastern sandy or acidic soils by the African Network for Tropical Soil Biology and Fertility Institute.

Cities in sub-Saharan Africa generate huge amount of waste ranging from 91 to 531 kg/hd/year, containing 50–75% of organic waste (Table 13.1); therefore, if the produced organic waste could be composted, hundreds of tons of compost could be produced daily and the policy of substituting the use of mineral fertilizers by compost should be implemented. For an integrated and sustainable solid waste management, any solid composting project should not just include technical aspects but also various key elements of sustainability.

13.5 Composting

Composting is the biological process of converting organic substrate, in the present of oxygen, into CO_2, H_2O, heat, new cell generation and humus. The end product of the composting process is compost, a product that is high in nutrients, rich in humic acids, and contains a diverse community of microorganisms. Compost can be utilized as organic fertilizer or soil additive, compost land application completes a circle in which nutrients and organic matter which have been removed by harvesting of different agricultural products are replaced (Nwaga et al. 2010). Centralized composting processes for MSW primarily fall into two categories: windrow composting and in-vessel composting, windrow composting being the most suitable for large quantities of diverse biodegradable waste (Last 2006).

13.5.1 Microbial Process of MSW Composting

Composting is a controlled self-heating, aerobic microbiological process that is facilitated by bacteria and fungi under the direct influence of some physical and chemical parameters. Microbial biodegradation predominately occurs in the thin liquid films (biofilms) on the surface of the organic particles. Temperature, moisture content, oxygen content, material particle size and nature of the feedstock with particular importance to carbon over nitrogen (C:N) ratio are the five key factors that to be monitored for optimal composting process (Evans 2001; Last 2006).

Under optimal conditions the composting process can be divided into four phases (Fig. 13.1a): (i) a first mesophilic phase, (ii) a thermophilic phase, (iii) a second mesophilic phase and, (iv) a maturation and stabilization phase (Hoitink and Boehm 1999; Tuomela et al. 2000; Insam and de Bertoldi 2007; Ryckeboer et al. 2003)

i. **The initial mesophilic phase** (10–42 °C) last for only few hours or a couple of days; it is characterizing by the proliferation of mesophilic bacteria and fungi which degrade primarily the readily available nutrients; that activity raises the temperature to about 45 °C (thermophilic condition) leading to the death and lyse of the vegetative cells and hyphae, only heat resistant spores survive.
ii. **The thermophilic phase** (temperature 45–80 °C) last few days, several weeks (particularly in food wastes) or even months (particularly in wood wastes); it is characterized by the development of a thermophilic microbial population comprising some bacterial species, actinomycetes and fungi. High temperatures support the degradation of recalcitrant organic materials such as lignocelluloses and wood, as well as the elimination of pathogenic and allergenic microorganisms which are generally mesophilics.
iii. **The second mesophilic phase** is a stationary period without significant changes of temperature because microbial heat production and heat dissipation balance each other; mesophilic microorganisms (thermophilic bacteria, actinomycetes, and fungi), often dissimilar to those of the first mesophilic phase, recolonize the substrates.
iv. **The maturation (or curing) and stabilization phase** last several weeks to several months, it is characterized by a gradual temperature decrease and a recolonization of the material by mesophilic microorganisms having survived the thermophilic phase and an invasion of microorganisms from outside, these microorganisms will pursue the degradation process and compounds such as lignin-humus complexes are formed that are not further degradable.

Bacteria, actinobacteria and fungi are present and active during MSW composting (Fig. 13.1b) and the most efficient composting process is achieved by mixed communities of bacteria and fungi (Waksman et al. 1939; Gray et al. 1971). The bacterial community during MSW composting includes both Gram-positive and Gram-negative bacteria, grouped into the six major bacterial classes or genera: Actinobacteria, *Thermoactinomyces*, *Acetobacter*, *Lactobacillus*, *Bacillus*, and *Clostridium*; from the phylum Actinobacteria, Bacteroidetes, Firmicutes, Proteobacteria

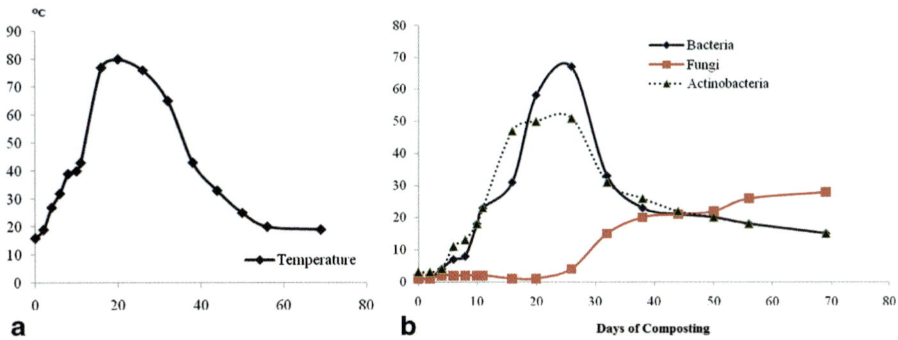

Fig. 13.1 **a** Temperature variations and **b** relatives abundance evolution of the different microorganisms (Bacteria, Fungi and Actinobacteria) observed during composting of MSW

and Deinococcus-Thermus (Danon et al. 2008; Partanen et al. 2010; Chandna et al. 2013). The diversity of fungi during composting is lower than the one of bacteria, fungi recovered during composting of MSW belong to the groups Mucorales, Pezizomycotina, Basidiomycota, Saccharomycetaceae and Dipodascaceae, from the phyla Ascomycota, Basidiomycota and Zygomycota (Ryckeboer et al. 2003; Hultman et al. 2010).

Compost should not contain pathogens or viable seeds both in terms of phytohygiene and human diseases, and it should be stable and suitable for use as biofertilizer or soil conditioner (Epstein 1997; Ryckeboer et al. 2003). Several microorganisms (both bacteria and fungi) present in compost could bc uscd as biocontrol agents of soilborne plant pathogens, e.g., *Flavobacterium balustinum*, *Bacillus subtilis*, *Bacillus cereus* and *Pseudomonas fluorescens*, *Trichoderma* spp., *Gliocladium virens* and *Paecilomyces* spp. (Raupach and Kloepper 1998; Fuchs 2002; Noble and Coventry 2005).

13.5.2 Advantages of Composting and Compost Application

The conversion of organic waste to compost results in a large quantity of mass reduction, which depends on the nature of the substrates and the composting process (Amlinger et al. 2008; Zhang and Matsuto 2010). Addition of compost to soil modifies considerably its biological, physical, and chemical properties in the short term as well as in the long term (Ryckeboer et al. 2003). the use of compost in agriculture aids in replenishing and maintaining long-term soil fertility by providing optimal conditions for soil biological activity and a slow flow of nutrients adapted to the needs of the crop (Gobat et al. 2003).

Composting is very well implemented in both developed countries as a way of dealing with the biodegradable fraction of the municipal waste; the benefits of composting are directed toward three directions:

Economically; composting can be started with very little capital and operating costs, it is easy to implement at different levels (household to large scale composting) and can generate greater income from the avoidance of methane emissions to the substitution to agrochemicals (Hoornweg et al. 1999).

Environmentally; composting allow the reduction of organic content within the local disposal site, less greenhouse gases (GHG) emissions as compared to landfilling, recovery of organic materials and production of soil enhancer/organic fertilizer, reduction of chemical fertilizer application within the municipality, reduction of leachate production and potential ground and surface water pollution, odour, potential hazardous air emissions, visual intrusion and litter nuisance (Bogner et al. 2008).

Socially; composting increases employment opportunities in the cities, integrates additional livelihood opportunities, supports poorer farmers with bio-fertilizers supply, reduces health risks to the neighbouring communities and promotes the awareness of the potential value of waste, which normally is perceived as odorous und useless (Perla 1997). Adding compost in tropical acidic or sandy soils may increase water and nutrient holding capacity and storage; and also reduce soil acidity and toxic metallic ions such as aluminium and manganese.

13.5.3 Factors Constraining Composting in Developing Countries

Despite the numerous advantages of composting they are many constraints to the development of this technology, obstacles that could be overcome. Municipal managers first take into consideration the cost of any change in MSW management policy, so when evaluating composting promotion programs, they first check the cost of composting versus landfilling. Dumping or landfilling is inexpensive and not subject to effective environmental controls, while composting is relatively expensive, therefore, the cost of composting should be competitive with the cost of landfilling.

The biggest difficulty faced by most composting programmes in generating income is finding a market for the compost; it is difficult to secure finances since the revenues generated from the sale of compost rarely cover processing, transportation and application costs, a suitable marketing plan for the produced is necessary. e.g., in Yaoundé/Cameroon, 700 t of municipal wastes are collected daily, and technically, 150–200 t of compost could be produced daily, so the main worry for any centralized composting plant is to know how this compost could find a suitable place in the city and in the surrounds. The raised concern about the cost of compost could find solution in the way that the end-users have to be located closer to the composting plant; the solution should be many decentralized small composting plants, targeting specific farmers. Compost substrate for nurseries (horticulture, fruit trees: cocoa, coffee, oil palm) of specific interest is the potential market of integrating compost for increasing soil microbial activities and large scale production of beneficial microorganisms such as nitrogen fixing bacteria and mycorrhizal fungi inocula (Nwaga et al. 2012).

Land tenure and ownership is sometime restricted, any investment in time and resources in community composting within urban areas is risky, the municipalities and local governments should ease land access to NGOs, Cooperatives and Common Initiative Groups (CIP) promoting composting.

In sub-Saharan Africa, there is no policy to separate household wastes at the origin and in the different waste dumps, all kind of waste are collected, including organic waste, metal, glass, hospital waste, and liquid waste. Compost made from organic waste originated from these dumps will be highly contaminated are will expose the workers and consumers health.

Establishment of compost quality that meets the international standards is required, and the government and local authorities should take suitable laws for waste segregation at the origin, which would improve the quality of the compost to be produced.

13.6 Conclusion

A holistic approach of soil fertility recommended for Sub Saharan African soils is based on a sustainable management which integrated biological, chemical and socio-economic considerations (Swift and Woomer 1994; Bationo et al. 2004). Land scarcity and soil fertility decline are major obstacles to the intensification of urban agriculture consequent to the increase demand for food supply. With regard to the cost and adverse effects that the agrochemicals have to the environment and the public health, and considering the quantity of organic waste that is produced daily, composting represents a viable option. Composting has therefore to be careful monitored in other to ward off any sanitary risks in a confined environment and to find a suitable place in the market. Municipal managers and all involved stakeholders in urban planning have to evaluate composting programs within an integrated municipal waste management system.

References

Achankeng E (2003) Globalization, urbanisation and municipal solid waste management in Africa. Proceedings of African studies association of Australasia and the Pacific: African on a global stage

Amlinger FF, Peyr SS, Cuhls CC (2008) Green house gas emissions from composting and mechanical biological treatment. Waste Manage Res 26:47–60

Armar-Klemesu M (2001) Urban agriculture and food security nutrition and health. Thematic paper 4. In: Bakker N et al (ed) Growing cities, growing food: urban agriculture on the policy agenda. DSE, Fefdafing

Bationo A, Kimetu J, Ikerra S, Kimani S, Mugendi D, Odendo M, Silver M, Swift MJ, Sanginga N (2004) The African network for soil biology and fertility: new challenges and opportunities. In: Bationo A (ed) Managing nutrient cycles to sustain soil fertility in sub-Saharan Africa. Academy Science, Nairobi

Bogner JE, Lee CA, Chanton J, Spokas K (2004) Reducing greenhouse gas emissions through landfill gas recovery in South Africa. In: WasteCon 2004 Seventeenth waste management conference and exhibition, Johannesburg, South Africa

Bogner J, Pipatti R, Hashimoto R, Diaz C, Mareckova K, Diaz L, Kjeldsen PS, Faaij A, Gao Q, Zhang T, Ahmed MA, Sutamihardja RTM, Gregory R (2008) Mitigation of global GHG emissions from waste: conclusions and strategies from the Intergovernmental Panel on Climate Change (IPCC) Fourth assessment report. Working Group III (Mitigation). Waste Manage Res 26:11–13

Bougnom BP, Mair J, Etoa FX, Insam H (2009) Composts with wood ash addition: a risk or a chance for ameliorating acid tropical soils? Geoderma 153:402–407

Bougnom BP, Knapp B, Etoa FX, Elhottová D, Koubová A, Insam H (2010) Designer compost for ameliorating acid tropical soils: effects on the soil microbiota. Appl Soil Ecol 45:319–324

Bougnom BP, Knapp BA, Etoa FX, Insam H (2011) Possible use of wood ash and compost for improving acid tropical soils. In: Insam H, Knapp BA (eds) Recycling of biomass ashes. Springer, Berlin, pp 87–103

Camara O, Heinemann E (2006) Overview of the fertilizer situation in Africa. Background paper prepared for the African Fertilizer Summit. Abudja, Nigeria, p 21

Chandna P, Nain L, Singh S, Chander Kuhad R (2013) Assessment of bacterial diversity during composting of agricultural byproducts. BMC Microbiol 13:99

Cohen M, Garrett J (2009) The food price crisis and urban food insecurity. IIED, London

Collivignarelli C, Sorlini S, Cavallari S, Vaccari M (2007) Waste management and recovery in developing countries. In: Diaz LF, Eggerth LL, Savage GM (eds) Managing solid wastes in developing countries. CISA, Padova, pp 33–40

Couth R, Trois C (2010) Carbon emissions reduction strategies in Africa from improved waste management: a review. Waste Manage 30:2336–2346

Couth, R, Trois, C (2011) Waste management activities and carbon emissions in Africa. Waste Manage 31(1):131–137

Couth R, Trois C (2012) Cost effective waste management through composting in Africa. Waste Manage 32:2518–2525

Danon M, Franke-Whittle IH, Insam H, Chen Y, Hadar Y (2008) Molecular analysis of bacterial community succession during prolonged compost curing. FEMS Microbiol Ecol 65(1):133–144

Diaz LF, Savage GM, Eggerth LL (2007) The management of solid wastes in economically developing countries. In: Diaz LF, Eggerth LL, Savage GM (eds) Managing solid wastes in developing countries. CISA, Padova, pp 17–29

Dossaa LH, Abdulkadirb A, Amadoua H, Sangarec S, Schlechta E (2011) Exploring the diversity of urban and peri-urban agricultural systems in Sudano-Sahelian West Africa: an attempt towards a regional typology. Landscape Urban Plann 102:197–206

Drechsel P, Dongus, S (2010) Dynamics and sustainability of urban agriculture: examples from sub-Saharan Africa. Sustainab Sci 5:69–78

Earthtrends (2008) Population, health and human well-being: country profiles. Earthtrends. http://earthtrends.wri.org/country_profiles/index.php?theme=4. Accessed July 2008

Egal F, Valstar A, Meershoek S (2001) Urban agriculture, household food security and nutrition in Southern Africa. Mimeo, FAO, Rome

Epstein E (1997) The science of composting Lancaster. Technomic Publishing Company, Lancaster

Evans G (2001) Biowaste and biological waste treatment. James and James, London

FAO (2008) Urbanization and food security in Sub Saharan Africa. Information paper for the FAO 25th African regional conference. ftp://ftp.fao.org/docrep/fao/meeting/012/k1915e.pdf. Accessed 31 Jan 2014

Floret C, Pontanier R (2001) La jachère en Afrique tropicale: de la jachère naturelle Ã la jachère améliorée. Le point des connaissances. CTA, IRD, John LibbeyEurotext. Actes d'un séminaire. Dakar, Sénégal, p 356

Freeman DB (1991) A city of farmers: informal urban agriculture in the open spaces of Nairobi, Kenya. Mc Gill-Queen's University Press, Montreal and Kingston

Fuchs JG (2002) Practical use of quality compost for plant health and vitality improvement. In: Insam H, Riddech N, Klammer S (eds) Microbiology of composting. Springer, Heidelberg, pp 435–444
Gobat JM, Aragno M, Mattey W (2003) Le sol vivant, 2e édition revue et complétée. Presses Polytechniques et Universitaires Romandes, Lausanne, p 319
Gray KR, Sherman K, Biddlestone AJ (1971) A review of composting—part 1. Process Biochem 6:32–36
Henry RK, Yongsheng Z, Jun D (2006) Municipal solid waste management challenges in developing countries—Kenyan case study. Waste Manage 26:92–100
Hoitink HAJ, Boehm MJ (1999) Biocontrol within the context of soil microbial communities: a substrate-dependent phenomenon. Annu Rev Phytopathol 37:427–446
Hoornweg, D, Thomas, L, Otten, L (1999) Composting and its applicability in developing countries. World Bank working paper series, 8
Hultman J, Vasara T, Partanen P, Kurola J, Kontro MH, Paulin L, Auvinen P, Romantschuk M (2010) Determination of fungal succession during municipal solid waste composting using a cloning-based analysis. J Appl Microbiol 108:472–487
Hunger G, Stretz J (2006) Proof of Service—an innovative multifunctional tool for sustainable solid waste management strategies in developing countries. In: WasteCon 2006, Eighteenth Waste Management Conference and Exhibition, October 2006, Cape Town, South Africa
Insam H, de Bertoldi M (2007) Microbiology of the composting process. In: Golueke C, Bidlingmaier W, de Bertoldi M, Diaz L (eds) Compost science and technology. Elsevier, UK, pp 25–48
Integrated Soil Fertility Management (ISFM) (2012) Integrated soil fertility management in Africa: from microbes to markets. Nairobi, Kenya (22–26 October 2012)
Kulaba SM (1989) Local government and management of urban services in Tanzania. In: Stren RE, White RR (eds) African cities in crisis: managing rapid urban growth. Westview Press, Boulder, pp 203–245
Lachance A (1993) Cultiver en ville Le CRDI Explore 21:8–9
Last S (2006) An introduction to waste technologies. Waste Technologies UK Associates
Lock K, van Veenhuizen R (2001) Balancing the positive and negative health impacts. Urban Agric Mag 3:1–5
Maxwell D (2003) The importance of urban agriculture to food and nutrition. In: Annotated bibliography on urban agriculture. Sida and ETC, Leusden, pp 22–129
Maxwell D, Levin C, Csete J (1998) Does urban agriculture help prevent malnutrition? Evidence from Kampala. Food Policy 23:411–424
Memon PA, Lee-Smith D (1993) Urban agriculture in Kenya. Can J Afr Stud 27:25–42
Monni S, Pipatti R, Lehtilä A, Savolainen I, Syri S (2006) Global climate change mitigation scenarios for solidwaste management, VTT Publications 603, Julkaisija Publisher, Finland. Available from: http://www.vtt.fi/inf/pdf/ publications/2006/P603.pdf. Accessed Sept 2010
Mougeot LJ (2000) The hidden significance of urban agriculture. 2020 Vision. Focus 3, Brief 6 of 10, IFPRI, Washington
Mougeot L (2006) Growing better cities: urban agriculture for sustainable development. International Development Research Centre, Ottawa, Canada, 97 pp
Muniafa M, Otiato E (2008) Solid waste management in Nairobi, Kenya. A case for emerging economics. In: WasteCon 2008, Nineteenth Waste Management Conference and Exhibition, Cape Town, South Africa (October 2008)
Ndoumbé NH, Ngnikam E, Wethe J (1995) Le compostage des ordures ménagères: L'expérience du Cameroun après la dévaluation du Franc CFA. Bulletin africain 5:4–10
Noble R, Coventry E (2005) Suppression of soil-borne plant diseases with composts: a review. Biocontrol Sci Technol 15:3–20
Nwaga D, Jansa J, Abossolo Angue M, Frossard E (2010) The potential of soil beneficial microorganisms for slash-and-burn agriculture in the humid forest zone of Sub Saharan Africa. In: Dion P (ed) Soil biology and agriculture in the tropics. Springer, Berlin, pp 81–107
Nwaga D, Diaw C, Ntede NH, Sarasin G, Bilong P, Tsala NG, Etoa F-X, Njomkap JC (2012) Microbial inoculants for eco-agriculture: Potential market development in Central Africa. Inte-

grated Soil Fertility Management in Africa: from Microbes to Markets, Nairobi, Kenya (22–26 October 2012)
Obudho RA, Foeken, DWJ (1999) Urban agriculture in Africa: a bibliographical survey. African Studies Centre, Leiden
Okot-Okumu J, Nyenje R (2011) Municipal solid waste management under decentralisation in Uganda. Habitat Int 35:537–543
Page B (2002) Urban agriculture in Cameroon: an anti-politics machine in the making? Geoforum 33:41–54
Parrot L, Sotamenou J, KamgniaDia B, Nantchouang A (2009) Determinants of domestic waste input use in urban agriculture lowland systems in Africa: the case of Yaounde in Cameroon. Habitat Int 33:357–364
Partanen P, Hultman J, Paulin L, Auvinen P, Romantschuk M (2010) Bacterial diversity at different stages of the composting process. BMC Microbiol 10:94
Perla M (1997) Community composting in developing countries. Biocycle 38:48–51
Potts D (2009) The slowing of sub-Saharan Africa's urbanization: evidence and implications for urban livelihoods. Environ Urban 21:253–259
Raupach GS, Kloepper JW (1998) Mixtures of plant growth-promoting rhizobacteria enhance biological control of multiple cucumber pathogens. Phytopathology 88:1158–1164
Ravallion M, Chen S, Sangraula P (2007) New evidence on the urbanization of global poverty. Policy research working paper 4199. World Bank, Washington
Rotich HK, Yongsheng Z, Jun D (2006) Municipal solid waste management challenges in developing countries: Kenyan case study. Waste Manage 26:92–100
Ryckeboer J, Mergaert J, Vaes K, Klammer S, De Clercq D, Coosemans J, Insam H, Swings J (2003) A survey of bacteria and fungi occurring during composting and self-heating processes. Ann Microbiol 53:349–410
Sanchez PA (2002) Soil fertility and hunger in Africa. Science (Washington) 295(5562):2019–2020
Satterthwaite D, McGranahan G, Tacoli C (2010) Urbanization and its implications for food and farming. Philos Trans R Soc Lond B Biol Sci 365:2809–2820
Sawio CJ (1993) Urban agriculture research in East and Central Africa: record, capacities and opportunities. Cities Feeding People Report 1. IDRC, Ottawa
Seng B, Hirayama K, Katayama-Hirayama K, Ochiai S, Kaneko H (2013) Scenario analysis of the benefit of municipal organic-waste composting over landfill, Cambodia. J Environ Manage 114:216–224
Smit J, Ratta A, Nasr J (1996) Urban agriculture: food, jobs and sustainable cities. UNDP, New York
Swift MJ, Woomer PL (1994) The biological management of tropical soil fertility. Wiley, New York, p 243
Tuomela M, Vikman M, Hatakka A, Itävaara M (2000) Biodegradation of lignin in a compost environment: a review. Bioresour Technol 72(2):169–183
UNDP (1996) Urban agriculture. Food, jobs, and sustainable cities. United Nations Development Programme. Publication series for habitat II, vol 1. UNDP, New York
UN-Habitat (2006) State of the World's Cities 2006/7. The millenium development goals and urban sustainability: 30 years of shaping the habitat agenda. Earthscan, London
United Nations (2008) World urbanization prospects: the 2007 revision, CD-ROM edn. United Nations Department of Economic and Social Affairs, Population Division, New York
Van Veenhuizen R (2006) Cities farming for the future. Urban agriculture for green and productive cities. RUAF Foundation, IDRC, Ottawa
Waksman SA, Cordon TC, Hulpoi N (1939) Influence of temperature upon the microbiological population and decomposition processes in compost of stable manure. Soil Sci 47:83–114
Yeung YM (1988) Agricultural land use in Asian cities. Land Use Policy 5:79–82
Yhdego M (1995) Urban solid waste management in Tanzania issues, concepts and challenges. Resour Conserv Recycl 14:1–10
Zezza A, Tasciotti L (2010) Urban agriculture, poverty, and food security: empirical evidence from a sample of developing countries. Food Policy 35:265–273

Zhang HH, Matsuto TT (2010) Mass and element balance in food waste composting facilities. Waste Manage 30:1477–1485

Zougmoré R (2003) Integrated water and nutrient management for sorghum production in semi-arid Burkina Faso. Tropical resource management papers, no.°45. Wageningen University, Wageningen, p 205

Index

D. K. Maheshwari (ed.), *Composting for Sustainable Agriculture*, Sustainable Development and Biodiversity 3, DOI 10.1007/978-3-319-08004-8

Printed by Printforce, the Netherlands